"Simon Fairlie provides us with an unusual and extremely important gift in his new book, Meat: A Benign Extravagance. *Everyone interested in how their food choices can affect the ecological, social, and economic health of the communities in which they live should read this book."*

Frederick Kirschenmann, Distinguished Fellow,
Leopold Center for Sustainable Agriculture, and
President of Stone Barns Center for Food and Agriculture

"This is a tremendous and very timely book: the world's meat consumption is rapidly rising, leading to devastating environmental impacts as well as having long-term health implications for societies everywhere. Simon Fairlie's book lays out the reasons why we must decrease the amount of meat we eat, both for the planet and for ourselves. This brilliant book is essential reading for anyone who cares about food and the environment."

Rosie Boycott, Founder of Spare Rib and Virago Press,
ex-editor of the *Independent, Independent on Sunday, Daily Express*
and *Esquire* magazine, broadcaster, writer and campaigner and
currently Food Advisor to the Mayor of London

"Meat, animals and dairy have been in the firing line for so long that in some circles, the assumption is taken for granted that there is no case, ever, anywhere, to be made for the role of animals in farming, landcare or diet. This book by Simon Fairlie is a wonderful and challenging correction. As a former Welsh Black breeder who farmed upland wet English hills but who gave up meat years ago (but takes dairy produce), I found this book a riveting read. As an academic who grapples with what land is for and what a sustainable diet might be, I assure you that this book is essential reading. Fairlie's beautifully written, practical yet erudite book covers the terrain that policy-makers now realise needs to be addressed. Fairlie makes the case for 'not throwing the baby out with the bath water,' or should that be 'don't demonise the animal before you know its function and value'?"

Tim Lang, Professor of Food Policy, City University London

"No one has ever analysed the world's food and agriculture more astutely than Simon Fairlie — an original thinker and a true scholar. Here he shows that while meat is generally a luxury it is often the best option, and could always be turned to advantage — if only we did things properly; but this, with present economic policies and legal restrictions, is becoming less and less possible. Everyone should read this book — especially governments, and all campaigners."

Colin Tudge, Biologist and author

Some hae meat and canna eat,
And some wad eat that want it;
But we hae meat, and we can eat,
Sae let the Lord be thankit.

The Selkirk Grace

MEAT

A Benign Extravagance

Simon Fairlie

Chelsea Green Publishing
White River Junction, Vermont

Chelsea Green Publishing is committed to preserving ancient forests and natural resources. We elected to print this title on 30-percent postconsumer recycled paper, processed chlorine-free. As a result, for this printing, we have saved:

22 Trees (40' tall and 6-8" diameter)
7 Million BTUs of Total Energy
2,074 Pounds of Greenhouse Gases
9,987 Gallons of Wastewater
606 Pounds of Solid Waste

Chelsea Green Publishing made this paper choice because we and our printer, Thomson-Shore, Inc., are members of the Green Press Initiative, a nonprofit program dedicated to supporting authors, publishers, and suppliers in their efforts to reduce their use of fiber obtained from endangered forests. For more information, visit: www.greenpressinitiative.org.

Environmental impact estimates were made using the Environmental Defense Paper Calculator.
For more information visit: www.papercalculator.org.

Meat: A Benign Extravagance was originally published in 2010 in the United Kingdom by Permanent Publications, The Sustainability Centre, East Meon, Hampshire GU32 1HR, UK. www.permaculture.co.uk

Printed in the United States of America
First Chelsea Green printing December, 2010
10 9 8 7 6 5 4 3 2 1   10 11 12 13 14

Our Commitment to Green Publishing
Chelsea Green sees publishing as a tool for cultural change and ecological stewardship. We strive to align our book manufacturing practices with our editorial mission and to reduce the impact of our business enterprise in the environment. We print our books and catalogs on chlorine-free recycled paper, using vegetable-based inks whenever possible. This book may cost slightly more because we use recycled paper, and we hope you'll agree that it's worth it. Chelsea Green is a member of the Green Press Initiative (www.greenpressinitiative.org), a nonprofit coalition of publishers, manufacturers, and authors working to protect the world's endangered forests and conserve natural resources. *Meat* was printed on Joy White, a 30-percent postconsumer recycled paper supplied by Thomson-Shore.

ISBN: 978-1-60358-324-4
Library of Congress Cataloging-in-Publication Data on file with the publisher

Chelsea Green Publishing Company
Post Office Box 428
White River Junction, VT 05001
(802) 295-6300
www.chelseagreen.com

CONTENTS

Foreword

Simon Fairlie's *Meat: The Benign Extravagance* is the sanest book I have read on the subject of how the human race is going to feed itself in the years ahead. Its main attention is given to the pro-meat versus no-meat debate, but it really involves an intense scrutiny of what we know and don't know about the entire food chain. Fairlie's search to ferret out the truth in these matters is awesomely thorough—he leaves no stone unturned.

I have been tempted to try writing a book like this one, waving a flag of caution before all the fervent advocates of one diet or another. But I doubt that I could endure the displeasure, even wrath, that would be visited upon me if I pointed out, as Fairlie does, that neither vegan nor meat glutton, nor factory farmer nor, horrors, even my own favorite food-production system—pasture farming—has all the answers. Fairlie is made of sterner stuff.

Somebody had to write this book, and thank heavens it was a writer with the wit and wisdom of Simon Fairlie. The reason his book is so important is that what it addresses—food security, first and foremost—is being undermined by well-intentioned people of all persuasions who are demanding rules and regulations in food matters without enough knowledge. We all have very firm convictions about what we want to eat and don't want to eat, but the only direct contact with the food chain that most of us experience is what we see when we sit down at the table. We have only foggy notions of how all forms of life interact in the food chain, how we are all seated at an unimaginably vast table, eating and being eaten. In fact, even scientists who make these matters their lifelong study know only a little, and the honest ones readily admit it.

But because we have zillions of reams of information about food production and endless columns of numbers to pick from in support or denial of whatever we want to believe, we think that human intelligence has analyzed the subject well enough to start dictating public policy about what we should eat and how it should be produced. This book is most valuable because it will convince the open-minded reader that when we start making grand statements about the earth's food-carrying capacity, more than a little humility is in order. No matter how fervently we support the no-meat or the pro-meat point of view, or how much allegiance we have for any particular dietary bible, or what kind of farming we think best serves humanity's food purposes, or what we think about carbon footprints, global warming, greenhouse gases and any of the other trendy phrases with which the news batters us, or what economic religion we think best serves the purpose of providing food for all, I challenge anyone to read this book and *not* realize that no one has all the right answers, because neither science nor ideology knows all the right answers yet.

Humility is a wonderful asset in the pursuit of knowledge, and Simon Fairlie gives the reader plenty of opportunity to acquire some. He addresses every aspect of the food and agriculture debate with unrelentingly thorough research and unswerving, sometimes almost ruthless, logic. Wherever he finds financial self-interest prevailing over objective data, wherever he spies ideology undermining science, he does not spare his rod of criticism. That is the beauty of this book; it does not take sides. It asks for more knowledge and objective thinking from everyone, and in the meantime, seeks compromise. In these days of polarization in almost everything, this book is a benign gift of clear thinking. The author also brings a sharp sense of humor to the debate, adding to the pleasure of reading the research.

The no-meat versus pro-meat camps might ponder the lessons of Prohibition days. No doubt trying to make old demon alcohol disappear was a noble idea, but we learned the hard way that it just isn't going to happen. And so it is with eating meat. As Fairlie argues, allowing for moderation works better for overall food security than trying to make farm animals disappear, and just might make it easier for vegetarians to follow their diet preferences, too.

I am constantly amazed at how many people quoted in this book seem to believe that they have the one and only answer to food issues even though they apparently have not one farthing of experience in farming. Simon Fairlie has spent considerable time actually doing farm work, which undoubtedly has informed his research to great end. I offer this to you, reader—within, find an important, well-written, and absolutely crucial addition to the agricultural canon, which I trust will have a much-needed impact on the future of sustainable agriculture.

GENE LOGSDON

ACKNOWLEDGMENTS

I would like to thank all who have looked through and commented on various sections of this text. To cite some by name would be to overlook others.

Special thanks are due to Tara Garnett, whose Food Climate Research Network provides an online library of information on these issues; to Marnie, of South Petherton Public Library, for her patient and adept renewal of overdue books; and to the anonymous bookhounds in Interlibrary Loans, who rarely fail to deliver.

Dedicated to Mary, without whose inconstancy this book would never have been written.

I think I could turn and live with animals, they are so
placid and self-contain'd,

I stand and look at them long and long.

They do not sweat and whine about their condition,

They do not lie awake in the dark and weep for their sins,

They do not make me sick discussing their duty to God,

They do not write environmental life cycle analyses.

Not one is dissatisfied, not one is demented with the mania
of owning things,

Not one kneels to another, nor to his kind that lived
thousands of years ago,

Not one is respectable or unhappy over the whole
earth.

Walt Whitman, *Leaves of Grass* (adapted)

1

INTRODUCTION

The most fertile districts of the habitable globe are now actually cultivated by men for animals, at a delay and waste of aliment absolutely incapable of calculation. It is only the wealthy that can, to any great degree, even now, indulge the unnatural craving for dead flesh.

With these words, written in 1813, Percy Shelley introduced a fresh, socialist argument into a centuries-old debate which had hitherto mainly focussed on the brutality to animals and the damage to human health caused by meat eating. Not only did the consumed and the consumer suffer – so too did those excluded from consumption.

As the world's population has surged, causing increasing pressure on land and natural resources, the concerns raised by Shelley have come to the fore. In a 1976 debate in *The Ecologist* on the role of meat in an ecological society, Peter Roberts of Compassion in World Farming argued:

> There is starvation because of poverty and because of greed and because we devote the major part of the world's food resources and its expertise to the feeding of animals instead of children. If we continue along these lines famine will increase on a scale never before seen ... The answer is plain. We must get rid of the animal in the food chain.[1]

A quarter of a century later he was echoed by George Monbiot in *The Guardian*:

> The world produces enough food for its people and its livestock, though (largely because they are so poor) some 800 million are permanently malnourished. The number of farm animals on earth has risen fivefold since 1950: humans are now outnumbered three to one. Livestock already consume half the world's grain, and their numbers are still growing almost exponentially ... As a meat-eater, I've long found it convenient to categorize veganism as a response to animal suffering or a health fad. Faced with these figures, it now seems plain that it is the only ethical response to what is arguably the world's most urgent social justice issue.[2]

The state of affairs that Monbiot describes is unambiguously iniquitous; if there is one single injustice for which our economic system could be held responsible above all others, it is the fact that it has, time and again over the last 200 years, diverted food out of the hands of the hungry and funnelled it

---

1   The Ecologist (1976), 'Should an Ecological Society be a Vegetarian One?' *The Ecologist* 6:10, December.
2   Monbiot, George (2002), 'The Poor Get Stuffed', *The Guardian*, 4 December.

down animals' throats to provide meat for the rich. Nothing written anywhere in this book should be taken to imply that depriving the poor of food to provide luxuries for the rich is anything but, as Jeremy Rifkin puts it, 'a form of human evil'. Nowhere in this book do I put the case for eating lots of meat, because there isn't one. Meat is an extravagance.

However, to conclude that veganism is the 'only ethical response' is to take a big leap into a very muddy pond. The fact that some people get *all* the meat, while others starve, is not in itself an indictment of meat, any more than the fact that some people can afford their own car while others have to walk is an argument against buses (though it is an argument against private cars). The vegan response brings to mind the Tupamaros' slogan: 'Everybody dances or nobody dances'.[3] This is fair enough; but is it not better that everybody should have the opportunity to dance, rather than nobody? And if there is not enough space on the floor for all to do so at once, why not take it in turns?

This book is concerned with the environmental ethics of eating meat. The central question it asks is not whether killing animals is right or wrong, but whether farming animals for meat is sustainable. From this springs a range of secondary questions: Is meat-eating a waste of resources? Is meat a way of robbing the hungry to fatten the well-fed? Does meat-eating cause disproportionate levels of global warming? Does the rearing of animals for meat deprive wild animals of habitat and the world of wilderness? These are charges that many vegans and vegetarians have levelled against meat-eaters and there is substance in them. They deserve addressing, and when I began this book I was not aware that anyone had ever tried to do so very comprehensively – though now, five years later, it is a hot topic.

Meat: A Benign Extravagance began as a personal enquiry, grew into a research project, and has ended up as a collection of essays with pretensions to the coherence of a book. I embarked upon it because I like eating meat and keeping livestock, and I wanted to address doubts I had about the sustainability and environmental justice of my way of life. The conflict between vegans and animal farmers has loomed large in my life: as an agricultural worker, smallholder, environmental journalist and hippie, I have frequently come into contact with both. Too many farmers have a narrow perspective of the social and environmental issues that confront us; and too many vegans have an equally limited understanding of the way nature works. It helps me, and I hope others, to get the issues down on paper.

I also embarked upon it because a gap appeared in my life where I had time to spare, but no opportunity to get away. The evidence I produce is not, alas, derived from splodging around farms in Wellington boots or trekking across savannah with nomadic herders. It consists of a trawl through what academics pompously call 'the literature', though whether it is peer-reviewed or not is not my main criterion. The farms and nomadic herds I view through the window of other people's accounts. The only time I get my hands dirty is when I try to sift out the bullshit.

This is not an easy matter. Livestock farming is a subject, I have discovered, where every answer uncovers two questions, and every statistic cloaks an ideological assumption. Shelley was perhaps right in 1813, but today the extravagance of livestock farming is no longer 'incapable of calculation'. It can be calculated, though with difficulty. Substantial sections of this book are an

---

3   Words written on the walls of a night club in Uruguay by the Tupamaros, left wing revolutionaries, in 1970.

attempt to pick a path through a crop of statistics that stretches in all directions as far as the mind's eye can see, and which I have done my best to render palatable to those who find a proliferation of percentages and decimal points indigestible.

The response to the problems addressed in this book, Roberts and Monbiot both claim, is 'plain': stop eating meat. But the role that animals and meat play in the ecology of human food production is too complex to allow for any instant ideological solution. The vegan answer is not plain, it is simplistic. My purpose in writing this book is not to expose the inadequacy of the vegan case, since only a minority are convinced by it anyway, but to hold it up as template, a clean sheet, against which the impacts and ambiguities of our role at the top of the food chain can better be studied.

This book is therefore both a critique of veganism and a tribute to vegans and vegetarians from Shelley onwards who deserve credit for initiating and widening the debate and persuading us to look at animals in different ways. They have radically influenced people's diets, and demonstrated that people can live happily without meat products; and they are starting to demonstrate that people can farm successfully without animals. Above all they have shown that meat-eating is dispensable, and that is a handy thing to know. However the fact that something is dispensable does not necessarily mean that we have to dispense with it. As I suggest in Chapter 10, the dispensability of meat is, paradoxically, a good reason for carrying on eating it.

This is not to say that I am impartial or have no ideological position. For six years in my early adulthood I was vegetarian, but I was faced with its inconsistency when I started keeping goats. What was to be done with the male kids? Being poor, I chose to eat them, and became a born-again carnivore (the worst kind). Now I favour the consumption of a modest amount of meat and dairy foods (not much more than what the FAO calls 'default' livestock production). I like keeping livestock and I support small farmers and peasants in their struggle against agribusiness. It would be foolish to pretend that these preferences haven't influenced my choice of subject matter or coloured my own interpretation of the statistics. I feel instinctively that the world would be much the poorer without domestic livestock and I want to work out why.

There are a number of other matters relating to animals which I have tried to steer clear of. I am not overly concerned with questions of dietary health, nor do I take any interest in the diet and dentition of our remote ancestors. I do have views about the ethics of killing animals and animal welfare, but as far as possible I have tried to keep them out of this book. However I see no reason to disguise the fact that I find factory farming and laboratory experiments upon animals a good deal more reprehensible than killing an animal competently with a knife.

Finally a word of advice: the environmental impact of meat and livestock is a complex matter, and this book is denser and heftier than I originally intended, possibly too much so for some readers. Although the book pursues a line of argument, each chapter is designed as far as possible to stand up as an essay in its own right, and can therefore be read as such. Four chapters (chapters 4, 5, 6 and 9) are revised versions of articles already published in *The Land* magazine. Sometimes, to avoid repetition, I have referred back to material covered in earlier chapters. But in general a reader daunted by the volume of facts, figures and references in this volume, can safely cherrypick the chapters that he or she finds most intriguing.

2

Sedentary Pigs,
Nomadic Cows, Urban Chickens

*The emergence of the great Western cattle cultures and the
emergence of world capitalism are inseparable.*
Jeremy Rifkin

*It could be said that European civilization – and Chinese
civilization too – has been founded on the pig.*
Jane Grigson

In respect of their environmental performance, carnivores didn't get off to a very good start. Round about 11,000 BC as the glaciers of the last ice age receded, five Eurasian species of large mammal (the woolly mammoth, the woolly rhino, the giant elk, the musk ox and the steppe bison) were hunted to extinction. In the Americas and Australia the record was even worse: by 7,000 BC, 32 genera of large animals, including horses, giant bison, oxen, elephants, camels, antelopes, pig, ground sloths and giant rodents had disappeared from the future New World. Nobody is quite sure to what extent their extinction was due to climate change, but there is little doubt that human predators had a hand in the matter.[1]

The problem for humans at the time was that they didn't have much else to live off (though they did appear to pick off one preferred species before moving onto the next). Once humans had learnt how to practise agriculture, and how to domesticate animals – two processes which happened more or less at the same time because the crops attracted the animals – hunting pressure could be reduced, and demand for extra meat focussed upon increasing the domesticated stock rather than killing off the wild. Domestication of sheep is thought to have first occurred in Mesopotamia around 9,000 BC, while domesticated cattle appear first in southeast Europe three millennia later. Darwin thought that the domestic pig probably first emerged in China, and there is uncertain evidence of its presence there around 8,000 BC.[2]

There are a few elementary ecological facts which explain why cows and pigs have since followed different patterns of domestication, and why pig culture can be viewed as 'Eastern' while cow culture is more 'Western'.

Cows, sheep and goats are ruminants; their digestive system, with mutiple stomachs, is designed to extract carbohydrates and proteins from low quality fibrous vegetable material, in particular grass. Horses are not ruminants, but they have a long colon which enables them to subsist on a diet of coarse grass. Though cattle can subsist in woodland, their preferred terrain, and the terrain

---

1   Ponting, Clive (1991), *Green History of the World*, Sinclair-Stevenson; Harris, Marvin (1977), *Cannibals and Kings*, Random House, p 34.
2   Watson, Lyall (2004), *The Whole Hog*, Profile Books, p 125.

where they excel, is savannah. Grassland, when it occurs naturally, is usually dry, and a relatively small percentage of the nutrients are available above ground in the form of leaves. Large grass-eating mammals have to migrate over large distances to locate sufficient food and water, and because the terrain is open, they prefer to move in large herds as a defence against predators, which in turn tend to hunt in packs.

Pigs on the other hand are omnivores, and like humans they are monogastrics (equipped with only one stomach). They cannot survive on a high fibre diet of leaves and grass, they need higher quality foods such as grains, nuts, roots, insects and carrion. Chickens are much the same. Neither animal is adapted for living in open grassland (except the specialized warthog) and pigs, who have no sweat glands and sparse fur, are not fond of prolonged sunshine. Neither pigs nor chickens migrate over long distances or move around in herds.

For these reasons, cows and horses were the animals favoured by the Kurgans, Aryans, Mongols, Huns and the other tribes who emerged out of the East and swept in waves across Russia and Europe, down into the Indian subcontinent and through the North of Africa over a period of 5,000 years until, in a sense, they were finally stopped by Charles Martel's cavalry at Poitiers in 732. We have only to imagine, briefly, the picture of Genghis Khan and his followers trying to herd thousands of swine across the Asiatic steppes, to appreciate why nomads didn't mess with pigs.

The pig, on the other hand, became the dominant domestic animal of those cultures which preferred to stay put. For the sedentary civilizations of China, and the forest dwellers and farmers of South East Asia and Polynesia, the pig was a much more sensible choice. Where there were trees and little grass it was more at home. And in areas where forests had been substantially replaced by intensive agriculture, the pig had a great advantage over the cow: because its digestive system is geared towards high value foodstuffs, a pig is about twice as efficient as a cow at turning substandard grains, waste foods and faeces into meat. Most animals are good at mopping up waste, but the omnivorous pig is king of the midden, and it is sedentary human societies which require scavengers. Nomads move on and leave their rubbish behind.

It is possible to pursue the distinction between pig-loving sedentary cultures and herbivore-dependent nomads a good deal further: for example it can hardly be a coincidence that it is Judaism, Islam, and Coptic Christianity, religions of nomadic herders, which forbid the eating of pig meat. 'Whatsoever parteth the hoof and is cloven-footed and cheweth the cud among the beasts that shall ye eat ... The swine, though he divide the hoof, and be cloven footed, yet he cheweth not the cud; he is unclean to you. Of their flesh shall ye not eat. ' These dietary rulings in Leviticus 11 are a clear endorsement for eating ruminants and little else. (One can take these observations further, and note that, by and large, the nomadic, cow-eating cultures whose people lived under a big sky, veered towards monotheism, whereas the pig-eating forest dwellers of South East Asia remained faithful to pantheistic cosmologies that saw spirits and ancestors lurking behind every tree. But that is beyond the scope of this book.)

The distinction between sedentary pig cultures and mobile cow cultures starts to become muddied when the nomads reach the point where they can go no further and are forced to settle. The Aryans who colonized India, pushing the Dravidians down to the bottom third of the subcontinent, brought their cow culture with them, and as one account puts it, they got rid of all the pigs.[3]

---

3   Spencer, Colin (1993), *The Heretic's Feast: A History of Vegetarianism*, Fourth Estate. No support is given for this assertion.

According to an interpretation best explained by Marvin Harris (which I have yet to see challenged), the population became sedentary and increased to the point where beef eating on any widespread scale was unsustainable. Pigs were not viable because Indian agriculture, much of it on dry rainfed soils, was more dependent upon oxen for ploughing than Chinese paddy, and so a cattle population had to be maintained in a country which is notoriously short of grazing lands. The solution, which developed as a result of Buddhist influence, was to keep the cow, but to forbid eating it, not because it was 'unclean' like the Semites' pigs, but because it was 'sacred'.[4]

In Africa, these problems do not seem to have occurred because there were wide areas of uncultivable grazing land and populations rarely attained the level where there were severe restrictions upon meat consumption. Prior to the arrival of European colonialists, nomadic and semi-nomadic cow-herding tribes followed established patterns of movement and observed customary rights and regulations which maintained some degree of ecological stability – but the cows, the tribes, and with them Islam, never permeated sub-Saharan Africa, probably because tsetse fly prevented their advance. Pig-rearing was practised patchily amongst sedentary populations, perhaps because Islam and the Saharan belt prevented them spreading from Asia, and perhaps because settled tribes could acquire animal protein either by hunting bushmeat, or by trading with nomads.

It was in the temperate climate of Europe that pig culture and cattle culture were to meet and combine. Much of Europe was wooded, and therefore highly suitable for pigs. Oak, beech, hazel and chestnut trees all provided nutritious mast (nuts) to fatten them on. But where the woodland was cleared away, grass grew – extremely good grass in many places, especially where the Atlantic Ocean washed the land with regular doses of light rain. Cattle thrived on these pastures, as well as sheep, which provided the wool to keep people warm through cold damp winters. And oxen were needed to plough the arable fields, many of which were heavy and clayey.

For two or three thousand years pigs, cattle and sheep (as well as horses, goats and chickens) seem to have thrived side by side with each other in Europe, and often complemented each other. Sheep, which grazed the rougher land, manured arable land that the oxen ploughed; pigs were fed on the whey that resulted from the peculiarly European method of making hard cheeses to keep through the cold winters when there was little milk. Towards the end of the Middle Ages when populations began to outstrip carrying capacity, there were signs of destructive competition between different elements of this mixed farming system. The shortfall in food production was exacerbated by a gradual deterioration in the fertility of the arable land, due to an insufficiency of nitrogen. Ploughing up more land was no remedy, since it decreased the amount of pasture available to provide manure.

Population pressure was relieved by the Black Death in 1350, which meant that much of the exhausted cornland could be turned back to pasture. As the population grew back to its former levels, one solution to population pressure was found in new crops of beans, turnips and nitrogen-bearing fodder varieties such as clover and sainfoin, which enabled more animals to be kept over winter, without diminishing the area devoted to corn. The other was the colonization of the New World which opened up new lands for emigrants who grew food and fibre to send back to the home country. European farming was to undergo further ups and downs, more the consequence of capitalist opportunism than of

---

4   Harris, Marvin (1986), *Good To Eat*, Allen and Unwin, pp 47-66.

ecological constraints. But in the interim more momentous events were taking place overseas. The cattle nomads whose westward drift had been held up for 1000 years on the Atlantic seaboard, were on the move again.

As we have seen, the Pleistocene extinctions had been more ruthlessly carried out in the Americas than in Eurasia. Virtually every large mammal that was potentially domesticable had been made extinct, and after the disappearance of the bridge of land across the Bering Straits there was no means of getting them back. In the Andes the llama and alpaca were domesticated by the Incas. In North America, if it was possible to domesticate the bison without the aid of the horse, native Indians had shown little interest in trying. As far as domestic animals were concerned, the New World presented itself as a tabula rasa to the European colonialists, upon which they could stamp whatever sort of animal economy they saw fit.

Jeremy Rifkin, in his flamboyant polemic against the cattle industry, *Beyond Beef*, expounds what might be called the 'bovine prerogative' theory of American history. The manifest destiny of the pioneers, he suggests, was the proliferation of an atavistic cattle culture whose roots lay in the rituals of the 'neolithic cowboys' who had migrated across the Steppes, in the 'Mithraic blood sacrifices' of ancient Rome, and in the Celtic 'warrior bull cult' (not, I might add, to be confused with 'the boar cults of the Celts, which identify fearsome warriors as boars with giant tusks'.[5] Rifkin writes:

> Centuries before Melville's Captain Ahab battled with the great white whale, Spanish matadors were already apprenticing for man's new role on the world scene, facing down the 'forces of nature' in dusty arenas in scores of small village towns on the Iberian Peninsula. Spanish explorers transported the ancient Iberian cattle complex to the shores of America in the 16th century. The Spanish conquerors of the New World bore a striking resemblance to the fierce nomadic tribesmen of the Eurasian steppes who had set out to conquer Europe over 5,000 years earlier.[6]

Columbus, Rifkin notes, on his second voyage, unloaded in Haiti '24 stallions, ten mares and an unknown number of cattle', and Cortes introduced longhorn cattle to Mexico, as if that was an indication of what was to follow. But Rifkin neglects to mention that on the same voyage, Columbus also brought 'eight sturdy Iberian pigs', while Cortes entered Mexico City at the head of a cavalcade which included a drove of Spanish swine.[7] The Plymouth Brethren arrived without cows, but with six goats, 50 pigs, and many hens – not surprising, since they were arriving on a well-wooded seaboard. Cattle didn't arrive until the following year. Ten years later, the Massachusetts Bay colony boasted '1500 cattle, 4000 goats and innumerable swine'. Imported pigs were allowed to run loose over most of northern Manhattan Island, and the stockade which kept them away from the farms later gave its name to Wall Street.[8]

Cattle spread up through the drylands of Mexico, but the predominance of pigs over cattle was to persist on the East coast, while the Southern states, from Virginia to Louisiana, became famous for their hams and fatback. A Southern physician called Dr John Wilson complained that people in the USA ate three times as much pork as Europeans: 'The United States of America might properly

5   Watson, *op cit*. 2, p 186.
6   Rifkin, Jeremy (1992), *Beyond Beef*, Dutton, p 41.
7   Watson, *op cit*. 2, p 140.
8   *Ibid*, p 143.

be called the great Hog-eating Confederacy, or the Republic of Porkdom.'[9] By the second half of the 18th century, the eastern states were salting surplus pork and shipping it to Europe, a century before chilled beef was able to make the same journey. In the early 19th century, up to half a million pigs a year were driven from as far away as the Ohio river to New York and Philadelphia, along 'well-worn hog trails as clearly marked and as famous as the cattle trails of the Southwest'[10] – although Hollywood has yet to come up with a hogman movie. Even during the cattle bonanza years of the late 19th century, the pig industry held its own: 'It was the runaway success of homesteaders' hogs that stopped them from dreaming about ranches in Oklahoma or gold in California'.[11] Eventually, as the midwest began producing increasing surpluses of corn, the hog frontier and the salt-pork barrelling industry advanced to Cincinnati and finally Chicago. In the United States, just as in the Old World, the folks who settled in the east stuck with pigs, and they outnumbered the pioneers who moved west with their cattle.

Right up until the 1950s, US citizens ate more pork than they ate beef, though you would never have guessed this from reading Rifkin. The two factors that finally tipped the balance in favour of beef had nothing to do with land use, still less a macho cow cult, but reflected the needs of the meat packing industry. The first was the introduction of refrigeration which allowed cattle carcases to be trimmed and packed in Chicago slaughterhouses, rather than having to be distributed on the hoof in cattle trucks to butchers in the East. Refrigeration took away the main advantage of pork – that, being fatty, it tasted better when salted, and so could be packed and preserved without refrigeration; and it meant that Chicago packers could take over and centralize the lucrative beef byproducts industry.[12]

But it was the rise of the hamburger that finally tipped the scales in favour of beef. As far as the meat industry was concerned the hamburger carried out one crucially important function. Because it consisted of ground beef, meat from different cows could be mixed together. Lean beef, from the Western rangelands, or from Central or South America, could be made more palatable by mixing it with fatty beef fed on grains in feedlots. There was no reason why the lean beef could not have been mixed together with a small amount of fatty pork; this would have been more efficient in terms of grain use and probably cheaper. But such a hybrid would not have conformed to the statutory definition of 'hamburger'. Marvin Harris, after making unsuccessful inquiries to the US Department of Agriculture as to how the Federal Code's definition was decided, concluded:

> The exclusion of pork and pork fat from hamburgers suggests that beef producers had more influence in government than pork producers.
> If true, this would be the natural outcome of a basic difference in the organization of the two industries which has persisted since the late 19th century. Beef production has long been dominated by a relatively small number of very large ranches and feedlot companies, while pig production has been carried out by a relatively large number of small to medium farm units … To sum up, beef achieved its recent ascendancy over pork through the direct and indirect influence of all-beef hamburger.[13]

---

9   Cited in Adams, C (2000), *The Sexual Politics of Meat*, Continuum Publishing, p 169.
10   Watson, *op cit*. 2, p 166.
11   *Ibid.*, p 166.
12   Ross, E (1980), 'Patterns of Diet and Forces of Production: An Economic and Ecological History of the Ascendancy of Beef in the US Diet', in E Ross (ed), *Beyond the Myths of Culture: Essays in Cultural Materialism*, Academic Press.
13   Harris, Marvin (1987), *The Sacred Cow and the Abominable Pig*, Touchstone, p 126.

If Harris is right, the definition of this one word is responsible for one of the biggest ecological cock-ups in modern history, the beef feedlot industry, which pumps vast amounts of corn and forage from irrigated pastures down the throats of animals that are least able to process it efficiently. One wonders also whether the Jewish lobby did not have some influence upon the definition, since Harris adds:

> Ground pork can be eaten, ground beef can be eaten; yet to mix the two together and call it a hamburger is an abomination. It all sounds suspiciously like a rerun of Leviticus … In mediating the age-old struggle between pigs – consummate eaters of grain – and cattle – consummate eaters of grass – the USDA had followed ancient precedents.

Over the last 20 years, the grain-fed meat industry has spread from the USA and other wealthy countries to emerging countries in the Third World, but it has not taken much beef with it. Brazil is producing increasing quantities of feedlot chicken, much of it for export to Europe. China's consumption of pork and poultry has shot up, and she is having to import grain in order to meet demand. India has outstripped the USA as the world's largest producer of dairy products, a feat that has been achieved without a massive increase in feed-grain input. Where developing countries, such as Argentina, Botswana or Brazil, have developed beef industries, they have been based on grazing, rather than on grain. The USA's excessively grain and hormone fed beef is not the norm, but an aberration, reflected in the fact that both Canada and the EU refuse to buy it.

The 6,000 year long movement of cowboy culture westward has not yet come to an end. Its last frontier is in the Amazon where it is probably wreaking more havoc than anywhere previously. But it is 'atypical' and hopefully soon on the wane. The world's livestock consumption is currently broadly split into four quarters: beef, dairy, pork and poultry. Beef is declining in popularity, while pork, and more especially poultry, are on the ascendant. 'Ruminant production' the FAO remarks, 'both meat and milk, tends to be much more rural-based,' because 'ruminants' higher daily fibre requirements entail bulk movement of fodder.'[14] As human civilization urbanizes, and more people are crammed into larger and denser concentrations, so our livestock become urbanized, and they too are crammed into larger and denser concentrations. The pastoral steer and the rural cow yield to the agrarian and proto-urban pig, which in turn is now yielding to the megapolitan broiler hen.

Nonetheless the rivalry between ruminants and monogastrics is far from exhausted. The 'white revolution', which has quadrupled milk production in India and made her the world's largest dairy producer, vies for importance with China's booming pork and poultry industry. In overdeveloped countries, battle lines are hardening between those who advocate 'grass farming' as the most ecological and humane approach to animal husbandry, and those who hold that the world's appetite for meat can only be met by feeding grain to pigs and chickens in 'confined animal feeding operations'. This is a separate conflict from the ethical dispute between carnivores and vegans. Yet beyond factory-farmed chicken lie species of lab-cultured meat that the writers of Leviticus never even dreamed of, and that is where the interests of agribusiness and of vegans may one day converge.

---

14   Steinfeld, H *et al* (2006), *Livestock's Long Shadow*, FAO.

THE LAND REQUIREMENTS

OF LIVESTOCK

3

AN ACRE A MEAL?

'The monopolising eater of animal flesh would no longer destroy
his constitution by eating an acre at a meal ...'
Percy Shelley, A Vindication of Natural Diet

T he primary environmental objection to meat-eating has always been that it is inefficient: in other words, it uses disproportionate amounts of the world's land and resources. Since 2006, this matter has become somewhat eclipsed by concern about carbon emissions from animal production causing global warming. However, livestock are responsible for a relatively small proportion of total greenhouse gas emissions, whereas they occupy well over half of our agricultural land. This chapter, and the following four, are about the land requirements of livestock.

It takes anything from three to twenty kilos of vegetable nutrients fed into an animal to produce one kilo of nutrients in the form of meat. This, it is argued, is both unsustainable and unjust in a world of six billion people, moving towards nine billion, where many people already go hungry. There is not enough land to feed all the world's population to the standard enjoyed by the carnivorous industrial countries, and so the meat industry is a way of diverting food from poor people to rich. A shift to a vegan diet would release large amounts of protein and food resources.

All of this is substantially correct, and it is a charge which every meat eater should consider. However it is a charge which could be levelled against a good many vegetable products. The production of strawberries, baby sweetcorn, and asparagus, not to mention coffee, tea, wine and chocolate, may be just as nutritionally inefficient, probably more so – and, as third world development agencies have often testified, industries of this kind often take land away from the poor which could otherwise be used for growing staples. All luxuries are, by definition, extravagant. There is a suspicion here that some who single out meat for environmental stricture on this account do so to add weight to previously held moral objections to killing animals.

Nonetheless, meat can be distinguished from other luxuries in a number of ways. Firstly it is, potentially, a staple food, providing protein, fat, vitamins and carbohydrates. You can live entirely on meat but you can't live entirely on strawberries or coffee. It is not just a luxury, it is a luxury staple – and indeed is promoted as such by advocates of the Atkins diet. It therefore poses a direct and extravagant alternative to grain based diets.

Secondly, meat is distinguishable by the sheer scale of the industry. According to an analysis carried out in the 1990s by Friends of the Earth, out of the 3.9 million hectares abroad which provide the UK with imported food, 2.3 million supply meat or animal feeds – against 1.6 million hectares supplying oil, fruit, coffee, chocolate etc.[1] About 50 per cent of arable land in the UK is devoted to meat.[2] If the UK reduced or eliminated its meat consumption, it would have ample land on which to produce its own food, and no need to import grain and meat from other countries, releasing land there to produce more food for poorer people. If everybody in the world stopped eating meat, there would be enough land for everybody to obtain sufficient quantities of protein, carbohydrates and fat, and to enjoy coffee and strawberries as well.

Put in these simple terms, the case against meat seems irrefutable. But the way the world operates is more complicated than that: inefficiencies often have a role to play in the greater scheme of things. The next few chapters are an attempt to unravel that complexity, to find out how inefficient meat really is, and also whether that inefficiency serves any purpose.

Measuring Efficiency

The carnivorous human chooses to eat at the top of the food chain. In *Beyond Beef*, Jeremy Rifkin illustrates this with a hierarchy consisting of grass, grasshoppers, frogs, trout and humans.

> At each stage of the food chain, when the grasshopper eats the grass, and the frog eats the grasshopper, and the trout eats the frog and so on, there is a loss of energy. In the process of devouring the prey, about 80-90 per cent of the energy is lost as heat to the environment. Only between 10 and 20 per cent of the energy that was devoured remains within the tissue of the predator for transfer to the next stage of the food chain. Three hundred trout are required to support one man for a year. The 300 trout, in turn, must consume 90,000 frogs, that must consume 27 million grasshoppers, which live off 100 tons of grass.[3]

In one respect this doesn't seem too reprehensible. It may take several acres of grass to keep a human alive, but the grass also supports millions of grass-hoppers, thousands of frogs and hundreds of fish. By choosing to eat fish and large herbivorous mammals, humans are assuming a position which otherwise is occupied by bears, large cats and wolves. There would be nothing intrinsically wrong or unnatural in this but for the fact that there are a lot of humans, while the pyramidal nature of the food chain requires that there can only be a limited number of predators and scavengers. There is currently not enough meat around for every human to eat at the level at which it is currently consumed in the USA and other wealthy countries.

Even worse, an increasing quantity of the food consumed by cows, pigs, sheep and poultry consists of grains and pulses – high quality vegetable protein and carbohydrates which could be consumed directly, and therefore less wastefully, by humans. Since there are some 800 million malnourished people in the world who do not get enough grain to eat, let alone meat, food is effectively being taken out of the mouths of these people to feed to domestic animals for the inefficient nourishment of a much smaller number of wealthy people.

Rifkin's observation, in the paragraph cited above, that 'between ten and twenty per cent of the energy that was devoured remains within the tissue of the predator for transfer to the next stage of the food chain' is a measure of feed

---

1   McLaren, D *et al* (1998), *Tomorrow's World*, Earthscan, p 144.
2   *Annual Abstract of Statistics* (2009) 21.3 'Agriculture Land Use'; DEFRA (2009) *Agriculture in the UK*, Table 5.2.
3   Rifkin, J (1992), *Beyond Beef*, Dutton, p 60, citing G Tyler Miller. Quote slightly edited.

efficiency. This percentage can also be expressed as a ratio – the 'feed conversion ratio'. So, for example, 20 per cent is the equivalent of 5:1, meaning that five units of nutrition fed to an animal produce only one unit of nutrition when that animal is eaten as meat.

This seems simple enough, but the matter soon becomes riddled with variables and complexities, such as the following:

(i)   Different animals have different digestive systems, and so perform differently from each other.

(ii)   The same animal can perform differently under different circumstances.

(iii)   'Feed' can mean all food ingested by the animal, including grass, tree leaves and crop residues – or else only the food which could otherwise have been eaten by a human.

(iv)   The feed conversion ratio is expressed in a number of different ways, which employ different units of measurement, for example:

- protein in the feed compared to protein in the meat;
- energy (calories) in the feed compared to energy in the meat;
- weight of standard feed compared to weight of the live animal;
- weight of standard feed compared to the weight of edible meat in an animal;
- amount of land required to grow a given quantity of vegetable nutrients compared to the amount of land required to produce meat of the same nutrient value (expressible either as proteins or as calories, or an amalgam of both).

Of these, it is the last that really matters, because land is the limiting factor. If there were boundless acres of land for everyone, then it would make relatively little difference to the hungry and poor of the world what the metabolic conversion rate of livestock was. Unfortunately land is in short supply, or at least subject to strong competing demands, and the metabolic conversion rates can be helpful, though rarely decisive, in telling us what we need to know – how frugally or extravagantly land is being farmed. What we are ultimately after is a 'land-take ratio' telling us, for example, that a hectare will produce x times as much human nourishment when put down to corn as it will when given over to livestock.

A further problem is that different commentators, for no obvious reason other than their ideological allegiance, regularly come up with different figures. Almost every item of literature on the environmental impact of meat cites a typical feed conversion ratio, or else a range, but the ratio given can be anything from 1:1 to 20:1. Anyone who dares to research the matter further finds themselves facing a mass of conflicting statistics, incompatible units, shifting variables and ill-defined terms. Understandably, popular writers and members of the public like to alight on one safe, handy figure for assessing the extravagance of a meat diet in comparison to a vegan diet, and frequently they veer towards the ratio of 10:1.

The earliest (and most elegantly expressed) use of the ten-to-one figure I have located is in Shelley's *A Vindication of Natural Diet* where he observes:

The quantity of nutritious vegetable matter consumed in fattening the carcase of an ox, would afford ten times the sustenance … if gathered immediately from the bosom of the earth.[4]

Here for example is an early modern statement, in respect of protein, from Peter Singer:

> Assume we have one acre of fertile land. We can use this acre to grow a high-protein plant food, like peas or beans. If we do this, we will get between 300 and 500 lbs of protein from our acre. Alternatively we can use our acre to grow a crop that we feed to animals, and then kill and eat the animals. Then we will end up with between 40 and 55 lbs of protein ... Most estimates conclude that plant foods yield about ten times as much protein per acre as meat does.[5]

The attraction of this rule of thumb, irrespective of which units of measurement and methodology are employed, is so marked that I have made a collection of examples, a number of which are given in the panels overleaf, to demonstrate the variety of approaches that can be taken.

Although common, use of the 10:1 formula is by no means universal. Many writers take pains to provide a range of conversion ratios specific to different species of animal, or in order to distinguish between protein content and energy content. The trouble is that different writers come up with different figures, which are often a reflection of their ideological position.

For example Marvin Harris, a pro-carnivore anthropologist, writes: 'it takes four grams of protein in the grains to produce one gram of protein in the meat.'[6] Ingrid Steen, a Danish agronomist, states that 'the cereals to meat conversion ratio in intensive animal husbandry is 6 to 1 for red meat'.[7]

On the other hand Frances Moore Lappé, author of *Diet for a Small Planet*, reports: 'For every 16 pounds of grain and soy fed to beef cattle in the United States, we only get one pound back in meat on our plates ... 16 pounds of grain has 21 times more calories and eight times more protein – but only three times more fat – than a pound of hamburger.'[8] So her protein conversion ratio for beef is 8:1 while her energy ratio is 21:1. Jeremy Rifkin, author of *Beyond Beef*, considers that 'Cattle have a feed protein efficiency of only 6 per cent [more than 16:1].'[9] And the Canadian expert on nitrogen fertilizers, Vaclav Smil, is vague about the efficiency of beef: 'Protein conversion efficiencies [for beef] are just 5 to 8 per cent ...' which is between 12.5:1 and 20:1. In an accompanying visual chart, more noticeable than the sentence in the text, he classifies beef at the top figure of 20:1.[10]

The average of all these deviants is still around 10:1, and there is no doubt that this figure acts as a strange attractor for commentators seeking a convenient rule of thumb; and, more importantly, it has been lodged in the mind of members of the public, from Shelley onwards, who take an interest in these matters. It is not surprising that the easiest and most memorable number should slip into common public usage. What is less clear is why the weight for weight figure for grain-fed meat should vary from 6:1 (Steen) and 9: 1 (Rifkin) to 16:1 (Moore Lappé), while the protein ratio for beef ranges from 4: 1 (Harris) and 8:1 (Moore Lappé) to 20:1 (Smil).

---

4   Shelley, Percy (1813), *A Vindication of Natural Diet*, London, p 20.
5   Singer, Peter (1990), *Animal Liberation*, Pimlico, 2nd edition, p 165.
6   Harris, Marvin (1987), *The Sacred Cow and the Abominable Pig*, Touchstone, p 21.
7   Steen, I (1988) 'Phosphate Recovery', *Phosphorus and Potassium* 217, Sept-Oct 1998, http://www.nhm.ac.uk/research-curation/projects/phosphate-recovery/p&k217/steen.htm
8   Lappé, Frances Moore (1982), *Diet for a Small Planet*, Ballantine.
9   Cited in Rifkin (1992), *op cit*.3, p 161.
10   Smil, Vaclav (2004), *Enriching the Earth*, MIT Press, p 165.

TEN TO ONE — A STRANGE ATTRACTOR

Below are some examples of the 10:1 ratio, as it appears in available literature. The first demonstrates the use of a 10:1 protein ratio in a highly charged vegan polemic.

> *The key to this extraordinary industry is the inefficiency of meat production – about 10 kg of prime vegetable protein to produce just 1kg of meat protein. Times that by the 45 billion animals barbarically slaughtered last year and the equation begins to take on an obscene dimension.[1]*

Here is a figure provided by a website page entitled 'How to Win an Argument with a Meateater':

> *Percentage of protein wasted by cycling grain through livestock: 90.[2]*

Michael Pollan gives a similar ratio for meat-eating, but focussed on calories:

> *It's true that prodigious amounts of food energy are wasted every time an animal eats another animal – nine calories for every one we consume.[3]*

This is British permaculturist Patrick Whitefield, who refers to land use:

> *Typically only 10 per cent of the food consumed by farm animals is available as edible meat when they are killed. The ratios for milk and eggs are somewhat better, but similar. This is obviously inefficient. A hectare of land can support ten times as many people on a vegan diet than on a purely carnivorous one.[4]*

In fact the ratios for milk and eggs are not at all similar to that of beef, they are universally agreed to be a great deal better than 10:1. Here is another example of the 10:1 ratio expressed in units of land, which in this case explicitly refers to grazing:

> *A given area of suitable land can produce up to ten times more food by growing plants, than by grazing animals on it.[5]*

In the next example, Jeremy Rifkin, who is usually scrupulous about facts, makes it clear he is talking about feedlot beef :

> *It takes nine pounds of feed to make one pound of gain in a feedlot steer, with six pounds of this consisting of grains and by-product feeds and three pounds of roughage. Only 11 per cent of the feed goes to produce the beef itself, with the rest either burned off as energy in the conversion process, used to maintain normal bodily functions, or excreted or absorbed into parts of the body system that are not eaten – like hair or bones.[6]*

---

1   Wardle, Tony (n.d.), 'Hype, Hypocrisy and Hope', *Growing Green International* 9. Cf also: 'because crop production yields ten times more food than animal farming …' Dave from Darlington, 'Horse Power ', *Growing Green International* 5.
2   Found at: www.vegsource.com/how_to_win.htm
3   Pollan, Michael (2006), *The Omnivore's Dilemma*, Bloomsbury, p 199.
4   Whitefield, Patrick (2004), *Earthcare Manual*, Permanent Publications, p 258.
5   Dave from Darlington (n.d.), 'Sustainable Agriculture', *Vohan News International* 3.
6   Rifkin (1992),*Beyond Beef*, Dutton, p 160.

Nick Fiddes gives a range of values:

> *The conversion of grain into animal flesh requires about ten calories for every calorie provided for human consumption, or five grams of protein input for the production of one gram of meat protein; for beef the ratio is more like twenty to one.*

But when he is required to generalize, he comes out with this:

> *it makes about ten times more sense in efficiency terms to eat the grain, than feed it to the cows and then eat them.[7]*

Even the advocate of meat-eating, Marvin Harris, opts for 10:1 in respect of human energy input:

> *The net calorie return on each calorie of human effort invested in plant production is on the average about ten times greater than the net calorie return obtainable from animal production.[8]*

And elsewhere he gives 10:1 as the conversion factor for calories in grain:

> *It takes nine additional calories to provide one calorie for human consumption when grain is converted into animal flesh.*

Although the 10:1 figure is infectious, this Professor of Political Science at the University of Maryland took a bit of convincing:

> *In 1990, when I first read that ten people could be fed with the grain that you would feed a cow that would be turned into the food for one person, I was impressed. But I was not moved … I thought, if I give up meat, it won't have that impact: it probably won't have any impact on anything at all, except me. I was wrong.[9]*

But here is a student journalist who has no hesitation about pasting the ten to one estimate into her copy, and appending some rather unconvincing scientific authentication:

> *Scientifically, it's clear that a carnivore diet is less energy-efficient. Each step on the food process has a ten percent efficiency rate. The first step is the food producers, or the plants. The next phase on the meat-eating ladder is feeding the plants to the livestock. Then, that livestock is fed to humans. If that middle step were cut out, the world's crops could conceivably support 10 times more humans. 'We've been aware of this for a while' said Johannes Feddema, associate professor of geography at the University of Kansas.[10]*

---

7   Fiddes, Nick (1991), *Meat a Natural Symbol*, Routledge, p 211.

8   Harris, Marvin (1991), *Cannibals and Kings*, Vintage, p 193.

9   Boyan, Steve (n d) *How our Food Choices can Help Save the Environment*, http://www.earthsave.org/environment/foodchoices.htm

10   Ashley Thompson, *Beef: It's What Affects Global Warming*, Multimedia Reporting (Bradford-Utsler) 10 May 2006, http://reporting.journalism.ku.edu/spring06/bradford-utsler/2006/05/beef_its_what_affects_global_w.html

One way to settle this matter is to turn to the feed calculations in an agricultural handbook, since writers of these pricing manuals for farmers do not have an ideological axe to grind. Nix's *Farm Management Pocketbook* has tables which show that for UK grain fed beef, raised from 115 kilos at 12 weeks to a slaughter weight of 530-550 kilos, the feed conversion ratio of grain to live animal is 5:1 in an average enterprise, and 4.6:1 for a high-yielding outfit.[11] Assuming that 55 per cent of the beast ends up as meat that is a conversion ratio of feed to meat of between 9.1:1 and 8.4:1. Bearing in mind that we are now comparing weight of feed and meat, rather than protein content, Rifkin appears to be about right.

However – and this is where room for error or manipulation creeps in – that accounts for only the final 415 kilos of the animal's growth. The first 115 kilos of growth, much of which occurs in the mother's womb, is a more complicated matter. Since a cow usually has only one calf per year, it is necessary to attribute to the calf's feed budget the upkeep of its mother over an entire year, plus a fraction of what it cost to rear the mother to calf-bearing age, minus a fraction of the mother's eventual meat value. If the mother is a suckler beef animal, then she will probably have been living on grass; but the inefficient feed conversion ratio can still be invoked, and a zealous analyst can quite easily find a way of determining the feed to live weight ratio of the 115 kilo calf to be anything up to 30:1 – which in turn makes the feed to meat ratio of the finished animal around 19:1.

However a 12 week old calf weighing 115 kilos delivered to a feed lot is unlikely to come from a beef suckler herd, it is more likely to be a byproduct of the dairy industry (Nix specifies that his figures are for dairy offspring). The amount of energy a cow puts into its calf when it is in the womb is relatively insignificant, the main expenditure is when the calf is suckling. Even so, a dairy cow can rear to three months a good deal more than one calf. According to a dairy manual written in 1946, when dairy yields were about half what they are now, 'when milking freely a foster-cow may be suckling as many as four calves at any one time, and during her lactation she may suckle eight or even a dozen'.[12] Rearing a calf up to the weaning age of three months certainly represents no more than a tenth of the upkeep of a modern high yielding dairy cow, plus 150 kilos of solid feed – in total probably somewhere around 700 kilos of grain equivalent. This suggests that the first 115 kilos of a beef calf emanating from the dairy industry are produced at a feed to live weight ratio of about 6:1, which is a feed to meat ratio of about 11:1. Adding this to the 9:1 ratio for feeding of the animal after weaning, we arrive at an overall figure of ... 10:1.[13]

The same calculation for pigs, based on Nix again, is a lot easier. An average sized pig yielding 64 kilos of meat requires 257 kilos of feed, including a proportion of the food given to the sow that nursed it and the boar that fathered it. This works out at a weight for weight feed to food ratio of almost exactly 4:1.

---

11   Nix, John (2007), *Farm Management Pocketbook*, Imperial College London, p 96.
12   Garner, F (1946), *British Dairying*, Longmans, p 143.
13   That figure is for fattening a calf derived from the dairy industry. Fattening a calf from a dedicated beef herd is arguably more likely to produce meat at a ratio of anything up to about 20 to 1 – but much of the feed is likely to be rangeland grass that humans cannot eat. Nix (2007), *op cit. 11*, p 89; Belyea, R *et al*, *Using NDF and ADF to Balance Diets*, University of Missouri Extension, 1993, http://muextension.missouri.edu/explore/agguides/dairy/g03161.htm; *Dairy Cattle Nutrition*, Penn State, http://www.das.psu.edu/dairynutrition/nutrition/tables/

CAST versus CIWF

The ten-to-one rule therefore seems to be about right for beef. Shelley was apparently correct when he pronounced that eating plants 'gathered from the bosom of the earth' was ten times as efficient as eating the meat from an ox. But beef represents less than 20 per cent of all the animal protein consumed by humans, and all the other major sources of animal protein – pigs, poultry, dairy and fish – are universally agreed to have feed conversion ratios markedly lower than beef.[14] This is in part because a sow has up to 20 piglets a year, and hens and fish are even more prolific, whereas a cow normally only has one calf, and the costs of maintaining the mother are reflected in the meat. However, this same lack of fecundity also helps to make the cow more efficient than the pig in another respect. Whereas pigs have at least a dozen teats, cows only have four (and goats only two) making them a good deal easier to milk – and milk production is more efficient than meat production, largely because you don't have to kill the animal to get it.

During the course of my research, I opted to take as a starting point a set of figures for feed conversion which I found in a report entitled *The Global Benefits of Eating Less Meat*, written by Mark Gold for Compassion in World Farming (CIWF). The table states that ten kilos of standard animal feed are required to produce one kilo of edible beef; 4 to 5.5 kilos to produce a kilo of pork; 2.1 to 3 kilos for poultry; and 1.5 to 2 kilos for farmed fish (which are more efficient because they are cold-blooded). Unfortunately no figures are provided for dairy.[15]

I chose them, first because they include the standard 10:1 figure for beef which appears to be correct in the UK; second, because the figures given for other animals are around a broadly acceptable median of all the figures available; and third, because the figures were referenced in the CIWF report as coming from a document called *Contribution of Animal Agriculture to Meeting Global Human Food Demand*, from the Council for Agricultural Science and Technology (CAST), published in 1999. CAST represents a large sector of the United States food industry, including organizations with an interest in meat production such as the American Meat Science Association, the American Forage and Grassland Council and the American College of Poultry Veterinarians – and hence at the opposite end of the ideological spectrum from CIWF. These are therefore figures that are accepted by authoritative writers on both sides of the vegan/carnivore divide.

Or so I thought. However, I was unable to get hold of the CAST document in Britain, either through the library system or through CIWF. After some delay, I contacted CAST, who informed me that a document of that name didn't exist, at least not published by them; but they had published another report in the same year with a similar title: *Animal Agriculture and Global Food Supply*.[16] Assuming this to be the same document under a revised title, I ordered a copy, fully expecting the CIWF chart to be in it. It wasn't. After further enquiries, to CAST and to CIWF, I am still unable to establish where the figures come from. CIWF have told me that they cannot work out where they got them from. I know the feeling!

---

14   Seré, C and Steinfeld, H (1996),*World Livestock Production Systems*, Animal Production and Health paper 127, FAO; cited in Council for Agricultural Science and Technology (CAST) (1999), *Animal Agriculture and Food Supply*, 1996, p 25.

15   Gold, Mark (2004), *The Global Benefits of Eating Less Meat*, Compassion in World Farming (CIWF), p 23.

16   Council for Agricultural Science and Technology (CAST) (1999), *Animal Agriculture and Food Supply*.

The CAST document provides a slightly different and more detailed set of figures (see table 1) for six different countries, including the USA. These relate to levels of nutrients in feed and meat (whereas the CIWF table doesn't tell us how much protein or energy there is in the feed, or in the meat); and, surprisingly perhaps, they show meat production to be slightly more inefficient than CIWF alleges. The mean figure for beef of 13.4:1 reflects a US beef industry which performs less well than the UK husbandry examined in Nix's agricultural handbook.

Table 1: Conversion Ratios of Animal Feed to Human Edible Food in the USA

Meat	Energy	Protein	Mean
Beef	14.3: 1	12.5:1	13.4:1
Pork	4.75:1	5.25:1	5:1
Poultry	5.25:1	3.2:1	4.25:1
All three meats	7.2:1		
Eggs	5.9:1	4.2:1	5:1
Milk	4:1	4.75:1	4.4:1

(The average for all three meats is about 7.2:1; if we were to include eggs, milk and fish the average ratio would be lower still. Since 1993 the feeding of grain to pork and poultry has increased much faster than the feeding of grain to beef so the average feed conversion ratio for all meat is dropping. Sheep, goats and other ruminants are omitted because they represent a comparatively small proportion of the global meat industry, and their performance is broadly similar to that of cattle. Source: Figures derived from tables 4.17 to 4.24 in CAST, Animal Agriculture and Food Supply, 1999.)

Although CAST's feed conversion ratios are higher than CIWF's, it is what each does with them that makes the difference. Both sides use them to support their ideological positions, and in both cases they misrepresent the real state of affairs.

CIWF present their figures in a table with a comment in the main text, that 'from table 3 it is evident that to promote vast increases in meat production as an answer to world hunger has one overriding limitation – it depends upon an inefficient (and also a relatively expensive) product.' While this is true, the report fails to note that there are a number of other factors which combine to lower the ratio between meat and grain, to the point where it is a good deal more slender than their table suggests.

CAST, on the other hand, adjust their conversion ratio downwards by whatever means they can find, fair or foul. Mostly these means are quite reasonable; but by a statistical sleight of hand, which I shall come to in due course, they manage to conclude that there is no practical difference in performance between meat and grain consumption, and 'thus diverting grains from animal production to direct human consumption would, in the long term, result in little increase in food protein.' This is a flagrant lie.

To understand how both sides have applied the figures, it is necessary to examine in some detail the main ways by which the feed conversion ratios in the table above have to be adjusted downwards, and under what circumstances. There are three subsidiary reasons for reducing the ratios (relating to the nutritive value of meat, the byproducts of meat, and the superior yields of some animal feed crops) and one overriding reason – the ability of animals to consume food that humans can't eat.

Nutritional Value

The nutritional value of meat is different from that of grain, and varies according to the proportion of muscle to fat. Weight for weight, lean meat tends to have more protein than grain, but slightly less energy, while very fatty meat can have more energy than grain, but less protein.

The quality of most meat protein is generally regarded as higher than that of some vegetable protein, for example rice which is low in lysine. According to Pimentel, meat protein is '1.4 times more nutritious for humans than the comparable amount of plant protein,' and CAST agrees 'the biological value of protein in foods from animals is about 1.4 times that of foods from plants.'[17] Since Pimentel and CAST are on opposite sides of the argument, that much seems agreed, and if we applied it to our beef protein ratio of 12.5:1, that would bring it to close to 9:1.

However the 1.4 figure refers only to protein, not to energy, and in most diets energy is the critical factor, not protein. Anybody deriving sufficient energy from a diet based on grains and pulses, is likely to be ingesting adequate quantities of proteins. However, when the staple is a root crop, such as cassava or sweet potatoes, is there a strong possibility that a diet supplying sufficient calories may be deficient in protein. Phenomenal quantities of potatoes have to be eaten to provide a protein rich diet, and where potatoes are the main staple for poor people, milk or some other animal food is usually consumed as a supplement. A relatively small amount of meat or dairy can help to ensure adequate intake of nutritive elements such as vitamins A and B, calcium, iron and zinc, in circumstances where it is hard to obtain these through a balanced vegan diet.

There is also value in meat and dairy produce because they add variety to what might otherwise be a boring diet. In circumstances where good quality vegetable oil is not easily obtainable – for example Papua New Guinea or northern Europe – the fat content of meat or dairy is particularly highly valued. A vegan society, unless desperately short of land, would normally devote some of it to producing low yielding nuts or other delicacies to add variety to its staple diet, and some value must be accorded to meat for providing the same service. This is an important factor but there are diminishing returns. A society which eats meat only on occasional feast days clearly rates the variety factor very high indeed, whereas the average North American eating over half a pound of meat a day barely notices it. The provision of variety is arguably not a matter which should be factored into the equation, but one that should be given weight after comparing conversion factors, as in this imaginary example:

> Under prevailing circumstances in the community, it takes four kilos of feed to produce one kilo of pork: however villagers view that consigning 20 per cent of all their grain towards the production of a volume of pork which represents barely 6 per cent of their total food intake is worth it in terms of the variety added to the diet.

Bearing all this in mind, I am inclined to take the nutritional value of animal food as being about 1.2 times as great as that of grains.

---

17   CAST (1999) ibid., p 42. And Pimentel, D and Pimentel, M, 'Sustainability of Meat-based and Plant-based Diets and the Environment', *American Journal of Clinical Nutrition*, Vol. 78, No. 3, 660S-663S, September 2003. Neither author gives any substantiation for this figure. The Pimentels refer the reader back to Pimentel, D and Pimentel, M (1996) *Food, Energy and Society*, Colorado University Press, but I couldn't locate the figure in this book via the index.

Byproducts

In one of the 'ten-to-one' quotations cited in the panels on pp 16 and 17 Jeremy Rifkin mentions animal byproducts – and discounts them:

> Only 11 per cent of the feed goes to produce the beef itself, with the rest either burned off as energy in the conversion process, used to maintain normal bodily functions, or excreted, or absorbed into parts of the body system that are not eaten – like hair or bones.

But these parts of the dead animal 'like hair or bones' which are not eaten – 50 to 55 per cent in the case of beef and 25 to 30 per cent in the case of pork – are not wasted.[18] The motto of meat processors everywhere, whether industrial packers or smallholders in their kitchen, has always been 'everything is used except the grunt' (or 'moo' or 'baa', depending upon the species).

The most obviously useful byproduct of a cow is its hide, which once cleaned and tanned is of considerable value. And every other part has its use. Collagen, an element of the connective tissues, is used in glues, wallpaper, plasterboard and sausage casings. Gelatin is used in various foodstuffs and plastics. Fat is used in soaps, waxes, margarine and cosmetics. In 1982, animal fat represented 68 per cent of all of the fat and oil used in UK soap manufacture,[19] but now a good deal of that has been replaced by oil from palm plantations which are held responsible for deforestation in Malaysia and Indonesia.[20] The hooves and horn are used as an ivory substitute. Hair is used in paintbrushes. Insulin and glucagon from the pancreas, blood plasma, trypsin, thrombin, bone marrow, intestines and ACTH [a hormone] from the pituitary gland are all used for medical purposes.[21] Bones can be used for glue. Waste meat can be turned into pet food, or it can be rendered and fed to pigs or fish. Anything which isn't eaten or otherwise used can be turned into a high value fertilizer. The fact that some of these products are currently wasted is not a fact of nature: it is a consequence of globalization and of incompetent animal health policies (a matter which is covered in more detail in Chapter 5).

Moreover, 'edibility' is a subjective concept. Under duress, human beings can probably eat 90 per cent of a cow. The skin of a pig is normally eaten, but can be turned into leather; the skin of a cow is normally turned into leather, but it can also be eaten. There is a moment in *The Goldrush* when Charlie Chaplin boils up his boots and tucks into them with a knife and fork. Things haven't got that bad in Nigeria, but leather producers there are concerned by the growing popularity amongst poor consumers of *pomo*, a delicacy made by boiling and softening cow hides. 'These pomo eaters have decided to walk on bare feet because they have decided to eat their own shoes,' the provost of the Zaria College of Leather Technology was reported as saying.[22]

---

18   Ross, Eric (1980), 'Patterns of Diet and Forces of Production: An Economic and Ecological History of the Ascendancy of Beef in the US Diet', in Ross, Eric (ed.), *Beyond the Myths of Culture*, Academic Press, p 202.

19   Monopolies and Merger Commission, *Animal Waste: A Report on the Supply of Animal Waste in Great Britain*, MMC, 1985, p 12, www.competition-commission.org.uk.

20   Friends of the Earth *et al*, 'The Oil for Ape Scandal: How Palm Oil is Threatening Orang-utan Survival', Friends of the Earth research report, 2005. www.foe.co.uk/resource/reports/oil_for_ape_full.pdf

21   The above list comes from Rifkin (1992), *op cit.*, p 274.

22   BBC News, 'Nigeria Eats its Shoe Leather', BBC News Report, 14 November 2000. For a tempting recipe for Nigerian cassava salad with pomo, see www.nigerianbloggers.com/Nigerian_Blog_featuring_delicious_recipes/?media=rss; Nov 5, 2008

Which parts of a carcase are edible, or not, depends on social factors: how rich people are, what alternatives are available, and nowadays, what advertisers, health geeks and celebrity chefs happen to be propagating. In 1986 British people ate 1.1 kilos of offal per year out of 54 kilos of meat. In 2008 they consumed even more meat – 58 kilos – but a pathetic 250 grams of offal. Probably most offal eaters are now over the age of 60.[23] However chefs such as Hugh Fearnley-Whittingstall and Jamie Oliver are now propagating the virtues of 'the fifth quarter', so we may see the decline in offal consumption reversed.

In the days when there were still horse-drawn rag and bone men providing a door-to-door recycling service, my mother used to boil up beef or chicken bones and scraps for stock which became the foundation for a quality of soup that many people under the age of 40 have never tasted. It is not hard to find stories about impoverished 18th or 19th century families who supplemented their vegetable ration with a few bones or scraps. By comparison to what landowners were eating, of course, this gruel appears pitiful; but assuming that there were enough dried peas, onions, potatoes and carrots and so on, there is no reason it should be any different from the excellent split pea and meat stock soup that my mother served me. This is exactly the sort of dish that vegans claim we should now be living from – except that the added benefit of meatstock or scraps increases palatability, whilst imposing a minute environmental toll. Much of this low grade meat, which would no doubt be welcomed by the hungry of the world, currently ends up in pet food. Go to any slaughterhouse in the UK and you will see huge vats full of beef-fat waiting to be incinerated. But in posh delicatessens you can also find liquid stock, emulating the kind my mother made, for sale in jars; and there are still a few of us who buy dripping from the butchers.

What is the ecologic value of all these byproducts? It is not particularly helpful to compare the protein or energy content of a cow's hide, tallow, pancreas etc, with the protein and energy in the parts that we eat. However, about two thirds of the 45 per cent of a cow which is officially human-inedible can be rendered and fed to pigs at a conversion rate of around 4:1 (or to fish even more efficiently) and that alone suggests that the feed conversion ratio for beef should be reduced by one sixth. By placing a ban on the rendering of animal carcases, the EU has managed to turn a valuable resource into a disposal problem, but that is a matter I discuss in Chapter 5.

The only other guide to go by is their economic value, which is subject to political and economic influences. According to a UK Monopolies and Mergers Commission report, up until the 1970s, the byproducts of a beef carcase paid for the slaughtering costs (which in the 1980s were around eight per cent of the value of the animal) plus the profits, so a slaughterhouse could sell the dressed carcase of a beef cow for the same price that it bought the animal. However, 'the balance became unsettled towards the end of the 1970s', presumably because alternative products derived from oil, or from imports such as palm oil, were undercutting the market: by 1981 byproducts only brought in 3.5 per cent of the value of the carcase.[24] In slaughterhouses, vats full of tallow wait to be incinerated, because nowadays we prefer our soap and cooking fat to be made of vegetable oil.

The hide represents seven per cent of the weight of the carcase, and even today, when we could if we wanted manufacture all our leather products out of

23   Annual Abstract of Statistics (1998), 9.15 and (2009), 21.16. Liver was consumed once a week, by 18 per cent of the population in 1979, but by 2 per cent in 1996. Other kinds of offal (heart, lights, tripe, etc) were consumed once a week by six per cent of the population in 1979 but only by one per cent in 1996. I can no longer locate the reference for these statistics.

24   Monopolies and Merger Commission (1985), *op cit. 19.*

synthetic materials, its value when cured is weight for weight higher than that of cured beef. Globally, the value of leather in relation to beef appears to be a good deal higher than eight per cent. According to FAO figures supplied by the International Council of Tanners the value of the world trade in raw hides and skins ($4.4 billion) is over a quarter the size of the global market in beef, lamb and goat meat ($17 billion).[25]

Tara Garnett has argued that the full leather value of a carcase should not be taken into account 'since some of the leather goods available' – expensive leather sofa upholstery for example – 'are not needed and might not be produced at all'. Whilst she has a point, a lot of uses for leather are not frivolous: 56 per cent of all light bovine leather is used for shoes, and at least a proportion of these are nowadays regarded as necessary.[26] Some uses of leather, like some meat or vegetable dishes, are brazen luxuries; but that luxuriousness can be adequately accounted for by weighing the expenses (ie disproportionate land-take) against the use value (which in respect of pâté de fois gras or asparagus tips is expressed in calories and protein, and in the case of sofas by the value attached to something like standard cotton upholstery). Attaching no value at all to leather seems unduly puritanical; and anyway it suggests that the Nigerians are right and we ought to be turning our cow hides into *pomo*, or in the UK, beef scratchings.

In the case of beef, the value of the leather, plus all the other products and the rendered meat is very considerable indeed. If these are to be taken into account, it seems reasonable to reduce what we can now call the 'feed to animal produce ratio' by 15 per cent in the case of beef. The gains from byproducts are much more slender in the case of pork and dairy, while for poultry they are negligible.

Beyond these manufactured products there are a number of more intangible services that animals provide, including traction, nutrient accumulation in the form of manure, clearing land, grazing for conservation purposes, companionship and heat. Often these are not byproducts, but the main reason for keeping the animal, and it is the meat that is a byproduct. It is clearly fruitless to try to fold these benefits into any feed conversion ratio, and they are dealt with elsewhere in this book.

Crop Yields

A further reason given by the CAST scientists for reducing the conversion ratio is that often higher yields can be obtained from animal feeds than could be obtained from human food grown on the same land:

> A fact often overlooked in the feed grain/food grain debate is that
> the most important feed grain, maize, yields substantially more per
> hectare than the most important food grains, wheat and rice. Maize
> is consumed directly by humans but wheat and rice are strongly
> preferred in much of the world.

And again:

> The amount of meat, milk or eggs produced from a unit of land is a
> function of crop yield as well as conversion rate by the animal. A crop

25   www.tannerscouncilict.org/statistics.htm; Tara Garnett gives a figure of 7.7 per cent taken from a 2002 Catalan life cycle analysis. Garnett,Tara (2008) *Cooking Up a Storm*, Food Climate Research Network, p65, http://www.fcrn.org.uk/frcnPubs/publications/PDFs/CuaS_web.pdf
26   Garnett (2008), *ibid*.

such as alfalfa (lucerne), for example, yields much more tonnage than any food crop that might replace it ... As shown by the table below, in California, more human food energy and protein (of higher quality) is obtained per hectare from growing alfalfa and feeding it to dairy cows than by growing wheat. The alfalfa requires more water but less nitrogen fertilizer.[27]

Although they certainly have a point, I am a bit sceptical about the CAST scientists' examples, because the yields they give for wheat are so low. Their yield for maize on irrigated land in Nebraska is 7.41 tonnes per hectare, while the yield given for wheat on the same land is 3.49 tonnes. But an average yield for wheat in the UK is 8.25 tons or 7.5 tons milled, which is comparable with the Nebraskan maize. Either the Nebraskans aren't very good wheat farmers or else Nebraska isn't a particularly good place for growing high yield wheat, but is part of the corn belt – in which case perhaps the Nebraskans should jolly well learn to eat corn, like Mexicans do. There is a sense in which CAST's ostensibly scientific analysis reflects a mission to persuade all the world to eat wheat, in the same way that the world is being induced to eat burgers and wear trainers and baseball caps. On the other hand, I was brought up on wheat, and if I moved to Omaha, Nebraska, I'd no doubt continue to eat it; so maybe that's not a very good line of argument.

The case for lucerne is also suspect, because its abundant yield depends upon large amounts of water, which in California cannot be provided without levels of irrigation which many environmentalists predict are unsustainable. According to Marc Reisner, in 1986 California used as much water for cow pasture as all 27 million people in the state used, including for their swimming pools and lawns.[28] And here again, the CAST scientists are comparing these rather artificial lucerne harvests to wheat yields of only four tonnes per hectare.

In Britain, where rainfall is abundant and wheat yields are twice as high, lucerne or clover yields can reach 14 tonnes of dry matter per hectare without irrigation. Lucerne has a 20 per cent protein content, nearly twice as high as wheat, so a hectare of high yielding lucerne would produce about 2.8 tonnes of protein, over three times as much as a hectare of high yielding wheat or barley, while red clover is only slightly less productive.[29] Since both lucerne and red clover are legumes which add nitrogen to the soil, their use in a well managed dairy farm can produce nearly as much protein for human consumption per hectare as a hectare of grain, economizes on fossil fuel derived fertilizers, and being a perennial probably secures more carbon in the soil.

Also of interest to British observers is the productivity of our wide expanses of wheat compared to that of dense forests of fodder maize. It looks as though the fodder maize produces more nutrients: and indeed it does, but only just, because a lot of that maize is water. An average field of fodder maize in the UK provides 900 kg of protein and 130 gigajoules of energy, while an average field of wheat produces 825 kg of protein and 108 gigajoules of energy. Since a wheatfield also produces several tonnes of useful straw, it is clear that in the UK feeding wheat to humans is far more efficient than feeding silage maize to cows.[30]

27   CAST, *op cit.*16, p 35.
28   Reisner, Marc (1986), *Cadillac Desert*, Penguin.
29   See for example: Welsh Institute of Rural Studies, *Final Report to the Milk Development Council*, 5 August 2000, randd.defra.gov.uk/Document.aspx?Document=LS0301_723_FRA.doc
30   Nix (2007), *op cit. 11*; and *The Keys to Maximizing Maize Silage Profitability*, Pioneer, www.pioneer.co.nz

We should also bear in mind that in the rainy South West you can easily get a high yield from maize, but rather less easily from wheat; whereas in East Anglia bumper yields of wheat are easily achievable, while fodder maize is less common. In other words the relative productivity of animal feed crops and human food crops varies according to the locality. Although the differences in yield, on aggregate, are marginal, they can sometimes be significant, and I'm inclined to allow livestock farmers an advantage here, because they have greater ability to adapt to the local situation. In some situations animal fodder grows more abundantly than human food, and the carnivore farmer has a choice, whereas the vegan farmer doesn't.

It is also apparent that the animal feeds which are more likely to outperform human food crops are bulky whole-plant forages such as maize silage, brassicas, root crops, lucerne, or even grass, and these are more accessible to cattle than either to pigs and poultry, which do best on grain. As we have seen, cattle are also the animals which provide most in the way of leather and other byproducts. Both these factors serve to lower the feed conversion factor for beef in particular (at a guess to round about 7:1 or 8:1) and narrow the difference in performance between cattle and the non-ruminant species.

The Global Pig Bucket

There is a further reason why higher yields are obtained from feedcrops: animals are less fussy. Humans, unless they are underfed, tend to discard substandard seeds, roots and fruits; animals, unless they are overfed, have much lower standards. Humans like to mill grain, and milling reduces the yield of wheat by nine per cent – animals aren't so bothered.[31] Of the 24 tonnes of potatoes that might be produced on a hectare, a proportion might be substandard, a further proportion might go off in storage, and 22 per cent of the remainder might be discarded as peel,[32] but a pig would eat the lot. The food rejected by humans is often fed to animals, as bran, cake or in a compound feed, and this brings us on to the matter which threatens to make a complete nonsense of all these feed conversion ratios: that animals eat things which humans don't eat.

Human-inedible biomass can be loosely divided into two kinds: the stuff that we can't eat and the stuff that we won't eat. The vegetable matter we can't eat consists mainly of fibrous materials such as grass, straws and stalks, and it is fed mainly to cows and other ruminants who have gastric systems designed to turn fibrous matter into protein.

The food we choose not to eat consists mainly of (a) surplus and spoilt grains and roots (b) residues arising from food processing (c) kitchen waste, and (d) slaughterhouse waste. These usually contain high levels of nutrients and are better fed to pigs or to poultry whose gastric systems have evolved to digest highly concentrated food. Pigs, like humans, are omnivores who have difficulty digesting significant amounts of fibrous matter and require high concentrate foods to thrive. Cows are able to eat high protein foods such as grains but they metabolize them inefficiently compared to pigs or poultry. However they can make up for this by providing milk more efficiently than meat. That is why in Britain during the Second World War, the small amounts of animal feed at the country's disposal were directed into increasing the number and performance

---

31   Nix (2007), ibid., p 17.
32   Elferink, E V *et al* (2007), 'Feeding Livestock Food Residue and the Consequences for the Environmental Impact of Meat', *Journal of Cleaner Production*, pp 1-7.

of dairy cows, while the number of grain-fed pigs and poultry were drastically reduced.[33]

Some idea of the amount of waste that is or could be going into the global pig bucket can be gained from a more detailed examination of these four categories:

(a) Spoilt produce, and crops that fail to sell often never leave the farm so it is difficult to make any estimate of their importance. Substandard produce is normally fed by mixed farmers to whatever stock they have on the farm – whereas specialist arable farmers do not have this option to recoup some of their losses.

(b) According to California University food analyst J G Fadel, the processing byproducts of six major crop classes (vegetable oil, sugar beet pulp, grain, citrus fruits, almonds and cotton) in 1993 amounted to about a quarter of a billion tonnes of dry matter, which if fed to animals would supply enough energy to produce 435 million metric tonnes of milk – more than the entire world's milk supply at the time, or about a quarter of a litre per person per day.[34] Milk provides over a quarter of all the nutrients provided by livestock in the world.[35] These figures do not include some other significant sources of waste such as potato or banana residues.

However, about a third of Fadel's figure consists of residues left over after making vegetable oils – mainly from soybeans but also from groundnuts, sunflower and sesame seeds. These seeds can all be eaten whole by humans, and in the case of soybeans the value of the meal for animal feed is as high as the value of the oil. There is therefore doubt as to whether these oilseed meals should be regarded as a co-product, or a by product – a matter which is so convoluted that it requires coverage in a separate chapter (Chapter 6). Even so, the other residues and co-products identified by Fadel represent a prodigious quantity of nutrients which for the most part are already consumed by animals. The production of oil and ethanol for energy purposes also leaves behind a substantial volume of meal or of distillers' waste that can most profitably be used for animal feed; but using grains and oilseeds for biofuel is highly inefficient and there are widespread calls for it to be discontinued.

(c) In the industrial countries, oil seed residues are probably dwarfed by the domestic waste stream. According to a report from the government sponsored organization WRAP, UK households now throw away around a third, by weight, of all the food they buy, somewhere between 6.7 and 8.3 million tonnes. A further 8.7 million tonnes is thrown away by manufacturers, retailers and restaurants.[36] Quite a high proportion of this is high value foodstuffs – for example

33   DEFRA, Agriculture During World War II, 2008, http://www.defra.gov.uk/esg/work_htm/publications/cs/farmstats_web/History/WWII/WWII_stats.htm

34   Fadel, J G (1999), 'Quantitative Analyses of Selected Plant Byproduct Feedstuffs, a Global Perspective', *Animal Feed Science and Technology* 79, pp 255-268.

35   Very roughly 25 per cent of the protein, and 29 per cent of calories, according to calculations derived from 1996 production figures in Seré and Steinfeld (1996), op cit. 14.

36   At the time of writing two different figures are given in WRAP (2009a) *Household Food Waste*, and WRAP (2009b) *Non Household Food Waste*, http://www.wrap.org.uk/retail/food_waste/nonhousehold_food.html, and http://www.wrap.org.uk/retail/food_waste/index.html

about 25 per cent of all bread and bakery products are wasted.[37] The Labour government's food adviser, Lord Haskins, has claimed that altogether Britain throws away about 20 million tonnes of food every year, 80 per cent of which is from homes, shops, restaurants, schools etc, while the rest is lost between the farm gate and the shop shelf. United States consumers are even more wasteful: their overall food supply is equivalent to 3600 kcals per person, but surveys show that they only consume 2000 kcal of this.[38] If these figures are to be trusted, US citizens throw away 45 per cent of their food, which, if it were all fed to pigs at a 5 :1 conversion rate, would in theory supply pork equivalent to nine per cent of their diet.

Clearly, it would be more efficient to waste less food, than feed it to pigs. But even in poor countries (as in the UK during World War II) there is still a considerable amount of wastage. In the 1980s Cuba, with a population of ten million, was feeding about half a million pigs daily with a diet of 37 per cent food waste from schools, restaurants, hospitals etc, though not from domestic homes. The waste from these sources was providing roughly two kilos of pork per person per year for everyone in the country.[39]

(d) On top of this there are slaughterhouse wastes. In 1991, rendered meat and bonemeal (MBM) from UK slaughterhouses provided about 500,000 tonnes of high protein animal feed and constituted about four per cent of pig rations (as well as being fed to poultry and cows).

Today, if this extremely high nutrient MBM, together with Lord Haskins' 20 million tonnes of food waste were fed to pigs, it would produce at the very least 1,000,000 tonnes of pork, or about 70 per cent of all the pigmeat consumed in Britain – drastically cutting the UK aggregate feed conversion ratio for pigs from about 4:1 to not much more than 1:1. Unfortunately none of this food waste nor the MBM is being fed to pigs or any other animals in the UK – most of it is being inefficiently incinerated to make cement or provide energy. In 1996, MBM was banned in the UK as a response to the outbreak of BSE, in 2001 the ban was extended to Europe, and in 2002 using kitchen food waste to make pig swill was banned throughout the EU. How the mismanagement of the UK's livestock sector led to such a perverse waste of resources is described in detail in Chapter 5.

Thankfully the EU's neurotic approach to waste management has not yet spread to Third World countries where many peasants still rear their pigs on a diet consisting primarily of scraps, crop residues and whatever comes to hand. According to figures extrapolated from FAO statistics, in the early 1990s developing world pig-keepers fed their pigs 1.76 kilos of grain for every kilo of pork they produced, while pigkeepers in the developed countries fed 3.67 kilos. The global figure was 2.69 kilos of grain. The rich countries' figure of 3.67 kilos is high, compared to the much lower figures given by Peter Brooks for

37   DEFRA (2007), *Estimates of Household Food Waste, in Family Food*, https://statistics.defra.gov.uk/esg/publications/efs/2007/annex.pdf

38   Smil (2004), *op cit.*, p 166. Figures for food consumption are from USDA's 10th National Food Consumption Survey, see Cleveland L E *et al*, 'What We Eat in America', *Nutrition Today*, 32: pp 37-40.

39   Pérez, Rena (1990), *Feeding Pigs in the Tropics*, Ministry of Sugar, Havana, FAO Animal Production and Health Paper 132.

British pigs in the early 1990s; it reflects the large US pork industry concentrated mainly in corn growing areas, rather than spread out across the country to maximize access to waste. These figures suggest that well over half the feed given to pigs in the developing world is some kind of waste or residue; and that if the USA, the EU and other industrialized countries were to put their food recycling in order they could achieve a better performance.

All Flesh is Grass

Many of the high protein wastes fed to pigs, particularly oilseed cake, can be and often are fed to cows. They make cows more productive, in the sense that they produce more milk, or put on weight quicker, but they produce meat about twice as inefficiently as if they were fed to pigs.

On the other hand, cows and other ruminants are much more efficient than pigs and poultry at consuming fibrous waste products. In third world countries, crop residues are routinely fed to cows and goats which are hardier and better adapted to surviving on lowly fare than European Holsteins and Limousins. Rice straw, millet and sorghum stovers, yam peels, substandard roots and grains and so on are a significant component of third world ruminant diets. Very fibrous materials such as bagasse (sugar cane straw) and wheat straw are more easily fed after treatment with ammonia or alkali. In many cases, cows are folded on the stubble of harvested crops, which helps dispose of it, manures the land and assists in the sequestration of carbon. Nomadic herders sometimes make arrangements with sedentary farmers to take some of their crop residues – for example Fulani (Peul) pastoralists in Nigeria move out of their grazing areas in the dry season to feed their animals on cereal straw.[40]

According to the FAO, feeding rice straw to cattle is 'the most efficient way at present to recover the energy of the straw. Straw nitrogen can also be effectively recovered if the dung is subsequently fermented to produce gas'.[41] In 1999, Fadel calculated that the global crop residues from wheat, rice, barley, maize and sugar cane (bagasse) alone totalled three-quarters of a billion tonnes of dry matter, while a later study estimated that the same volume was potentially available as animal feed in Asia alone.[42] Being mainly fibrous straws, these are sometimes not fed to animals when better food is available or when they are needed for some other purpose such as litter, thatch or fuel. In Bangladesh and Thailand over 70 per of rice straw is fed to animals, whereas in South Korea the figure is only 15 per cent.[43] It should be borne in mind that when crop residues are fed to animals the bulk of the organic matter emerges at the tail end of the process in the form of manure, which either finds its way back to the soil or is used for fuel or building material. In theory, according to Fadel's calculations these residues provide enough energy to support the entire world's milk supply, and enough protein to support about a third of it.

40   Ayoola, G B and Ayoade, J A, *Socio-Economic and Policy Aspects of Using Crop Residues and Agro-Industrial Byproducts as Alternative Livestock Feed Resources in Nigeria*, University of Agriculture, PMB 2373, Makurdi, Nigeria /www.ilri.cgiar.org/InfoServ/Webpub/Fulldocs/X5519b/x5519b1g.htm
41   Jackson, M G, 'Rice Straw as Livestock Feed', *Ruminant Nutrition: Selected Articles from the World Animal Review*, FAO, 1978, http://www.fao.org/docrep/004/X6512E/X6512E07.htm
42   Devendra, C and Sevilla, C (2002), Availability and Use of Feed Resources in Crop-Animal Systems in Asia, *Agricultural Systems*, 71(1) pp 59-73; cited in Steinfeld, Henning *et al* (2006), *Livestock's Long Shadow*, FAO, p 41.
43   Steinfeld *et al* (2006), *ibid.*, p 41.

In the UK there is a similar range of crop residues available for cows, such as apple pomace, sugar beet pulp and various kinds of straw. As William Cobbett sourly observed, almost the only major human food plant whose residue cannot be eaten by animals is the potato: 'While the wheat straw is worth from three to five pounds an acre, the haulm of the potatoes is not worth one single truss of that straw'.[44] But collectively these crop residues provide a tiny amount of nutrition for the national cattle herd, partly because there is such an abundance of higher grade feed around, and also because wheat and other crops have been bred so that most of the goodness goes into the seed and very little into the straw. Oat straw is a good deal more nutritious and palatable to livestock than wheat straw; but few crops of oats are grown nowadays.

Happily, the most abundant fibrous material of all happens to be a food highly relished by the cow, precisely because it has not been bred for seed production – namely grass. Anyone who has kept a dairy cow cannot fail to have marvelled at the way she can graze on the grass that we tread underfoot and turn it overnight into prodigious quantities of milk, which within an hour or two can yield butter. How can solid fat be created so fast from such an unpromising material? What machine could do the same? Attempts to extract edible substances from grass and weeds with machines have so far only produced an unappetising gunge which requires mixing with sweeteners and nuts to make it palatable.

Despite the fact that she is dealing with such fibrous material, the cow does this both willingly and surprisingly efficiently, given that (unlike a machine) she is also keeping herself alive and having offspring at the same time. In a survey carried out on dairy cows in New York State, about 60 kg of protein was produced from 190 kg of vegetable protein, a conversion rate of just over 3:1. Of the 190 kg of protein, about half was from grains and about half from forage. If the forage was grass from land that could not reasonably be cultivated, then the protein conversion rate of potential human feed to dairy produce in this case was just over 1.5:1.[45]

The 3:1 ratio seen in New York can also be observed in the performance of British farms. A typical UK dairy farm, with black and white Holstein cows, produces roughly 7900 litres of milk per hectare per year, which is about 250 kilos of protein, whereas a hectare of land put over to winter wheat would produce about 750 kilos of protein – a ratio of 3:1. On an organic farm the protein ratio is nearer 2:1, because the yield per hectare of organic cows is about two thirds that of conventional dairy cows, while organic wheat yields are little more than half those of conventionally grown wheat. In respect of energy the ratio is 5.4:1 on conventional farms (if you don't factor in the fossil fuels required for the fertilizers that boost the wheat and grass yields) and 3.7:1 on organic.[46]

However inefficient the conversion ratio of any given animal may be, if if it is grazing entirely on land which could not otherwise be used for arable production or some other highly productive activity, then it cannot be said to be detracting from the sum quantity of nutrients available to the people of the

---

44   Cobbett, W. (1822), *Cottage Economy*, Peter Davies, 1926, p 51. The tomato plant, a cousin of the potato, is also disliked by animals.

45   Pimentel, D and Pimentel, M (1979), *Food Energy and Society*, Edward Arnold, p 55.

46   Nix (2007), *op cit. 11*; and Lambkin *et al* (2007), *Organic Farm Management Handbook*, University of Wales Aberystwyth. Calculations as follows: 0.5 ha per cow plus feed costs of 0.25 ha and replacement costs 0.18 ha, minus returns from calves and cull cows 0.06 = 0.87 ha; 6850 litres per cow = 7873 litre per ha = 250 kilos protein; wheat 1 ha 7.5 tonnes at 10 per cent protein = 750 kg = **3:1**. Energy = 7873 litres x 600 = 4.7 million kcals. wheat 7500 kg x 3400 = 25.5 million = **5.4 to 1**. Organic 0.67 per cow plus 0.3 feed, plus 0.2 replacement minus 0.066 calf = 1.1 ha. 6000 litres per 1.1 ha = 5500 per ha.; protein 5500 x 0.032 = 176 kg protein organic 400 x 0.91 (milling) = 3640 kilos x 0.1 = 364 = a little over **2 to 1**; energy 5500 x 600= 3.3 million kcals; 3640 x 3400 =- 12.3 million = **3.7:1**.

world, but is adding to them. When I kept Jersey cows, they produced 3500 to 4000 litres of milk per year on a diet consisting of about one tonne of bought-in organic lucerne, plus hay and pasture. The lucerne, a legume which introduces nitrogen into the ground, would normally be part of a long term arable crop cycle, and would require less than one tenth of a hectare. The grass and hay took up about one hectare of permanent pasture/meadow per cow, and some of that hectare was orchard. Theoretically the grassland could also have been folded into an organic arable cycle, or even cropped continously with the use of chemical fertilizers; but since it was on a significant slope there was no economic or ecological incentive to do so. The cows were effectively fed little or nothing which could have been human food.

The same is true of suckler cows and sheep, which, in the UK, to a very large extent, are grazed or fed on hay or silage from land which is either unsuitable for arable production, or not required for it. Even the archetypal American feedlot cow is not fed entirely on grain because cows need fibre. 'In the total feedlot system' says David Pimentel, 'the protein fed to the beef and breeding stock consists of about 42 per cent forage with the remainder being grain';[47] CAST provides a similar figure. Moreover, feedlot cows normally spend the first half, or even two thirds of their life on grass, where they have already gained 50 to 70 per cent of their weight.[48] If we round this off to 60 per cent it means that human edible food constitutes only 24 per cent of the animal's entire intake – which in practice brings a 8:1 feed conversion ratio abruptly down to about 2:1. However, much of the forage fed to US feedlot cows is lucerne derived from unsustainably irrigated arable fields: if we count this component of a feedlot cow's diet as equivalent to human food, then the effective ratio of human edible feed to meat and other animal products in US feedlot beef comes to about 3.2:1.

CAST v CIWF: The Result

The information provided in the foregoing pages is, hopefully, sufficient to convince the reader that whatever the metabolic feed conversion ratio may be for any breed of animal, in the real world there is a multiplicity of factors that can and regularly does lower the ratio significantly, sometimes to a point where the inefficiency of livestock rearing is negligible, or even negative.

But what of our rule of thumb figure, so casually passed off as ten to one by some commentators? Does this require revision, or should it be abandoned? Is it possible to arrive at a global aggregate efficiency ratio for livestock as a whole?

In 1979, David Pimentel estimated that more than 60 per cent of the world's livestock protein 'comes from animals fed grasses and forages that cannot be utilized by man. The remainder comes from livestock fed plant and animal protein that is suitable food for man.'[49] This puts the matter into context, but it does not really help us, not only because it is out of date, but also because we still do not know at what efficiency the 40 per cent was being converted into human food. In 1997, a major FAO census on livestock concluded:

> If it is assumed that all 966 million tonnes of cereals, roots and tubers
> used for livestock are edible for humans (in effect they are not, as there
> are considerable preparation losses in milling etc), then livestock gets
> 74 million tons of edible protein. On the positive side, the 199 million

47   Rifkin (1992), *op cit. 3*, p 161.
48   CAST (1999), op cit. 16, p 42.
49   Pimentel and Pimentel (1979) *op cit. 45*, p 53.

tons of meat, 532 million tons of milk and 53 million tons of eggs
produced globally in 1996 contain 53 million tons of protein.[50]

Assuming that this figure is correct, it means that the global aggregate
conversion ratio of human-edible feed to human edible meat is 74:53, or 1.4:1
(the figure has since been updated by the FAO and is now 77:58, closer to 1.3:1).[51]
Hopefully readers who have followed me in examining all the above mitigating
factors and circumstances are now in a better position to appreciate how the
feed conversion ratios that we began with (ranging from around 4:1 for poultry
and milk to around 13:1 for beef) translate into the FAO's global aggregate of
1.4:1. In fact the FAO authors go on to argue the ratio may be near enough one
to one: 'While input is higher than output, if cereal preparation losses on the
input side and improved protein quality on the output side are considered, a
reasonable balance emerges.' They might also have added leather and other
animal byproducts into the equation.

We are also now in a position to examine how both CIWF and CAST have
used the figures to support their particular causes. Mark Gold, for the CIWF,
in the section entitled 'The Inefficiency of Animal Foods – Food Conversion
Rates' simply omits to mention the superior nutritional value of meat, or the
value of animal byproducts, or the possibility that animal feed crops may be
more productive than human crops; and he fails to acknowledge the service
which ruminants and pigs provide by turning inedible vegetation and waste
into human food. Grazing is only mentioned in a favourable light in the section
entitled 'The Value of Livestock to Poor Communities', as if it were acceptable in
poor countries, but not in wealthy ones.

The authors of the CAST report do refer to all these factors, although (being
North American) they play down pig waste, and emphasise ruminant use of
grassland:

> Poultry and pork production are the most efficient on the basis of total
> food produced from total food intake but, on average, ruminants return
> more human food per unit of human-edible feed consumed because
> most of their feed is obtained from materials that cannot be consumed
> directly by humans. This fact has been overlooked in some assessments
> of the role. On a global basis, less than three kilograms of grain are
> required to produce a kilogram of meat from any of the species.

So far so good, but then a few lines later they continue:

> It has been estimated that, on a global basis, animals produce a
> kilogram of human protein for each 1.4 kilograms of human food
> protein consumed. The biological value of protein in foods from
> animals is about 1.4 times that of foods from plants. Thus diverting
> grains from animal production to direct human consumption would, in
> the long term, result in little increase in total food protein.[52]

These paragraphs come from the 'Interpretative Summary'— the only part
of the report which can be viewed on the internet. Perhaps it was written and
then grafted on to the report by someone other than the scientists who drew up

---

50   De Haan, C et al (1997), *Livestock and the Environment: Finding a Balance*, FAO 1997, www.fao.
org/AG/aga/lspa/LXEHTML/tech/index.htm
51   Steinfeld *et al* (2006), *op cit.* 43, p 271.
52   CAST (1999), *op cit.* 16, Interpretative Summary.

the main document. If the second paragraph seems to make sense to you, read it again, it is worth dissecting in detail.

The 1.4:1 kg of human food protein comes, not from CAST's own exhaustive analysis of conversion ratios, but from FAO's 1997 analysis, which I have just referred to. The fact that animal protein also happens to be regarded as 1.4 times as nutritious as vegetable protein is a coincidence too good to ignore (never mind that energy intake may be a more reliable indicator of nutritional value). What a perfect opportunity to suggest to the reader that these two ratios cancel each other out, without risking academic credibility by actually spelling it out. 'Thus', the summary continues 'diverting grains from animal production to direct human consumption would, in the long term, result in little increase in total food protein.'

It shouldn't take the reader who has been following the argument long to spot the mountain of deceit invested in that little word 'thus'. The final sentence is not a consequence of the first two sentences, because the conversion rate of 1.4 kg of human food protein to one kilo of edible meat protein is an average derived from the sum total of all livestock, some of whom are fed a great deal of grain in their diet, while others (as the authors have only just pointed out) are fed little or none. If we stopped feeding grain to animals we would still retain over half of our meat supply and also benefit from about three times as much nutrition in the form of grain as there was in the meat foregone. The last paragraph of the sentence is a cleverly disguised lie.

In fact the main text of the report comes to a slightly different conclusion: 'From the perspective of human protein, the use of animals does not decrease the amount of protein available for humans.'

This makes sense if it is taken to mean that *if everybody stopped eating meat completely, we would have no more protein than we do now*, which is a logical and correct conclusion to be drawn from the FAO's analysis. This is not the same as saying that *'diverting grains from animal production would result in little or no increase in food protein'*, because if everybody stopped eating grain-fed meat but carried on eating meat from animals that were not grain-fed then we would have more protein. In fact the main body of the report – the part not available on the internet – does briefly point this out: 'The fact that several of the input:output ratios are greater than 1:1 does mean that feeding less grain to animals would translate to somewhat more total food for humans.'

'Somewhat more total food for humans'? How much is somewhat? Is not this the main reason for CAST carrying out all its exhaustive 'input:output' analyses? What could be more relevant to a report entitled *Animal Agriculture and Global Food Supply*? But on this matter the scientists are completely silent, and hastily move off the subject by counselling the reader in the very next sentence: 'It would also mean a food supply with less variety and lower nutrient density.'

However, the figure is easy enough to work out from their tables: 400 million tonnes of cereal grain. This is the extra food you would get if the 600 million tonnes fed to animals at an estimated 3:1 efficiency were instead put over to human consumption. It is enough to feed about 1.3 billion people, over 20 per cent of the world's population, at least 300 million more than the number of people estimated to be malnourished.[53]

More to the point, two thirds of this 600 million tonnes of grain is used in the industrialized countries for the benefit of the 20 per cent of the world's population who live there. And even more to the point, given that the report

53   CAST (1999), *ibid.*, p 41.

is written by US scientists, a quarter of this grain is used to fatten animals for the benefit of the five per cent of the world's population who live in the USA. In effect, what the CAST report says is that it is not a waste of protein for North Americans and Europeans to use extravagant quantities of human food to fatten animals, because farmers in the Third World make up for it by rearing animals on feed which couldn't be eaten by humans. This is like saying 'it's no problem if we Americans burn barrels of oil and release tonnes of carbon, because hey! Those guys over in India and Africa hardly use any.'

Whilst a casual reader or slipshod journalist might easily overlook this flaw in CAST's logic, and its evasion of the obvious conclusion, it is hard to believe that the authors of a report endorsed by 38 scientific bodies were too stupid to spot them, so I am inclined to conclude that they were happy to mislead.

4

DEFAULT LIVESTOCK

Default: a course of action that a program or operating system will take, when the user or programmer specifies no overriding value or action.

Dictionary.com

In one of a series of studies on feeding food processing wastes to animals in the Netherlands, Dutch food analyst Sanderine Nonhebel and colleagues supplied the simple diagram which can be seen in Fig 1.[1] This 'hockey stick' curve shows that up to a certain point B, livestock can be fed on food residues, and so their environmental impact in respect of the land they require remains very low; but beyond this point, as people start to consume meat from animals fed on dedicated feed crops, the environmental impacts become much greater. The converse of this is that if the Dutch reduced their meat consumption from point C, where it is currently, down to point B, then substantial reductions in land use and environmental impact would be achieved; but further reductions in meat consumption below point B result in negligible environmental gains.

Figure 1. The Meat Consumption Curve

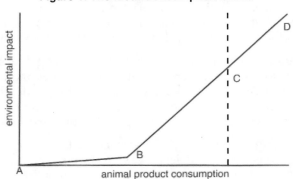

Credit: E V Elferink, S Nonhebel and H C Moll, "Feeding Livestock Food Residue and the Consequences for the Environmental Impact of Meat", *Journal of Cleaner Production*, 2007 20, (1-7)

For historical reasons the Dutch food industry is more highly specialized than any other, except perhaps for those of its neighbours Belgium and Denmark. During the later Mediaeval period, the Flemish nations, obliged to feed a high population from a small land base, but with financial resources from

---

1   Elferink, E V, Nonhebel S, and Moll, H C (2007), 'Feeding Livestock Food Residue and the Consequences for the Environmental Impact of Meat', *Journal of Cleaner Production*, xx, 1-7.

their trading activities at their disposal, became adept at intensive agriculture, and remain so to this day. Nowadays, the Netherlands imports large amounts of animal feeds, mainly soya and cassava, for its meat and dairy industries, which, together with potatoes, sugar beet and other crops, support a booming food processing industry, both for home consumption and for export. The residues from this industry are considerably more than would result from processing food purely for national consumption. Nonhebel calculates that the residues from the average Dutch consumer's diet, if fed to pigs, are sufficient to provide every inhabitant with 135 grams of pork per day, or 49 kilos per year – far more than the average world per capita consumption of meat.

This sounds too good to be true, and it is. The Dutch consume a lot of soya oil, and 40 per cent of these food residues are soybean meal. It is doubtful whether the meal extracted from soybeans can legitimately be regarded as a residue, since it is worth at least as much as the oil itself. This is a complex question which is addressed separately in Chapter 6.

However, in other countries there are other reasons why a certain amount of meat or dairy consumption has a low environmental impact, and hence comes within the AB section of the curve. Wherever animals are grazed (but not overgrazed) on land that is unsuitable for arable production, they will be relieving pressure on the arable land, and helping to retrieve otherwise inaccessible nutrients and bring them within the food chain. In less developed countries, the feeding of food wastes to pigs, poultry or fish has not been outlawed, and these resources can be recycled to provide human nutrition. There are other situations where the prime reason for keeping animals is not to provide food, but meat is a byproduct or a co-product of an integrated system – for example when animals are used for traction, or when fish or ducks are integrated into a complex cropping and fertility cycle.

In the FAO's most recent overview of the global livestock industry, *Livestock's Long Shadow*, the authors refer to what they call a 'default land user strategy' for livestock. It is an ungainly term, but the word 'default' – in its modern sense signifying what one automatically falls back on when nothing else is specified – does sum up the role that animals play in a food economy which is not expressly designed to produce meat or dairy. This is what the FAO have to say:

> Livestock are moving from a 'default land user strategy' (ie as the only way to harness biomass from marginal lands, residues and interstitial areas) to an 'active land user' strategy (ie competing with other sectors for the establishment of feedcrops, intensive pasture and production units). This process leads to efficiency gains in the use of resources.[2]

How growing corn and feeding it to livestock at a five or ten to one conversion ratio can be construed as an 'efficiency gain' is a mystery that only economists can unveil. But the FAO are correct to state that in traditional agricultural communities the main role of animals is to turn vegetation that cannot be eaten or economically harvested by humans into useful goods and services. In a default economy, animals are not often reared as the single product towards which everything else is subjugated, unless (as in some pastoral environments) there are no other available products. Peasants with access to arable land, unlike people who buy their meat from a supermarket, are only too aware from the exigencies of their daily life that livestock are an inefficient way of producing human food. Pouring hard-earned grain down the throat of an animal is not

---

2   Steinfeld *et al* (2006), *Livestock's Long Shadow*, FAO, p 76; see also p 31.

something any poor farmer is likely to do to any great extent, unless there is a large surplus and no market for it. The role of the animal in the traditional, peasant, default economy is to mop up wastes, residues, surpluses, and marginal and interstitial biomass, and to provide services such as traction and nutrient movement. Default livestock don't monopolize land that is required for other intrinsic elements of the agricultural ecology: the extent and nature of the animal foods and services available are determined by the agricultural system as a whole.

In its 2009 State of Food and Agriculture report (SOFA), the FAO again describe how traditional default livestock is under threat from industrial agriculture:

> The move towards modern production systems has implied a decline
> in integrated mixed farming systems and their replacement by
> specialized enterprises. In this process, the livestock sector changes
> from being multifunctional to commodity specific. There is a decline
> in the importance of traditionally important livestock functions, such
> as provision of draught power and manure, acting as assets and
> insurance, and serving sociocultural functions. Livestock production
> is thus no longer part of integrated production systems, based on local
> resources with non-food outputs serving as inputs in other production
> activities within the system.[3]

How big is the world's default livestock sector – how much meat and dairy produce does it, or could it produce? To arrive at anything close to a precise figure would require a accounting exercise that probably only the FAO are capable of carrying out – and to my knowledge they haven't done so. However we are in a position to make a broad estimate by summarizing some of the figures cited above:

(i) Food Processing Wastes Fadel estimates the processing byproducts of six industries in 1993 were sufficient to support the production of more than the entire world's milk supply, or roughly a quarter of all the nutrients provided by livestock in the world.[4] (However, of this, about a quarter of the energy and more than a third of the protein was derived from soybean meal, which is more a co-product than a byproduct.)

(ii) Crop Residues Fadel also states that crop residues from wheat, rice, barley, maize and sugar cane (bagasse) potentially provide enough energy to support the entire world's milk supply, and enough protein to support a third of it.

(iii) Food Waste In the USA and the UK around about a third of all food is wasted; Cuba in the 1980s was producing about two kilos of pork per person per year from institutional (not domestic) food waste.[5]

3   FAO (2009), *Livestock in the Balance*, State of Food and Agriculture Report for 2009 (SOFA 2009), p 29.
4   Fadel, J G (1999), 'Quantitative Analyses of Selected Plant Byproduct Feedstuffs, a Global Perspective', *Animal Feed Science and Technology* 79, pp 255-268. While milk production has increased about 10 per cent since then, crop yields have increased correspondingly so the estimate probably still applies.
5   Hall K D *et al* (2009), 'The Progressive Increase of Food Waste in America and its Environmental Impact', *PloS ONE*, November 2009, Volume 4, Issue 11.This paper calculates that the amount of food wasted in the US has grown by 50% since 1974 and amounts to around 1400 kcal/person/day.

(iv) Slaughterhouse Waste According to Dutch food analysts E V Elferink and Sanderine Nonhebel, prior to the BSE crisis 10 per cent of EU animal feed consisted of meat and bone meal, a figure which seems high.[6] Peter Brooks' estimates that animal products comprised about four per cent of commercial pig rations in the UK in the 1990s, which seems low by comparison – could that be because the British were putting so much into their cow rations? [7]

(v) Grazing It is hard to obtain a total figure for the amount of meat and dairy produce raised on grass, because so much grazing takes place on mixed farms where livestock are fed a mixture of grass, crop residues and feed.

> **(a) Ranching and Grass Farms** The FAO state that about 24 per cent of beef and 12 per cent of dairy are produced on farms where there is nothing but grazing land.[8] It can be assumed that most of this takes place on land which is unsuitable for cultivation, and where grazing is the best option for bringing biomass into the food chain.

> **(b) Land Reserved for Pasture** An FAO study from the 1990s estimates that 27 per cent of the world's milk and 23 per cent of the world's beef is produced on grasslands managed solely for that purpose – representing roughly an eighth of all animal protein. The figures for dairy in (a) and (b) differ because some of the land in (b) is found on mixed farms, and hence also included in **(c)**.[9]

> **(c) Mixed Farms** About 70 per cent of beef and 87 per cent of dairy is produced on mixed livestock and arable farms, where livestock often have some access to crop residues, and areas of grazing land that cannot be cultivated.[10]

> **(d) Fertility Building** On many mixed farms livestock fed on leys or rotated grassland are part of the fertility building cycle. Though they may not be an indispensable part, they often add to the complexity and hence the resilience of the ecosystem, at little added expense.

(vi) Traction Some meat and dairy comes from animals whose main role is to provide traction.

If you include all the above it is probably safe to say that default livestock farming provides between a third and two thirds of all the meat and dairy produce currently produced in the world. I wouldn't like to be more precise than that, not least because the situation is changing rapidly with the increasing adoption of factory farming in the more prosperous developing countries.

However, for convenience, let us suppose that default livestock produce half of all the world's supply of meat and dairy. In Fig 1 on page 35, the line AB represents the default half of the world's available animal products, and the line BC represents the livestock fed on human-edible crops grown specifically for animals. There are huge environmental gains to be made from reducing global meat consumption back to this level; it will release large amounts of feed

6   Elferink, E V and Nonhebel, S (2007), 'Does the Amazon Suffer from BSE Prevention', *Agriculture, Ecosystems and Environment*, 120, pp 467-9.

7   Brooks, P (c1993), *Rediscovering the Environmentally Friendly Pig*, Seale Hayne Agricultural College, unpublished.

8   Steinfeld *et al* (2006), *op cit.2*, p 53; FAO (2009), *op cit.3*, p 26.

9   Seré, C *et al* 1995, *World Livestock Production Systems: Current Status, Issues and Trends*, FAO 1995.

10   FAO (2009), *op cit.3*, p 26.

grain for human consumption, easily enough to feed all those who currently are underfed, and a substantial contribution towards feeding the two or three billion extra people expected to be alive on the planet in 40 years time.

Now if you are a privileged white middle class Briton like myself, who has veered away from the diet of meat and two veg, and even had spells as a vegetarian, then halving the amount of meat and dairy in one's diet seems quite tolerable, generous even. Although an enthusiastic carnivore, I can happily subsist on a less than half the average meat and dairy enjoyed by the residents of Western Europe, in fact within my social circle, that is pretty standard. The relentless diet of full English breakfast and meat and two veg pursued until recently in Britain – or the everlasting succession of charcuterie, steaks and fromage that many Frenchmen traditionally mop up with their baguettes – takes away much of the excitement that can be derived from slicing into a lump of bacon twice a week throughout the winter, cracking into a cheese you have waited three months to mature, or frying up the liver from the pig you have just slaughtered.

However, if you spread this 50 per cent equitably across the world's population, you have to spread it pretty thinly. In 2000, there were 37 kilos of meat and 78 kilos of milk available for every man, woman and child in the world.[11] Halve this, and each person gets about 18 kilos of meat per year (350 grams or about three quarter-pound hamburgers a week) plus 39 kilos of milk (about 1.33 pints of milk, or barely 75 grams of cheese per week).

That is a very rough estimate of the amount of animal nourishment each of us could derive as an animal byproduct from an otherwise vegan agricultural system. If it seems disproportionately low on milk and cheese, that is because global figures are skewed by the Chinese and other nations who don't touch milk. Nutritionally five kilos of milk are equal to about one kilo of meat (or half a kilo of hard cheese and 4.5 kilos of whey to feed pigs). But milk generally supplies more nutrients per hectare than meat, so I would hazard that four kilos of milk are the ecological equivalent of a kilo of meat. This would suggest that the vegetarian in a default livestock world would be entitled to 111 kilos of milk per year (slightly more than half a pint day); the cheese eater would get 11 kilos of cheese a year plus the occasional bonus of sausages from the whey.

These are slim pickings for those of us who like our meat and cheese. They will be even slimmer for our grandchildren if the world's population grows to the predicted nine billion in 2050. That brings the equitably distributed default ration down to 12 kilos of meat per year plus 26 kilos of milk.

If you are a meat or cheese lover and this looks like gloom to you, you have forgotten the premise that we started from: that this is all free. It is *default* meat and dairy, it is the surplus that we get from growing the grains and vegetables that are necessary to maintain a given population. So long as a human population can be maintained on this planet through the cultivation of vegetables, a certain amount of meat will be your birthright. I may have miscalculated the amount by a factor of perhaps 15 per cent either way, but the section AB of Fig 1 will always be there and it will always provide a measure of free meat. How much further up the hockey stick handle we can afford to go is a matter of debate.

Probably we can afford some level of grain production for livestock. The amount of grain currently produced is more than enough to feed the world's population adequately if only default meat and dairy products were consumed.

11   Much of this discussion is derived from Tara Garnett Garnett,Tara (2008) *Cooking Up a Storm*, Food Climate Research Network, p65, http://www.fcrn.org.uk/frcnPubs/publications/PDFs/CuaS_web.pdf

There would be a substantial surplus of grain – at least 150 million tonnes (the equivalent of an extra eight kilos of meat per person per year if we assume a 3:1 conversion ratio) – and there are other reasons for wishing to maintain a surplus to feed to animals, which I shall examine in Chapter 10.

By contrast, the premise of the UN Food and Agriculture Organization, as expounded in Livestock's Long Shadow, is that livestock production must move away from being a 'default land user strategy' to what they call an 'active land user strategy', but might more helpfully be termed 'industrial farming'. Elsewhere they state:

> In modern times, livestock production has developed from a resource-driven activity into one led mainly by demand. Traditional livestock was based on the availability of local feed resources ... Modern livestock production is essentially driven by demand, drawing on additional feed resources as required.

That demand, in their view, has to be met. It is not true to say that the FAO never mention the possibility of reducing or managing demand for meat or dairy products. In the SOFA report you can find the following sentence tucked away at the end of the section on climate change:

> In developed countries, reducing the consumption of meat and dairy products could lower emissions without harming human health.[12]

That's it. There is no further discussion, nor rebuttal, of a proposal which solves 90 per cent of the problems that their report agonizes over, and which many thousands of environmental commentators advocate. Somewhere in Livestock's Long Shadow there is a similar statement, but it is so deeply buried and so inconsequential that I can no longer find it.

The view of the FAO economists is that a rural population content to consume what their local environment provides is already outnumbered and soon to be superseded by an urban proletariat who expect to buy anything they want in a supermarket, or aspire to be rich enough to do so. The FAO never seriously suggests that it is anything other than the duty of capitalism to provide as sumptuous a spread as it can, and anticipate that between 1999 and 2050, global meat production and milk production will double.

Having accepted this premise, the FAO are correct in drawing the following conclusions:

> If the projected future demand for livestock is to be met, it is hard to see an alternative to intensification of livestock production. Indeed the process of intensification must be accelerated if the use of additional land, water and other resources is to be avoided. The principle means of limiting livestock's impact on the environment must be to reduce land requirements ... This involves the intensification of the most productive arable and grassland used to produce feed or pasture; and the retirement of marginally used land where this is socially acceptable and where other uses of land, such as for environmental purposes, are in demand.

Intensification will have a radical influence on the structure of farming, and, predictably, farms will get bigger. 'The shift to intensive production systems is accompanied by increasing size of operations, driven by economies of scale,'

---

12   FAO (2009), *op cit*.3, p 73.

which will include 'a relative expansion of concentrate-based production systems'. To supply these factory farms 'intensification also needs to occur in the production of feedcrops, thereby limiting the use of land assigned to livestock production, either directly as pasture or indirectly for feedcrops.' These feedlots will be focused mainly on monogastric animals – pigs and chickens – because these species convert plant nutrients into protein more efficiently than ruminants.

This shift to large scale and industrial farms, the FAO continue, 'is only achieved at the cost of pushing numerous small- and middle-scale producers and other agents out of business … Small family-based livestock producers will find it increasingly difficult to stay in the market.' Currently 1,300 million people, or 20 per cent of the world population are engaged either full-time or part-time in livestock production, of whom 987 million are classed as poor. Many of these people will have to be displaced:

> Many producers will need to find alternative livelihoods … The loss of competitiveness requires policy interventions, not necessarily to maintain smallholder involvement in agriculture, but to provide opportunities for finding livelihoods outside the agricultural sector, and enable an orderly transition…. This trend raises social issues of rural emigration and wealth concentration. Diversification within and outside the agricultural sector, and social safety nets are some of the policies developed to address these issues.

The consequence is a rapid 'urbanization of livestock' in which rural economies are undermined, and enclosed by a dominant urban and globalized economy:

> As a result of economies of scale, industrial livestock production generates substantially lower income per unit of output than smallholder farms, and benefits go to fewer producers. Furthermore, economic returns and spillover effects occur in the generally already better off urban areas. The shift towards such production has thus, a largely negative effect on rural development.

In the above, the FAO are describing a process of industrialization that has, in its own individual way, already taken place in the UK over the last 200 years and whose spread throughout the developing world they predict, endorse and promote. Today many people in Britain have misgivings about the industrialized agricultural system that we have inherited, and are trying to de-industrialize it through support for organic farming, local foods, real meat, community supported agriculture, animal welfare measures, and campaigns against GM, pesticides and junk food. To the FAO these are an indication that we in Britain, have reached the 'post-industrial' phase where, as they put it, 'environmental and public health objectives take predominance'. The poorest in the developing world have no choice but to progress through three prior stages of industrialization and urbanization before they arrive at our state of grace, and even if they had the choice, that is what they would choose to do.

In the SOFA report, the FAO take a more guarded approach towards the process of industrialization, largely because it threatens the livelihoods of hundreds of millions of poor peasants. Nonetheless, they still consider that the inducement for smallholders to abandon livestock farming:

> is an integral part of the economic development process and should not be viewed as a negative trend. Concerns arise when the pace of change

in the livestock sector exceeds the capacity of the rest of the economy to provide alternative employment opportunities.[13]

There is no shortage of reasons for challenging the trajectory laid down by the FAO. Why, we may ask, should the rest of the world follow the path we have trodden, can they not learn from our mistakes? What is to be gained by dispossessing peasants – the only people on this earth whose environmental footprint does not extend into other people's space – and herding them into megacities of jaw-dropping unsustainability? When the world is already facing a resource crisis, how can it make sense to promote a centralized food economy that pours good human food down animals' gullets and that relies so heavily upon fossil fuels for its fertility, processing, transport, refrigeration, packaging and commercialisation?

These are all crucial questions, but they cannot be posed with much integrity by those who enjoy a diet that is unattainable for others on the lower rungs of the FAO's ladder of urbanization. However commendable it may be in Britain to eat only free range chicken fed on local organic grain, it is difficult to dispel an odour of hypocrisy when the only opportunity millions of people in developed countries are ever likely to have to obtain chicken on a regular basis is through factory farms on the edge of sprawling conurbations. Those who have opted for a vegan diet have, to some degree, transcended this dilemma. But a vegan diet, laudable though it may be for the individual, is neither sensible nor attainable for society as a whole.

To address this dilemma, I propose, as an alternative, the 'default livestock diet', not because it is necessarily the most sensible or the most admirable, but because it is comparatively easily defined. A vegan diet is one without animal products; the industrial meat diet advanced by the FAO is one that allows for anything that the consumer can afford and the producer can supply. In the muddy spectrum between, there are not many secure footholds, but the default livestock diet can be reasonably clearly defined as one that provides meat, dairy and other animal products which arise as the integral co-product of an agricultural system dedicated to the provision of sustainable vegetable nourishment. It is the section AB, the lower end of the hockey stick. As such it provides, not an orthodoxy to which we should aspire, but a benchmark by which we can assess the sustainability and the environmental justice of what we eat.

However, the term 'default livestock' is an unprepossessing one. As one livestock farmer Bill Grayson told me, it has 'a pejorative feel about it', which is perhaps what the FAO intended. The original meaning of the term is 'failure of something, want, defect' and one particularly unfortunate early meaning was 'lack of food or other necessaries'.[14] Grayson adds 'I think we do need something more inspiring if we are to debate the issue successfully within the wider social context'.

Its true that advertising a pack of sausages as 'made from 100 per cent default pork' is hardy likely to be a selling point. Despite toying with terms such as 'intrinsic', 'convivial' and 'low impact', I struggled to find any word which succinctly conveys the meaning of animal foods and services whose extent and nature is determined by the agricultural system as a whole. While I was mulling this over, Tara Garnett, of the Food Climate Research Network developed the very similar concept of livestock raised on 'ecological leftovers'. Garnett's definition is tighter than mine, since she restricts it solely to 'pigs and poultry

---

13   *Ibid.*, p 4.
14   *Oxford English Dictionary.*

fed on byproducts' and livestock reared on 'land unsuited for other agricultural purposes', whereas I would opt for 'grassland not required for other agricultural purposes' (which allows for livestock raised on rotated grassland). 'Ecological Leftovers' is a more explicit and easily understood term than default; but on the other hand, a pack of sausages advertised as 'made from 100 per cent leftover pork' is likely to be even harder to sell.

In the absence of any inspiration, I have settled for the word default, because I believe it is now commonly understood in the meaning which it holds in the phrase 'default font' – ie what happens automatically when we don't specify anything different – and that is how the FAO intended it to be understood. A default is the natural outcome of a system when it is not being overridden for any specific purpose.

One other observation must be made here: that there is little difference between a default livestock strategy and a permaculture one. Permaculture, in the words of Bill Mollison who coined the term, 'is a philosophy of working with, rather than against nature; of looking at plants and animals in all their functions, rather than treating any area as a single-product system.'[15] I view permaculture as an agricultural, ecologic and social system whose efficiency, like Nature's, relies on a maximum of connectivity between different but complementary components, so that the byproduct of one is easily available to provide the feedstock of another. Monoculture does not exist in nature, nor does waste, for whatever is waste to one species is food or habitat to another. The object in permaculture is not to produce a maximum yield of one or two species, but to enhance the richness, the diversity and the interconnectedness of the entire system; under normal circumstances, livestock activity, nourishment and waste production play a contributory, but not a dominant part.

In a permaculture system the role of livestock is by definition a default one: the number and the species of animals is a function of the system as a whole; overemphasis upon one species, or upon livestock as a whole, will cause waste, inefficiency and stress. The FAO's preferred 'active land user' strategy (ie competing with other sectors for the establishment of feedcrops, intensive pasture and production units) is the direct antithesis of permaculture. On the other hand its description of traditional mixed farming cited earlier in this chapter could almost serve as a definition of permaculture:

> Multifunctional, integrated production systems based on local resource
> systems with non-food outputs serving as inputs in other production
> activities within the system.

From hereon, I will continue to use the term 'default livestock ' to specify 'animal products and services that arise as the integral co-product of a wider agricultural system', while I shall take 'permaculture' to refer to strategies or systems where livestock is entirely or largely limited to a default role.

---

15   About Permaculture (n.d.), http://www.permaculture.net/about/definitions.html

5

THE PLIGHT OF THE PIG IN THE NANNY STATE

A pig in almost every cottage sty:
that is the infallible mark of a happy people.

William Cobbett, Rural Rides, 12 September 1826

In March 2008, pig farmers from around Britain staged a rally at Whitehall, memorable mainly for reworking a Tammy Wynette song under the title 'Stand by Your Ham'. The pigmen were drawing attention to the second crisis in their industry within nine years. In 1999–2000 world pig prices crashed because of lack of demand, largely as a result of the economic downturn in SE Asia. In England, weaners weighing 20 pounds could be bought cheaper than hamsters. This time pig prices were more stable, but the cost of feed more than doubled. Producers claimed to be making a loss of £26 on every pig, and a piggy-bank motif on the campaign's website showed the industry losing £6 every second.

One can understand farmers' anger when the supermarkets are not paying them a fair price. But the problem goes further than this, and in many ways pig farmers have themselves to blame. Their piggy bank is leaching money because they have forgotten what pigs are for. Why do children put their pocket-money in piggy-banks? Because the role of pigs throughout history has been to accumulate resources, and to act as a hedge against the oscillating availability and price of grain.[1]

Pigs do this in two ways, the first of these being through their omnivorous appetite. Ever since the prodigal son turned his nose up at the pig swill and went back to daddy, the pig has been the main repository for nutrients that were unfit for human consumption in sedentary cultures. In his 1814 treatise on swine, Robert Henderson observed that pigs will 'feed on anything. They are a kind of natural scavenger: they thrive on the outcasts of the kitchen, the sweepings of barns and granaries, the offals of a market, and most richly on the refuse of a dairy … The facility of feeding them everywhere, at a small expense, is a national benefit.' In Flora Thompson's *Larkrise*, every house had its pigsty right outside the kitchen, ready to receive the 'little taturs, the pot liquor' and other detritus.[2] Engels commented on the innumerable pig pens in Manchester 'into which the inhabitants of the court throw all refuse and offal, whence the swine grow fat.'[3] Pigs fed on scraps during the Second World War made up for the lack of animal

---

1   Since writing this I have been directed to an article on wikipedia claiming that piggy banks evolved their porcine form because they were made out of a certain kind of clay, called 'pygg'. However this theory doesn't explain a 14th century piggy bank from Java pictured on the same web-page. http://en.wikipedia.org/wiki/Piggy_bank
2   Thompson, Flora (1948), *Larkrise to Candleford*, Reprint Society, p 22.
3   Engels, F (1845), *The Condition of the Working Class in England*, Panther Books, 1972.

feed. The beaches of Goa are (or at least were) kept spotlessly clean and the latrines emptied by pigs which end up as spare ribs in the beach bars.

The other main function of pigs is to use up surplus grain in a good year. This is not just an economy measure; it also serves to introduce elasticity into the grain market, through what is known as the 'livestock feed buffer'(see Chapter 10). Pigs are not the only recipient of feed-buffer grain – it is also fed to beef, dairy animals or poultry. But feeding it to cows is inefficient because the bovine gut is designed to digest fibrous materials such as grass, not high protein feeds like grain. Poultry convert grain into meat more efficiently than pigs; but the advantage of pigs is that they have a high percentage of fat – which is not only in short supply in temperate climates, but also makes it easier to preserve the meat (which is why we don't have chicken 'bacon' or chicken 'ham').

A farmer who feeds his (or her) stock entirely upon bought-in grain is putting himself in a risky position, since he is completely at the mercy of an inherently volatile market. A farmer who wishes to hedge against disaster should either be growing the grain himself, or else be partially reliant upon alternative feeds. The main alternative feed for cows is grass; for pigs it has always been, and it always will be, waste.

This is what the UK's pig farmers have manifestly failed to do – they are now largely dependent upon grain and market-linked products for their feed – and that is why their piggy bank has been losing money.

In the early 1990s Professor Peter Brooks of Devon's now defunct Seale Hayne agricultural college gave a lecture entitled 'Rediscovering the Environmentally Friendly Pig'. In it he provided figures showing that only 33 per cent of UK compound pig food consisted of grains fit for human consumption; 22 per cent was oil seed residues and the remainder consisted mainly of various other kinds of animal and vegetable food residues.[4]

However, he warned that the pig's role as recycler of waste food was under threat from a number of factors, including:

(i)   Changing animal feed legislation

(ii)   The concentration of the animal feed industry into national corporations too large to cope with intrinsically variable raw materials or those only available in small quantities.

(iii)   Lower cereal prices as a result of EU policies and trade liberalization.

(iv)   The influence of the supermarkets who were 'increasingly dictating methods of production … and imposing limits on the range of raw materials and dietary inclusions'.

Brooks concluded:

the Pig must not be allowed to become a competitor with Man for food products but must remain a converter of that which Man cannot eat, or rejects, into a product which he can and will eat.

Fifteen years later, Professor Brooks' fears have been fulfilled. The proportion of cereals in pig food has almost doubled. Mole Valley Farmers' grower mix contains 66 per cent cereals, while Simon Mounsey of Feed Statistics reckons that

---

4   Feed figures from *Feed Facts Quarterly* 1992, cited in Brooks, P (c1993), *Rediscovering the Environmentally Friendly Pig*, Seale Hayne Agricultural College, unpublished.

'off the top of my head the proportion of cereals is now around 60 per cent.'[5] The bulk of the remainder consists of oil seed residues – rape, sunflower, but mostly soya – international commodities whose price is linked to the global grain price.

All the trends which Brooks pointed to have played a role in making pigs dependent upon cereals. But much of the blame for the decline of the environmental pig must be laid on DEFRA and its predecessor MAFF who have done everything in their power to regulate this recycling industry out of existence. The pressure to do so has come, not from the pig industry, but from the requirements of the more influential beef and dairy industries. In the second half of the 20th century, 'improvements' in the yield of dairy cows were so great that cows could not physically eat enough grass and grain to produce the milk that their metabolisms were capable of generating, so it became necessary to find a more concentrated source of nourishment. Initially this was provided by fishmeal from the Peruvian anchovy fishery, but when the fishery collapsed in 1972, because of overfishing to meet European demand, another source of instant protein had to be found. The answer was to feed meat and bone meal (MBM), which for years had been rendered and fed to omnivorous pigs and poultry without any problem, to herbivorous cows, and the result was BSE.

The BSE scare made the British public both aware and fearful of the rendering industry and prompted a series of crackdowns on the use of rendered meat and catering waste. Having caused a health scandal by feeding dead meat to herbivores, the government reacted in 1996, by banning the practice of feeding meat and bone meal to omnivorous pigs, whose gut is just as well adapted to eating meat as yours or mine. The result is that now large quantities of slaughterhouse wastes, particularly rich in phosphorus (which the world is slowly mining to exhaustion), are being incinerated in the production of cement, or else at Glanford power station in Norfolk,[6] and then landfilled, instead of finding their way back into the food chain and thence into the soil via the more natural route of an omnivore's gut. Meanwhile, pigs and cows alike are now increasingly fed on GM soya from North America, or non-GM from the Amazon.

In 2000 the United Kingdom Renderers' Association was still petitioning the government to allow MBM to be fed to cross-species omnivores;[7] but in 2001 the ban was extended across the whole of the EU. The consequence has been a huge increase in the import of non-GM soya from Brazil: the extent of recent deforestation in the Amazon exactly matches the shortfall in animal feed that the EU has had to make up by importing non-GM soya.[8] The rationale for the ban is that there a risk of pig feed being fed to cows. However, the World Health Organization has guidelines for feeding MBM to monogastrics, even in countries where there is a history of BSE, and deforesting large swathes of the Amazon to provide soya poses a far greater risk to public health.[9] The EU seems to be aware of the problem and stated in 2005 that its strategic goal is to 'move towards relaxation of certain measures of the current total feed ban when certain conditions are met'.[10]

5   Personal communication.

6   Energy Power Resource (EPR) website; http://www.eprl.co.uk/profile/index.html

7   Lawrence A (2000), Letter from Alan Lawrence, secretary UKRA to Barbara Richards, Food Standards Agency, 15 Sept 2000. www.food.gov.uk/multimedia/pdfs/lawrence.pdf

8   Elferink, E V and Nonhebel, S (2007), 'Does the Amazon Suffer from BSE Prevention', Agriculture, Ecosystems and Environment, 120, pp 467-9 .

9   WHO *et al* (2007) *Joint WHO/FAO/OIE Technical Consultation on BSE: Public Health, Animal Health and Trade, Conclusions and Key Recommendations*; OIE Headquarters, Paris, 11-14 June 2001 http://www.fao.org/ag/aga/AGAP/FRG/Feedsafety/PDFs/BSEWGF81101.pdf

10   European Commission (2005), *The TSE Roadmap*, Brussels 2005, http://ec.europa.eu/food/food/biosafety/bse/roadmap_en.pdf

But worse was to come. When the 2001 foot and mouth epidemic was revealed to have originated on a farm which had been illegally feeding uncooked swill under slack MAFF inspection, the Government temporarily banned the feeding of all swill and animal residues to any animals, including pigs and poultry – a move conveniently obscured by the smoke from the burning pyres. The Irish and French governments did the same.

When the foot and mouth crisis was over, instead of rescinding the temporary ban, the UK government made it permanent, with the approval of the NFU and the British Pig Association, who were happy to disassociate themselves from such mucky practices, and pour nice clean grain down their pigs' throats. There was minimal public consultation, and no compensation for waste food recyclers who had invested thousands of pounds in new machinery to comply with the government's increasingly stringent regulations. The Association of Swill Users complained to the Parliamentary Ombudsman who, six years later, concluded that both MAFF and its veterinary service were guilty of maladministration, but still refused to grant compensation.[11]

On 1 November 2002, with support from the UK, swill feeding was banned throughout the EU, partly because it was held responsible for some outbreaks of classical swine fever (though this is also endemic in the wild boar population). Germany and Austria were given a four year derogation because the ban put an end to their highly efficient food waste recycling systems, which fed an estimated six million pigs on swill, without their ever suffering a disease outbreak even a hundredth as serious as the UK catastrophe. In the peasant strongholds of Eastern Europe the ban may be having less effect. Hungarian Gábor Miklósi reported:

> Although no cases of swine fever have been reported in Hungary for quite a while now, farmers resigned themselves to the ban on swill without a grumble. They probably doubt that any authority is going to carry out dawn raids on hundreds of thousands of farms to check whether the regulations are being adhered to.[12]

The European pig swill and rendering industries thus served as the scapegoat for the UK government's century-long mismanagement of foot and mouth, a disease which causes no harm to pigs and is about as serious to cows as measles is to European children. Over six million animals were slaughtered by MAFF in 2001, to protect a beef export industry whose existence creates a vacancy for the very imports which cause the problem. The Government pinned the blame on the pig industry and helped persuade the EU to shut down the food recycling system right across Europe.

The abrupt closure of two of the EU's largest and most profitable recycling industries at a moment in history when the public was being urged to recycle everything else was nothing short of extraordinary. The swill ban was imposed as a panic reaction to an epidemic which could not possibly have been contained by restrictions on feed, and then absorbed into EU policy without any public assessment of its usefulness or effectiveness. The public appears to have taken it for granted that the ban is necessary for public health; but an examination of the historic figures for outbreaks of the diseases in question suggest that the health concerns reflect certain ideological considerations.

---

11   Parliamentary and Health Service Ombudsman (2007), *The Introduction of the Ban on Swill Feeding*, Dec 2007, www.ombudsman.org.uk/improving_services/special_reports/pca/swill_feeding/complaint.html

12   Miklósi, Gabor (2004), 'In Search of Lost Fat Content', *The Hungarian Quarterly*, No 173, Spring, 2004; www.hungarianquarterly.com/no173/9.htm#aut

During the Second World War the quantity of imported animal feed fell from 8,750,000 tonnes in 1939 to 1,250,000 tonnes in 1943, and most of this was reserved for milk production. The number of pigs on farms was virtually halved, but the Small Pig-Keepers Council lobbied local authorities to allow people to keep pigs in their back yards. In January 1940 a national campaign to save scraps for pig swill was initiated. Neighbourhood pig clubs were formed whose members fed their pigs with scraps from home, cafés, bakeries and anything that came to hand together with small rations of feed. In towns, waste food was boiled up into a concoction known as Tottenham Pudding – Ernest Onians, known as the Pudding King, made a fortune from selling it, which he invested in a collection of over 400 paintings. The pig clubs were so successful, that the Government exacted half the proceeds, to prevent pig keepers becoming noticeably fatter than the rest of the population.[13]

It is therefore interesting to examine DEFRA statistics on the incidence of disease during this 'special period'. The two main diseases in question are Classical Swine Fever (CSF) which in its acute form is frequently fatal for pigs, but doesn't affect other animals, and Foot and Mouth Disease, which is virtually harmless to pigs, but causes unpleasant but short-lived symptoms in cattle.

CSF was endemic in Britain in the period leading up to the war with an average of about 1800 cases per year in the previous ten years. There was a rise in the incidence of the disease in 1940 to 5019 cases, making it the worst year since 1898. Whilst this rise was partly attributed to the use of swill, it was probably helped by the fact that there was already an unusual amount of the disease around: there were 3286 cases in 1939, making it the worst year since 1916. The government therefore embarked on a campaign to ensure that all pig swill was cooked, and by 1942 the number of cases had dropped to 451, the lowest since records began. Admittedly, there were fewer pigs than between the wars, but the volume of swill feeding was much higher.

Foot and mouth was already subject to a slaughter policy, yet was still common in the UK, having occurred in every year in Britain since 1917. The incidence of disease was on average higher during the war, but not significantly, except for 1942, which had 670 cases, the worst year since 1924. On the other hand 1943, with only 27 cases, had the lowest count since 1930. None of the war years anywhere near matched the epidemic years of 1922-24, 1967 or 2001.

It is hard to attribute any statistically meaningful rise in disease to wartime swill use, or any conclusion other than that pigkeepers new to recycling needed to be reminded by the government that it is wise to boil pigswill. Over the long term the figures suggest that the main danger to livestock comes not from swill, but from the veterinary police themselves. Between 1879 and 1892, when there was no CSF slaughter policy, there were 68,741 cases affecting 330,913 animals, some of whom would have recovered. At an average of 4.8 pigs per farm, these incidents were obviously distressing, but they were unlikely to be catastrophic for the owners.

In 1893, a slaughter policy for CSF was initiated, but abandoned in 1917 after the culling of over 600,000 animals had failed to make any impression on the disease. It was reintroduced in 1963 and this time, after the massacre of over 400,000 pigs in three years, the disease was apparently eradicated. However in

13   BBC2 (2005), 'Pig Clubs Supplementing Meat Rations', WW2 People's War /www.bbc. co.uk/ww2peopleswar/stories/89/a4464489.shtml; The Living Archive (2003), *Pigs on the Home Front*, Wolverton, www.livingarchive.org.uk/nvq03/tracey/animals_pig.html; BBC (2003) *Pig Swill Estate Wins Poussin War*, BBC News. Monday, 4 February, 2002, http://news.bbc.co.uk/2/hi/ entertainment/1800588.stm

2001 the disease broke out in a shed containing 1,200 pigs on a farm in Suffolk which reared 3,500 pigs in total. Over 74,000 animals were slaughtered by the state vets before the disease was brought under control.

That, of course, is nothing compared with the bloodbaths carried out to rid the UK of foot and mouth, a disease which is regarded as an irritation in countries where cattle have built up resistance — foot and mouth is endemic in India, but that has not stopped it becoming the world's largest dairy producer. In the UK over 273,000 animals lost their lives in the 1922-24 epidemic, 487,000 in 1966-68, and in 2001, six million (or ten million according to those who believe MAFF was minimizing casualties). Two years before, in 1999 Y Leforban, secretary of the European Commission on Foot and Mouth Disease had warned that 'if the virus did manage to invade, it was likely to spread faster and further than ever before. This was because the trend towards intensive farming had led to larger, more densely stocked farms, while the decline of international trade barriers had facilitated long distance animal trading.'[14]

The vulnerability of large farms to infection can be seen most clearly in the rising toll per outbreak from classical swine fever. Between 1893 and 1917, when there was a slaughter policy, fewer than ten pigs were slaughtered per outbreak. During the Second World War, from 1941 to 1945 the average number of animals put down per outbreak was just 1.5. Since the slaughter policy was reintroduced in 1963 there have been nearly half a million pigs slaughtered at an average of 275 per outbreak. In 2001, there were 16 outbreaks, in which 74,000 pigs were slaughtered – an average of 4,625 per outbreak.[15]

While the loss of four or five pigs on a mixed farm is distressing to a small farmer, infection in a shed containing over 1000 pigs is a catastrophe. As the FAO observe, 'classical swine fever is a problem for pig producers who want to trade on the international market, but at a very low level of incidence it is an accepted risk for small scale pig producers.'[16] It is the vulnerability of confined animal feeding operations to disease that has led, not just to the swill ban, but to the draconian health and safety legislation that has forced the closure of numerous small slaughterhouses and made life difficult for small farmers. The National Farmers Union does not champion small scale swill farmers, it represents the big boys, so, like the British Pig Association, it supported the swill ban.[17]

Nor were the swill recyclers and renderers helped by their public image. The Co-op had already banned swill-fed pork from their shops in 1996, stating that it 'was not a natural feeding practice';[18] and in the UK there was a conspicuous silence from the champions of recycling, Friends of the Earth, who were no doubt afraid that raising the swill issue might alienate some of their vegetarian supporters. One of the rare voices raised against the swill ban was that of Boris Johnson, then MP for Henley. At a parliamentary debate in 2004 he reported:

> To take one example, Phyllis Court Hotel in Henley must now pay an
> extra £1,000 a year to a licensed collector, whose responsibility is to
> remove wet waste that previously went to a pigswill feeder. Given that
> there is room for only three years' waste in our landfill sites, that is not

14   Woods, Abigail (2004), *A Manufactured Plague*, Earthscan.

15   DEFRA (n.d.) *Animal Health and Welfare: Statistics – Outbreak Data* www.defra.gov.uk/animalh/diseases/notifiable/csf/stats.htm ; and www.defra.gov.uk/animalh/diseases/notifiable/fmd/stats.htm

16   FAO (2009), Livestock in the Balance, State of Food and Agriculture Report for 2009 (SOFA 2009), p 82.

17   Bradshaw, Ben (2004a) *Hansard*, 19 May 2004, http://www.publications.parliament.uk/pa/cm200304/cmhansrd/vo040519/text/40519w03.htm

18   The Co-op (2005) *Animal Feed*, www.co-operative.co.uk/en/corporate/ourviews/animalfeed/

the cleanest and greenest solution. It is estimated that the ban on swill feeding is generating an extra 1.7 million tonnes of waste per year, and that which does not fill up our landfill sites must be going down our drains, clogging up the sewers and attracting vermin.[19]

Agriculture Minister Ben Bradshaw replied that when the ban was imposed, only a small percentage of the national pig herd were fed on swill and only a small proportion of our food waste was going to pigs. In other words, the once-thriving swill industry was already so decimated by regulations that one might as well kill it off.

Bradshaw didn't specify exactly how much food was being wasted through-out the country, and there still seems to be some uncertainty. The government-sponsored Waste Resources Action Partnership (WRAP) states 'about 6.7 million tonnes – about a third of all the food we buy – ends up being thrown away', while another report commissioned by WRAP gives a figure of 5.5 million.[20] But these figures only represent domestic waste. According to New Labour's favourite food consultant, Lord Haskins, quoted in *The Independent*, we throw away about 20 million tonnes, half of all the food produced. Sixteen million tonnes of this is from homes, shops, restaurants, schools etc. The rest is lost between the farm and the shop shelf.[21] 'The government take their recycling targets seriously,' Bradshaw continued in his 2004 speech, 'and are aware that the amount of biodegradable waste sent to landfill must be reduced, not increased. DEFRA strongly supports the option of composting and the biogas treatment of catering waste.'

Five years later most of the waste is still landfilled or incinerated, and the most profitable recycling strategy has so far come from car owners making biofuel from the sudden surplus of used chip oil. However, in February 2008, DEFRA announced £10 million funding for an anaerobic digestion plant to process waste into energy and compost. WRAP calculates that if the 5.5 million tonnes of organic waste were anaerobically digested 'between 477 and 761 gigawatts hours per year of electricity would be generated – enough to meet the needs of up to 164,000 houses.'

This is somewhat underwhelming. WRAP are proposing to take waste equivalent to half the food necessary to feed a nation of 60 million people and burn its energy content to provide electricity for between 260,000 and 420,000 people. At 10 p per kilowatt hour, that's worth between £48 and £76 million.

The alternative of feeding the waste to pigs would provide at the very least 230,000 tonnes of pork per year with a retail value over £1 billion, based on a 24:1 conversion ratio achieved in a plant in Cuba in the 1990s.[22] That's the equivalent in calories of all the food required to feed 800,000 people. To produce the same amount of pork from cereal-fed pigs would require over a million tonnes of grain, enough to feed more than three million people. And this is only the post

19   Bradshaw, Ben (2004b), *Westminster Hall Debate on Pigswill*, 16 March 2004, www.benbradshaw.co.uk/speeches/speeches.php?id=33.

20   WRAP (2008), *Household Food Waste*, 2008, www.wrap.org.uk/retail/food_waste/index.html; WRAP (2007) *Cutting the Carbon Impacts of Waste*, Press Release, 5 March 2007 www.defra.gov.uk/news/2007/070305c.htm

21   Mesure, Susie (2008), 'The £20bn Mountain', *The Independent*, 2 March 2008.

22   This calculation is based on a 24:1 feed conversion ratio derived from FAO figures for a swill operation in Cuba producing a kilo of meat for every 24 kilos of swill (dry matter content, 14 to 20 per cent). US sources from the early 20th century for raw municipal garbage (e.g. from college halls) give conversion ratios of around 10:1 which suggests that the UK food waste could produce substantially more pork than the amounts I have given. *Pérez, Rena (1990), Feeding Pigs in the Tropics, Ministry of Sugar, Havana, FAO Animal Production and Health Paper 132*; see also J C Miller, 'The Value of Garbage in the Ration of Growing and Fattening Pigs', *Journal of Animal Science*, 1936a, pp 101-104, http://jas.fass.org/cgi/reprint/1936a/1/101.pdf

consumer waste. If Lord Haskins' 20 million tonnes were fed to pigs that would produce at least 800,000 tonnes of pork, more than one sixth of our entire meat consumption. And at the tail end of the whole process you can still extract some energy by anaerobically digesting the manure.

Around the time that these estimates were first published in *The Land* magazine, Tristram Stuart carried out an independent calculation based on figures provided by the Greenfinch anaerobic digestion plant, which currently generates 255 kwh of electricity per tonne of food waste, and could also provide hot water equivalent to another 76 kwh, with a potential value of £37.50. Stuart adopted a 15:1 waste to meat conversion ratio, based on his first hand experience of food waste processing plants in the Far East, which means that the pork derived from a tonne of waste food would be worth £330 – nine times as much. The CO_2 savings from food waste would also be nearly twice as great, and closer to 100 times as great if you included deforestation emissions caused by feeding soya to pigs.[23]

Of course, it would be even more efficient if we wasted less food. But there will always be a fair amount of waste and the figures reaffirm what is common sense – that keeping food in the food chain is more sensible than burning it.

Fortunately the British animal-hygiene neurosis hasn't yet spread to the rest of the world. Over half the world's pork is now raised in China, much of it in peasant households and small farms. A 1998 US study of Chinese agriculture noted: 'With the dissolution of many collective farms and the institution of the Household Responsibility System in the early 1980s, backyard [pork] production increased to 92.9 per cent by 1982.'[24] The advantage of decentralizing pig production is that it is easy to find waste food locally, easy to dispose of waste, and easy to ensure that nutrients cascade back to the land in the form of manure. A pig in every backyard is like a living dustbin in every backyard. Another US study of small-scale pig production in China observed:

> Hogs in China frequently consume large amounts of green roughage such as water plants, vegetable leaves, tubers, carrots, pumpkins, and various crop stalks ... green feeds account for 18 per cent of total feed consumption in backyard hog operations. Grain byproducts, such as bran and hulls are also used to feed swine ... By products from restaurants and manufactured food processes, such as alcohol, tofu and bean and tuber noodle production averaged between two and six per cent of total feed in backyard production. Frequently the primary grain and grain byproducts used for feed are derived from the crops grown and processed on the farm. Swine rations on backyard farms contain approximately 36.1 per cent purchased concentrate feeds, and in some provinces, such as Jilin, Shandong and Shaanxi, the share of concentrates in total feed is less than 20 per cent.[25]

The authors found that on the smaller farms (under 200kg pork per year) grain was only 31 per cent of the ration, while on the largest farms (over 500 kg of pork) it represented nearly 40 per cent of the ration. The researchers also

23   Stuart, Tristram (2009), *Waste: Uncovering the Global Food Scandal*, Penguin, pp 238-240.

24   Food and Agricultural Policy Research Institute (1998), *Impacts of Chinese Swine Feeding Practices on Future Chinese Feed Grain and Livestock Trade*, FAPRI, Iowa, USA, www.fapri.org/bulletin/nov98/chineseSwine.htm

25   Cheng Fang and Fuller, Frank (1998), *Feed-Grain Consumption by Traditional Pork-Producing Households in China*, Working Paper 98-WP 2003 Center for Agricultural and Rural Development, Iowa State University, December 1998. www.card.iastate.edu/publications/DBS/PDFFiles/98wp203.pdf

discovered that the more educated a farmer was, the more concentrate he was likely to feed his pigs – which doesn't say a lot for education.

Nor is pig-keeping restricted to the countryside:

> The Chinese do not waste anything that could be used as food for domestic animals or fowls. Peasants in the suburbs will collect kitchen wastes to feed pigs and poultry. Nearly all Chinese institutions – factories, offices, schools, barracks – run dining halls for their workers. These canteens usually keep pigs in their backyard as they can feed the animals on kitchen wastes. This is a regular practice in small cities and towns close to rural areas. Hygienic considerations restrict this practice in large cities.[26]

However, the Chinese are beginning to catch the British disease. That was written in 1987, and since then things have changed. Pork production has roughly tripled, outstripping US production, and now more than half of all pork in the world is produced in China. By the mid 1990s, 15 per cent of this pork was being produced in factory farms sited near large cities, and that figure has undoubtedly increased over the last ten years.[27]

Even so, at the beginning of the 21st century, 60 per cent of pig production was still taking place in people's backyards. The Chinese Government Feed Office estimates that in the years 2003 and 2004 China produced around 47 million tonnes of pork, 18 million tonnes of it on commercial farms, using only about 30 million tonnes of commercial feed.[28] It seems likely that the 28 million tonnes produced in Chinese backyards – which is about 31 per cent of the entire world's pork production – is still produced at a low feed to meat ratio. Despite the drift towards factory farms, the Chinese show that it is entirely feasible to feed large numbers of pigs on a diet which is 50 per cent waste, just as Britain did up until quite recently.

But while the Chinese pig industry is thriving, the UK industry is in trouble. The Stand By Your Ham campaign was a reaction to something close to a collapse of the British pig industry. The number of pigs in the UK declined from 8.1 million in 1998 to 4.8 million in 2007.[29] The British are still eating just as much pork, but around 60 per cent of it is imported, primarily from factories in Denmark and the Netherlands, where they are fed on a diet high in imported feed, but also from France, Germany, Ireland, Spain, Belgium, USA and Poland.[30] UK pig farmers complain, probably rightly, that they are disadvantaged by more stringent animal welfare legislation than in other countries. If Lord Haskins' figures are correct we are importing roughly the same quantity of pork and bacon as we could supply from post-farmgate food waste.

However, this is not such an ill wind. Over the last few years there has also been a discernible rise in the number of backyard and small-scale pigkeepers.

26   Furedy, C and Sun Hongchang (1987), *Municipal Waste and Resource Recovery in Chinese Cities, York University*, http://www.yorku.ca/furedy/papers/wr/ASEP87.doc.

27   Lei Xion (2006), *China Needs a New Type of Livestock Revolution*, Worldwatch, 12 Dec 2006, www.worldwatch.org/node/4772/print

28   Pork production figure of 47 million tonnes from Campbell, R. (2006), *Global Pork Meat Trade and Customer Demands on Feed Supply*, paper presented to 2006 Australasian Milling Conference. www.porkcrc.com.au/publications/Feed%20Millers%20Conference%20Global%20trade.doc/ Other figures from USDA (2005), *People's Republic of China, Grain and Feed Annual, 2005*. http://www.fas.usda.gov/gainfiles/200502/146118716.pdf

29   Annual Abstract of Statistics (2007) 21.7.

30   Sloyan, Mick (2006), *An Analysis of Pork and Pork Products Imported into the UK, British Pig Executive*, 2006, www.bpex.org/technical/publications/pdf/An_analysis_of%20pork_imports-11Apr06.pdf

There are no figures for this, and if there were they would be insignificant; but the shift in attitudes towards pigs, influenced by the campaigns of celebrity chefs like Hugh Fearnley-Whittingstall and the TV pigman Jimmy Doherty, is gathering pace. No cottage pig-keeper has any problem finding buyers for non-industrial pork or home cured bacon.

Many of these small-scale producers will be giving whatever food waste they have to their pigs, because that is what pigs are for. Often this is a minimal quantity, compared to a pig's voracious appetite: the amount of food wasted by a normal family is nowhere near enough to fatten a hog. But if the pigs are part of a community, or kept on a mixed farm, or there are obliging neighbours, or there is a dairy, brewery, market, or even a supermarket with unlocked skips nearby, then the amount of waste food available can be quite significant.

This trend may well be a sign of things to come. If the fossil-fuel dependent industrial farming system gradually collapses under the weight of its own contradictions, we will have to reabsorb pigs into the life of the community. Pigs will be kept in small numbers on mixed farms, in woodlands, at schools and prisons, near hospitals, behind restaurants and pubs, on allotments, at city farms and by neighbourhood pig clubs. Waste food will be recycled by the shortest possible route, according to DEFRA's own 'proximity principle'.[31]

Welcoming pigs back into the community may sound dotty because it seems so at odds with the sanitized suburbia promulgated by the nanny state. But we know that recycling food through pigs works because we did it during the Second World War – and if the prophets of peak oil and global warming are correct, we may be headed for similar conditions. It also looks as though the business of making Tottenham Pudding could become a simpler and less energy intensive business. Tristram Stuart visited a firm in Japan which sterilizes food waste by pasteurizing it at 90 degrees for just five minutes, and then inoculates it with a *Lactobacillus*, that ferments the swill much as if it were yoghurt. The product keeps for two weeks, is cheaper than other feed and conforms with Japanese recycling laws.[32]

Local pigs are also necessary if we are to get the nutrients back to the land. Peter Brooks' paper on 'Rediscovering the Environmentally Friendly Pig' remained unpublished because the pigs which he urged should be fed with recycled waste were to be kept on large scale, industrial farms, where manure accumulates in a heap that causes pollution and disposal problems.[33] If we are to have industrial-scale food processing at all, then its waste nutrients need to be cascaded back to mixed farms and smallholdings across the expanse of our land, and the most viable way to do this is through pig-food. It can be done through selling bags of subsidized compost, but at the current price of £10 for a 25 kilo sack, organic pig-feed justifies the transport a lot more than organic compost.

The return of the backyard pig will not only sort out our food waste problems more importantly it will enhance the cultural life of the nation. The pig has always been a mainstay of the poor family's independence and of a community's food sovereignty. 'Pigs for health' was an English expression; 'the pig pays the rent' an Irish one. It is because of the centrality of the pig to the security and the aspirations of settled peasants – be they Papuan tribesmen, the labourers of Larkrise, or the negroes of Maryland – that its life, its death and its afterlife are

---

31   DEFRA (2006), *New Technologies: Demonstrator Programme for Waste Technologies*, Criterion 14, http://www.defra.gov.uk/environment/waste/residual/newtech/demo/selection.htm

32   Stuart (2009), *op cit. 22*, p 279.

33   It was submitted to *The Ecologist* where I was a co-editor at the time, and I rejected it on these grounds. I have regretted doing so and apologise to Professor Brooks.

invariably imbued with ritual or religious significance.[34] 'Stand by your Ham'
is a call that resonates more fully amongst a proud and independent peasantry
than a bunch of whingeing factory farmers, some of whom have probably never
cured a ham in their lives. 'A couple of flitches of bacon', said William Cobbett,
'are worth fifty thousand Methodist sermons and religious tracts.'[35]

Reintegration into human society will also vastly improve the cultural life of
pigs. 'Dogs look up to us, cats look down on us,' Churchill is supposed to have
said, 'but pigs treat us as equals.' Pigs have as much to contribute to human
society as the canine and feline races, and as much to benefit. Once freed from
the factory, they will again be rooting in nearby woodland, sharing the life of the
farmyard, snuffling around with chickens in the *basse cour*, or lying pampered in
a sty by the kitchen. The congenial existence of the household pig is almost the
first thing Flora Thompson describes in *Larkrise to Candleford*:

> The family pig was everybody's pride and everybody's business ...
> its health and condition were regularly reported in letters to children
> away from home, together with news of their brothers and sisters. Men
> callers on Sunday afternoons came, not to see the family, but the pig,
> and would lounge with its owner against the pigsty door for an hour
> scratching the piggy's back ... The children on their way home from
> school would fill their arms with sow thistle, dandelion and choice
> long grasses, or roam along the hedgerows on wet evenings collecting
> snails for the pig's supper.

In one of his *Just So Stories*, 'The Cat Who Walked by Himself', Rudyard
Kipling compares the contracts made by the dog, the cow, the horse and the
cat as they submit to domestication by primitive humans. Kipling neglected to
include the pig – but 8,000 years ago when herds of wild swine were attracted to
the settlements of early agriculturalists, an interspecies bargain was negotiated.
'You give us your waste food and a bit of that extra juicy grass seed you have,
and we'll keep your camp clean and let you eat our surplus offspring (of
which we have many).' We have broken that contract by forcing our pigs into
concentration camps, and feeding them on concentrates, and we are materially
and spiritually the worse for it.

34   Leighton, Clare (1943), *Southern Harvest*, Victor Gollancz, p 64.
35   Cobbett, William, (1822), *Cottage Economy*, Peter Davies, 1926, p 103.

6

THE FAT OF THE LAND

Most people today turn up their noses at lard, which is strange since the name comes from the Greek *laros*, meaning 'pleasant to eat' via the noun *larinos* meaning 'fat'. 'Fat' too is a dirty word now, but in the days when people coined expressions such as 'fat of the land', 'bringing home the bacon', and 'cream of the crop', anything well endowed with edible oils was highly desirable, since it provided a concentrated form of energy.

For that reason, fat is particularly relished in cold or harsh climates: the Inuit are renowned for their prodigious appetite for seal blubber, while Afghanis traditionally eat virtually everything in a pool of warm grease taken from the huge, bosom-like appendages of fat which flop around on the buttocks of their sheep. But it is also in demand in warmer countries from people who have to sweat for a living. HR Davidson, in a 1948 textbook on pigs, states:

> Fat can be consumed with relish more or less in proportion to the amount of physical labour being carried out by the consumer. This relationship is much more important than that between temperature and consumption of fat, which usually receives more prominence. Very large quantities of 'fat backs' from the American packing houses are consumed by the negro and other workers in the cotton producing southern states of the USA, where the labour is hard and incomes low, but the temperature high.[1]

William Cobbett put it more bluntly: 'The man who cannot live on solid fat bacon, well fed and well cured, wants the sweet sauce of labour, or is fit for the hospital.'[2] But a taste for cruder kinds of fat tended to be a working class preference not only because they provided energy but also because they substituted for more expensive cuts of lean meat. Davidson continues:

> A cheap and easily managed source of animal fat is of great value to poor agricultural workers, and even yet these form the greater part of the world's population. So long as animal fat is available, the necessary protein can be obtained from vegetable sources very much more cheaply than from animals … Until recent times the predominant demand has been for a lard-producing type of pig rather than for one which would yield a high proportion of lean meat. With the increase of urban population and the increase in standards of living however, there is an increasing demand for a leaner type of pig.

Sixty years later the same dietary preferences linger on: building workers still breakfast in 'greasy spoons', while food outlets for office workers offer a more 'Mediterranean' fare with delicacies such as ciabattas, pastrami and sun dried

---

1   Davidson, H R (1949), *The Production and Marketing of Pigs*, Longman Green.
2   Cobbett, William, (1822), *Cottage Economy*, Peter Davies, 1926

tomatoes. Nonetheless the sophisticated urban consumer who declines gobbets of streaky bacon, and bread fried in sausage grease, may still be absorbing a fair amount of better disguised vegetable oil.

Up until about 1948, in the UK and other northern countries, most fat was of animal origin – margarine was largely made of animal fat, including whale blubber, and you would be hard pushed to find a bottle of vegetable oil in what (because it was where the bacon was hung) used to be called the larder. This was because until well into the 19th century countries north of the Alps didn't produce any vegetable oil to speak of, and olive oil was the only edible vegetable oil available – at a price. There is a biological explanation for this. Animals grow a layer of fat to keep warm, whereas plants contain oil to stop their seeds drying out; therefore animal fats are found mostly in cold climates, while vegetable oils are abundant in warm climates. There are exceptions such as linseed and rapeseed, but until recently these were regarded as inedible and used only for industrial purposes. In the north, animal fats are local foods – the rise of vegetable oil is a symptom of globalization.

The first edible vegetable oil to reach the market in the US, at the end of the 19th century, was 'Cottolene', a blend of cotton seed oil (a byproduct of the cotton industry) with a proportion of beef dripping. Food writer Alice Ross observes that it was one of the first branded and advertised commodities, and was marketed to upwardly mobile urban consumers who were encouraged to view pork as 'a less expensive meat ... associated with "the poorer classes"'.[3]

Since the 1950s, other vegetable oil products have followed the same promotional path as Cottolene, and largely replaced animal fats in margarine, in soap and as a cooking medium. A few of these – grape seed oil for example, or corn oil which constitutes only 4 per cent of wet-milled corn – are commercially viable because, like Cottolene, they are byproducts of another industry. Others such as palm oil and ground nut oil are competitive, at least in part because they are farmed in developing countries with cheap labour. In the 1970s, Canadian researchers bred a variety of rapeseed low in indigestible erucic acid, with the result that rape oil came to the table. It is now the third most common vegetable oil after soya and palm oil, and the only vegetable oil produced on a large scale in the UK.

Whilst these vegetable oils have outcompeted pig fat both financially and culturally, it remains unclear whether their ecological performance – in terms of land requirements or pollutant emissions for example – is better or worse than traditional pig fat production. The Tanganyika groundnut scheme of the late 1940s quickly established itself as the archetypal development disaster, and groundnut oil production has never quite vindicated itself since.[4] The incursions of palm oil plantations into tropical forests are currently giving cause for concern. Corn oil represents a small byproduct of a massive industry whose baleful influence upon global fast food consumption is eloquently spelt out in Michael Pollan's *The Omnivore's Dilemma*. Rape production in the UK is buoyant but unpopular, especially with beekeepers. Some varieties of vegetable oil, for example grape pip and sunflower, apparently perform quite well.

However, the crop which produces by far the largest amount of vegetable oil is soya. Grown primarily in the USA and Brazil, it is said to represent 70 per

3   Ross, Alice (2002), *Cottolene: The Mysterious Disappearance of Lard*, http://www.journalofantiques.com/Feb02/hearthfeb.htm
4   Wood, Alan (1950), *The Groundnut Affair*, Bodley Head. On more recent deforestation for groundut production in Senegal, see Coura Badiane (2001), *Senegal's Trade in Groundnuts*, TED Case Studies 646, 2001 www.american.edu/TED/senegal-groundnut.htm.

cent of the world's supply and its influence upon global food and commodities markets is second only to wheat.[5] As such it has attracted considerable attention from food analysts, not least in the Netherlands where soya represents around three quarters of all the oil consumed.

In 2005, two Dutch food analysts, Winnie Gerbens-Leenes and Sanderine Nonhebel, published an analysis of the land-take of the typical Dutch diet, based on a methodology developed by them in an earlier paper.[6] The resulting figures, reproduced in Table 2, are useful for anyone wishing to assess their personal food print. However, it should be borne in mind (a) that they refer to intensive chemical agriculture, using inputs entirely from arable production, and (b) the figures for meat and dairy are in real life highly variable, because animals can be fed on permanent grass (which is often less valuable than arable land) or on crop residues and waste, and this muddies the picture considerably.

The relatively low amount of land required for cereal and vegetable production and the high amount for meat, especially beef, are no great surprise. The land required for eight kilos of coffee, representing about two cups of coffee a day, is more significant than one might expect.

But the most surprising element in this analysis is the fat element. First, the quantity consumed is impressive – 29 kilos a year of butter, margarine and vegetable oil, over half a kilo per week. On top of that the Dutch could be ingesting another 15 kilos or so in the form of animal fat (on the assumption that the pork consumed is 15 per cent fat, the milk four per cent fat and the cheese 20 per cent fat). That's 44 kilos a year, the equivalent of a pack of butter every two days. (The average Briton consumes 33 kilos of butter, marge and veg oil, but less pork and cheese than the Dutch.)

Is this excessive? Gerbens-Leenes and Nonhebel provide figures for Dutch consumption from 1950 through to 1990. In 1950, right after World War II, when the Dutch ate less than half the amount of meat they eat now, their consumption of marge, veg oil and butter was 25 kilos. It went up to 30 kilos in 1960 and has stayed around there ever since. The estimated total consumption of fat went up from 40 kilos in 1950 to 44 kilos in 1990. Most of the animal fat consumption in the 1950s came from drinking large quantities of whole milk.

In other words, Dutch people's consumption of fat, since shortly after World War II, has been fairly constant. One could argue that a need for fat in 1950, to fuel a manual workforce living in homes with no central heating, has been supplanted by greed and hence obesity. But the rise in fat consumption is negligible compared to the increase in wealth, in meat consumption and in fossil fuel consumption over the same period. The conclusion we may draw from this (assuming the British are not that different from the Dutch) is that if, for environmental reasons, we had to retreat to a level of consumption equivalent to that enjoyed in Europe in the 1950s, then we would not be likely to make many savings in the realm of dietary fat consumption. People like their fat when they are rich, benefit from it when they are poor, and aren't going to give it up very easily. This in turn suggests that a move towards a vegan or vegetarian diet would result in increased consumption of vegetable oils (unless regulation, price incentives or health concerns acted as a deterrent).

5   Schill, Susanne (2008), 'Sizing Up the Soybean Market', *Biodiesel Magazine*, http://www. biodieselmagazine.com/article.jsp?article_id=2973

6   Gerbens-Leenes, W and Nonhebel, S (2005), 'Food and Land Use. The Influence of Consumption Patterns on the Use of Agricultural Resource', *Appetite*, 45, pp 24-31; Gerbens-Leenes, W *et al* (2002), 'A Method to Determine Land Requirements Relating to Food Consumption Patterns', *Agricultural Ecosystems and Environment* 90, pp 47-58.

Table 2: Land Requirements for Various Foods as Consumed in the Netherlands

Food	Land requirement m² per kilo	Consumption kg per capita	Land per capita m² per capita
Beverages			
Beer	0.5	91	45.5
Wine	1.5	15	22.5
Coffee	15.8	8	126.4
Tea	35.2	1	35.2
Total beverages			229.6
Fats			
Veg oil	20.7	13	269.1
Margarine	21.5	10	215
Low fat spread	10.3	3	30.9
Total veg fats			515
Meat			
Beef	**20.9**	**20**	**418**
Pork	**8.9**	**45**	**400.5**
Poultry	**7.3**	**17**	**124.1**
Other	**?**	**3**	**?**
Total 3 meats			**942.6**
Dairy and eggs			
Full fat milk	1.2	42	50.4
Semi-skimmed	0.9	42	37.8
Butter	13.8	3	41.4
Cheese	10.2	15	153
Eggs	3.5	9	31.5
Skimmed milk	?		
Condensed milk	?		
Total dairy eggs			314.1
Cereal and veg products			
Flour	1.4	66	92.4
Sugar	1.2	37	44.4
Potatoes	0.2	87	17.4
Vegetables (av)	0.3	63	18.9
Fruits (inc citrus)	0.5	95	47.5
			210.6
Total			2211.9

Meat = 43 per cent, Fat = 23   Dairy = 14   Beverage =10   Cereal and veg = 10

Credit: W Gerbens-Leenes and S Nonhebel, "Food and Land Use. The Influence of Consumption Patterns on the Use of Agricultural Resources", *Appetite* 45: 24-31 (2005)

The second noticeable feature of this table is how expensive vegetable oil is in terms of land use. The production of vegetable oils requires 20.7m² per kilo, and margarine even more. This is about the same as beef, which is by far the most inefficient form of meat to produce. By comparison, butter requires 13.8m² per kilo; and pork (which can comprise anything up to about 50 per cent fat) 8.9m². Elsewhere, Nonhebel writes: 'In the course of researching these food systems … I was struck by the huge acreage required for growing the 30 kilos of oil crops. It is over three times as big as the area required for growing the feed crops for 33 kilos of pork.'[7]

---

7   Nonhebel, Sanderine (2004), 'On Resource use in Food Production Systems: the Value of Livestock as "Rest Stream Upgrading System"', *Ecological Economics*, 48.

However, these figures are misleading because, as the authors point out:

The assessment of land requirements for individual food items was often complicated by a joint production dilemma. For soybeans, all of the land required was classified as being for soya oil, and oil cakes used for livestock products were merely seen as being a waste product. The consequence of assigning all the land to the main product is that land requirements for oil products were relatively high, while land requirements for livestock products were low.

It is this 'joint production dilemma' which makes comparisons between livestock based systems and stockfree systems complicated, and it casts doubt over the extent to which oilseed residues can be regarded as 'ecological leftovers' and animals fed on them regarded as default livestock.

Moreover this is not a marginal matter. Seventy per cent of the feed in the Dutch livestock industry is derived from food processing byproducts, and about 60 per cent of these byproducts come from three commodities: potatoes, sugar and soybean oil. Every year the average Dutch consumer, from his consumption of these three foodstuffs, generates 18kg of dried sugar beet pulp, 12kg of molasses, 7kg of potato waste and an impressive 72kg of soybean pulp. Fed to productive pigs, this is sufficient to produce 81 grams of pork per person per day, or over 29kg of pork per year. If you add in the 40 per cent of residues derived from other commodities you can produce 135 grams of pork per person per day, the equivalent of 49kg of pork per year, more than the average Dutch consumer was eating in 1990.[8] (On top of this, if slaughterhouse, catering and domestic food waste were also fed to pigs, as indeed they were not so long ago, then the quantity of pork produced from waste would be higher still.)

Many of these food processing residues do not have much value for other purposes. There is not a lot you can do with potato peelings from the potato crisp industry, sugar beet pulp or rapeseed residue, other than feed it to animals. However there is one particular commodity that stands out: 53 grams out of the 135 grams of pork that can be derived from food processing waste are attributable to high protein soybean residues (soya meal), and this is in such demand as animal feed that it is manifestly not a waste product; indeed it is debatable whether it is the oil or the meal that is the main product. There is four times as much meal as there is oil in a tonne of soybeans, while the oil is worth three and a half to four times the price of the meal.[9] This means that slightly over half the value of the soya lies in the meal, rather than in the oil. Nonhebel observes:

The production of vegetable oil is strongly intertwined with the production of feed. Their value is more or less the same so that, based on these data, the soybean cake cannot be considered as a residue but as a co-product. This implies that part of the acreage attributed to oil production should be attributed to meat.[10]

Indeed, but precisely what part? What proportion of the soya crop, and of oilseed crops in general, should be attributed to animal feed and what proportion to vegetable oil?

---

8   Elferink, E V, Nonhebel S, and Moll, H C (2007), 'Feeding Livestock Food Residue and the Consequences for the Environmental Impact of Meat', *Journal of Cleaner Production*, xx, 1-7.
9   Nonhebel (2004), *op cit*. 7. Cf Schill, Susanne, 'USDA Raises Projections for Soybean Prices', *Biodiesel Magazine*, 11 June 2008; www.biodieselmagazine.com/article.jsp?article_id=2429
10   Nonhebel (2004), *ibid*.

One possibility is to allocate land-take by weight: in other words the same number of square metres is assigned to a kilo of oil as to a kilo of meal. However, this distorts the picture because once the meal is converted inefficiently into meat it is worth far less per kilo in terms of human nourishment than the oil. It means, for example, that sunflower oil requires the same amount of land as soya oil (for a given total crop weight) even though twice as much sunflower oil is produced as soya oil.[11]

It therefore makes more sense to allocate land-take to oil and meal through an assessment of their value as end products. If we assume that the conversion ratio of soya meal to meat is 4:1 (a more correct ratio is 4.4:1), and that a kilo of meat is of the same value as a kilo of oil then we can apportion the land-take of each commodity on that basis. Table 3 shows the very different results that follow from allocating land requirements to the oil alone, by weight, and by their market value.[12]

Table 3: Area of Land Required to Produce Soya and Sunflower Commodities by Various Allocation Methods

	by Oil only	by Weight	by Market Value
Soya oil	22.22	4.44	11.1
Soya meal	4.44	2.78	
Sunflower oil	10	4.44	8
Sunflower meal	4.44	1.6	

All values in m². The methodology for this table is taken from Elferink *et al* (2007), however the values have been calculated by the author.

Sanderine Nonhebel and colleagues have calculated that, if land-take was assigned on the basis of the market value of the co-products, the Dutch consumer could be provided with 10.6 kg a year (29 grams a day) of pork with negligible land-take, because it had been solely fed on low-value residues from sugar and potatoes; or 29 kg a year (81 grams a day) of pork fed on residues from sugar, potatoes and soya, with an average land-take of about 5.5m² per kilo; or 55 kg a year (150 grams a day) from these three byproducts supplemented with grain crops, at an average land-take of about 8.0m² per kilo.[13] Remember, this is just from 60 per cent of available residues, which suggests that if 55 kg of pork were fed on all available residues, then it would take considerably less than 8m²

---

11   Both crops have similar total crop yields of 2.25 tonnes, which represent the global average yield for soya beans in 2007 and the West European 2007 yield for sunflower (the sunflowers are net of hulls). But soybeans are only 20 per cent oil, whereas sunfowers seeds are around 40 per cent. Figures from FAOSTAT, http://faostat.fao.org/

12   This valuation can either be achieved through an assessment of monetary value or of nutritional value. Under normal market conditions the monetary value broadly reflects the value of the end product – for example the fact that soya oil costs about four times as much as soya meal per kilo, mirrors the fact that it takes roughly four kilos of soyameal to produce a kilo of pork. In the case of the figures given in Table 3, it is assumed that the soya oil is four times as valuable as the meal, whereas the sunflower oil is six times as valuable – a rough reflection of typical market conditions. However market conditions do not always remain normal. Between 2001 and 2007, the price of sunflower oil rose steadily from a 20 year low of $350 per tonne to $1280 per tonne, while the price of sunflower meal stayed broadly the same at around $100 per tonne. This was at least partly due to subsidies for biodiesel oil crops. The high price of the oil would increase its share of the land-take, and decrease that of the meal, although in ecological terms nothing had changed. Clearly, volatile prices like these can make any assessment of the land requirements through market value alone meaningless. Figures from FAOSTAT and National Sunflower Association, Sunflower Statistics, www.sunflowernsa.com/stats/default.asp?contentID=218

13   Elferink *et al* (2007), *op cit*. 8.

of land per kilo, and a lot less land than the area of around 11m² required to produce a kilo of soya oil.

In other words, the more pork (or meat) that is consumed, the more land per kilo is needed to produce it, because a small quantity of meat can be reared on more or less worthless byproducts, while larger quantities require high value oilseed meal, or dedicated grain crops. The relative environmental impact of pork fat and vegetable oil can be viewed schematically in Fig 2, where the land requirement for a kilo of pork increases as consumption increases, following the hockey-stick curve described on (p 35), whereas the land requirement for soya and sunflower oil is constant. Under current conditions, large amounts of pork (and hence pig fat) can be consumed before its land-take starts to approach that of vegetable oil – but if the consumption of pig fat increased, the amount of soya fat eaten and soya meal available would decline, and this would affect the land take of the pork.

Figure 2. Land Requirements of Pork, Soya Oil and Sunflower Oil

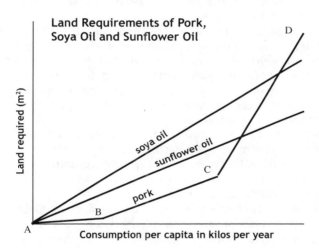

The steeper the gradient, the more land required per kilo consumed. The gradient AB shows the land requirement when pigs are fed on food wastes and low value residues. As per capita consumption increases, at BC, pigs must be fed on oilseed meal which uses more land (approximately the same amount of land as sunflower oil). CD shows land requirement when pigs are fed on grain, and so require more land per kilo than both soya and sunflower oil. In the Netherlands, point B is at 10.6 kg per year, and point C is at 29 kg per year (if the relative land take of meal and oil are allocated according to market value).

It is no doubt possible to get too pedantic about the land requirements for different oil and fat products. But the figures above also demonstrate how the vegetable oil industry and the livestock industry have become reliant upon each other – indeed they feed each other. The more vegetable oil we produce, the more meal there is to feed to livestock; and the more soya-fed meat we buy, the more vegetable oil there is to shove in our snacks. Is it any wonder that we are all becoming obese?

Which industry, if any, is driving this spiral of supply? In respect of the majority of oilseeds, the vegetable oil industry is undoubtedly the main driver because the value of the meal is relatively low. The byproduct constituency of sunflower, groundnuts, mustard seed, palm kernel, copra and sesame seed is in every case around 50 per cent or less, while the value of the oil is higher. Rape consists of 40 per cent oil and 60 per cent meal. Palm oil produces only 250 kilos of meal for every tonne of oil.[14] The soybean industry however follows a different pattern. Soya beans are only 20 per cent oil, and 80 per cent meal – yet the soybean industry is by far the largest, comprising around 60 per cent of the oilseed meal market and 32 per cent of the world's vegetable oil.[15]

Since the oil yield is so low, it might seem that it is the demand for meal rather than oil that drives the expansion of the soya bean industry. On the other hand, the yield of the meal is low as well, by comparison to conventional feed grains. In fact, in respect of land use, soya is an extravagant crop all round. A hectare (non-organic) produces about 2.25 tonnes of beans yielding 450kg of oil, enough to meet the total annual fat consumption of ten consumers in the Netherlands. Fed to pigs, the residues will produce 405kg of pork, which, with the oil, gives a total of 855kg of high value food.[16]

By contrast a hectare of non-organic mixed feed (wheat, beans and cassava) fed to pigs would produce 1,650kg of pork – nearly twice as much – of which 405 kg (25 per cent) could conceivably be fat.[17] Even a field of UK grown barley fed to pigs will provide considerably more food than a field of soybeans.[18] Feeding grain to pigs is wasteful, but it is not as extravagant in its use of land as growing soya beans for oil and feeding the residue to pigs.

So why grow soya? As a legume, it economizes on nitrogen, and fits into the US corn farmer's rotation. In the southern hemisphere it has proved effective as a pioneer crop on virgin land where yields of other crops would be low and labour is cheap – and hence is a major agent of deforestation in South America. And it is valued for feeding to overproductive dairy cows who cannot metabolize enough protein for their milk yield from less concentrated feeds. For these and other reasons soya comes onto the market at an attractive price.

Soybeans can be eaten whole by humans, and soya is clearly a much more efficient crop if the meal left over after making oil is consumed by humans, for example as soy milk or tofu. However, these products typically contain about four per cent fat, and to avoid creating a surplus of soymeal, a vegan society would probably have to reduce its consumption of soya oil.

Currently only about three per cent of soybean protein is consumed by humans, and some oil is used for industrial purposes.[19] The rest of the soybean industry is buoyed up by an unholy alliance of vegetable oil consumers and meat eaters, without either of whom cultivation would be unviable. Perhaps we should make it unviable. Soya oil consumers seeking to lower their food footprint would be advised to switch to higher yielding European-grown oils such as rape or sunflower oil, or else to lard; similarly motivated consumers of soya-fed meat would do better to chose livestock fed on genuine byproducts, lucerne or grass; or else go vegan.

14   Fadel, J G (1999), 'Quantitative Analyses of Selected Plant Byproduct Feedstuffs, a Global Perspective', *Animal Feed Science and Technology* 79, pp 255-268; the particularly low figure given by Fadel for palm oil meal is explained at the wikipedia entry for palm oil: http://en.wikipedia.org/wiki/Palm_oil.

15   FAO (2006), *Medium Term Prospects for Agricultural Commodities: Oilmeal*, www.fao.org/docrep/006/y5143e/y5143e0l.htm

16   Calculation derived from Nonhebel (2004), *op cit*.7.

17   Nonhebel (2004), *op cit*.7.

18   The UK average yield of 6.6 tonnes of barley per hectare in the UK, fed to pigs at a conversion ratio of 5 to 1, would provide 1,320 kilos of meat.

19   Steinfeld *et al* (2006), *Livestock's Long Shadow*, FAO 2006.

7

HARD TO SWALLOW

Of all the statistical clichés about livestock that are passed like a relay baton from one article or website to another, there is one that stands out in its enormity. George Monbiot, writing in *The Guardian*, tells us that 'every kilogram of beef we consume, according to research by the agronomists David Pimentel and Robert Goodland, requires around 100,000 litres of water to produce'. But it's unfair to single out Monbiot because the 100,000 litre figure pops up all over the place, as often as not preceded by the word 'staggering'. It can be found on dozens of websites representing groups as diverse as the Humanist Vegetarian Group and the Woodcraft Folk. Jonathan Porritt repeated it in *The Guardian* three years after Monbiot, in an article ambiguously entitled 'Hard to Swallow'.[1] There is also a variant, which appeared in the *Vegan* magazine, where Pimentel is quoted saying that 'it takes around 100 times more water to produce 1 kg of beef than it does to produce 1 kg of wheat' (which proves to be the same statistic, as we are also told that wheat requires a little less than 1000 litres per kilo).[2]

Every kilo of beef? 100,000 litres sounds like a lot of water, and it is a lot of water. It is the equivalent of an acre an inch deep in water. To put it into context, let's apply it to Bramley, an Angus/Jersey cross steer I had the pleasure of rearing. Bramley, when he was slaughtered, provided something over 125 kilos of meat. That means, according to Pimentel and Goodland, that over the course of his 16 month life he consumed 12,500 tonnes of water, which is the equivalent of an acre ten foot deep in water – or 25,000 litres on each of the 500 days of his life.

How he managed to achieve this feat this I am at a loss to explain, since all Bramley did while he was alive was hang out in a field and eat grass, without ever manifesting any unusual appetite for water. For the first six months of his life he drank about a gallon and a half of milk, and for the final ten months he ate maybe 50 kilos of fresh grass daily and drank perhaps 50 litres of water. Even if we include the water in the grass he ate, this cannot have been much more than 100 litres of water per day. And it's not as though he locked all this water up in his little body so that nobody else could use it. Most of it either went back to the land where it came from as urine or dung, or was else was transpired into the atmosphere to reappear as rain somewhere else – which is also what happens to the rather larger quantities of water drunk by the poplar trees that border the field he lived in.

So where could the other 24,900 litres per day that Pimentel claims Bramley consumed possibly come from? About 100 litres was used at the slaughterhouse. Perhaps one should include a proportion of what his mother drank; and what about the manufacture of plastic bags that the beef was packed in, and the eartag?

---

1 Porritt, Jonathan (2006), 'Hard to Swallow', *The Guardian*, 4 January.
2 Toms, Catriona (2003), 'Are Your Meals Costing the Earth?', *Vegan*, Winter 2003.

However many overheads one tries to summon up (and there aren't that many with suckler beef), it is impossible to arrive at any figure remotely resembling 25 tonnes per day or 100,000 litres per kilo. Bramley consumed at the very most 50 tonnes of water in his entire life which works out at about 400 litres per kilo of beef.

So how did Pimentel and his associates arrive at their figure? No doubt prodigious quantities of water are required for US feedlots, where thousands of animals are confined and fed upon lucerne and corn grown on irrigated land, and where mucking out must be an exercise of Augean proportions. But not all beef is feedlot beef, by a very long chalk, and even if the US feedlots' consumption bumps up the global average to 100,000 per kilo, which I doubt, it seems a pointless and misleading exercise to bunch free range cows and feedlot cows together under one integer.

I decided to check up the source material. In a book by Pimentel, Westra and Noss, I found a table which did indeed state that beef used 100,000 litres of water per kilogram.[3] There was not a word about how this figure was derived, but I was referred to an article in *Bioscience* by Pimentel and no less than nine other writers. I procured this and found that it gave the same table, but also an indication as to how the figure was arrived at:

> Producing 1 kg of beef requires approximately 100 kg of hay/forage and 4 kg of grain. Producing this much forage and grain requires approximately 100,000 litres of water to produce approximately 100 kg of plant biomass plus 5,400 litres to produce 4 kg of grain.[4]

This makes it clear that Pimentel is not just referring to grain fed beef. On the contrary, at 1000 litres per kilo of hay, against 1,350 litres per kilo of grain (which is more nutritious than hay), Pimentel's figures suggest that beef from a cow fed on grain requires *less* water than beef from a cow fed on hay or forage.

Anyone who has actually fed cows, and has a reasonable command of arithmetic, will observe that Pimentel's cow eats about twice as much as is normally necessary to produce a kilo of beef.[5] But this is a mere quibble compared to the question of how he manages to derive a figure of 1,000 litres of water per 100 kilo of hay. Hay, in regions where grass thrives, appears every year without the human application of any water whatsoever. Over large parts of Britain, it is difficult to stop hay growing, which is why people use lawnmowers.

The only possible way you can assign that much water consumption to hay is by taking into account every drop of rain that falls on the ground over a year. In the UK a hectare of average grass produces the equivalent of about 10,000 kilos of hay and receives about eight million litres of water in rainfall.[6] This is the equivalent of 800 litres of water per kilo of hay. The UK is blessed with high fertility and temperate rainfall, so 1,000 litres of water per kilo of hay equivalent is perhaps about right.

In other words, Pimentel's calculation takes into account every scrap of precipitation that falls upon the area of land that a beef cow might occupy. This appears to be corroborated by his statement in the same paper that 'on rangeland

---

3   Pimentel, D, Westra, L, and Noss, R F (eds) (2000) *Ecological Integrity: Integrating Environment, Conservation, and Health,* Washington DC: Island Press.

4   Pimentel *et al* (1997), 'Water Resources, Agriculture, the Environment and Society', *Bioscience,* 47:2, Feb 1997.

5   A growing cow puts on a kilo of meat in about 2.5 to 3 days, and requires about 17 kilos of hay per day, which works out at 40 to 50 kilos of hay per kilo of beef. See, for example, Lampkin, N *et al* (2004), *Organic Farm Management Handbook,* University of Wales, Aberystwyth, pp 152-3; or The Longhorn Cattle Society (n.d.), *Feeding Longhorn Cattle,* www.longhorncattlesociety.com

6   Leach, G (1976), *Energy and Food Production,* IPC Science and Technology Press.

more than 200,000 litres of water are needed to produce one kilogram of beef', since normally the only water which rangeland receives is rain. The reference for this figure was given as 'Thomas 1987'. When I got hold of 'Thomas 1987', a charming article about rangeland water by a farmer's son who eventually became President of New Mexico State University, it cited the figure but referred me to 'Thomas 1977'.[7] At this point I gave up the chase.

There is no doubt some virtue in calculating the amount of rain that falls from the sky upon the land which a beef cow occupies; but to suggest that this figure is a reflection of the toll that meat production exacts upon global water reserves is absurd. And to claim, as is currently now fashionable, that this figure represents 'embodied water' that can be exported to another country seems even more spurious.[8] Ninety-nine per cent of the rain would fall onto the ground, and do much the same thing, whether the cow were there or not. If the cow weren't there, the grass would still grow, and rabbits or deer or bison would graze it and consume the same theoretical amount of water.

Pimentel's figure is not the only one: half an hour on a search engine is sufficient to dig up a bewildering array of different estimates for the amount of water in a kilo of beef, Pimentel's just happens to be the largest. Marc Reisner gives between 16,000 litres and 166,000; the International Water Management Institute of Stockholm gives 70,000 litres; *New Scientist*, citing *Forbes* magazine, gives 50,000; Herb Schulbach of the University of California extension service gives 43,000 litres; *Diet for a New America* gives 20,000; Jim Oltjen of the Department of Animal Science at UC Davis California (supported by Gerald Ward at the Department of Animal Science at Colorado) comes up with 3660 litres; and the Alberta Beef Industry claims it is just 130 litres.[9] To borrow a comment made by George Monbiot about different figures given for the price of nuclear energy, the amount of water consumed by a beef cow appears to be a function of your political position.[10] The National Cattlemen's Beef Association cite Oltjen's figure – a man whose bias (we are told) is betrayed by the official University portrait of him in a cowboy hat. But whereas, in Monbiot's example, the New Economic Foundation's nuclear electricity costs eight times as much as the Nuclear Energy Institute's, in the field of bovine water consumption the ideological polarization is of a different order of magnitude: the green environmentalist's cow consumes 27 times as much water as the cowboys' cow, and 769 times as much as the cows in Alberta. In the face of such uncertainty, a *Newsweek* reporter in 1998 decided that the safest option was to state that 'the water that goes into a 1,000 pound steer would float a destroyer' and now there are 110 separate citations of this sentence on Google's search engine, mostly on vegan and vegetarian websites.

There is an attempt to explain the provenance of some of these figures in an article dating from about 2000, by John Robbins, heir to the Baskin Robbins ice

7   'Thomas, G W, 'Water: Critical and Evasive Resource on Semiarid Lands' in Jordan, W (ed.) (1987), *Water and Water Policy in World Food Supplies*, Proceedings of the Conference, May 26-30 1985, Texas A and M University, 1987; Thomas, G W (1977) 'Environmental Sensitivity and Production Potential on Semi-arid Rangelands' in Lubbock, *Frontiers of the Semi-arid World*, Texas Tech University.

8   E.g. The Water Efficiency Task Group (2006), *Virtual Water*, Victorian Water Industry Association Inc. http://www.vicwater.org.au/uploads/Water%20Restrictions/Virtual%20Water%20Final.pdf

9   Reisner, Marc (1993), *Cadillac Desert: American West and its Disappearing Water*, Penguin; Nelson, Jeff (2004), *8500 Gallons of Water for 1 Pound of Beef*, Vegsource Interactive Ltd, http://www.vegsource.com/articles2/water_stockholm.htm; Robbins, J (n.d.) *2,500 Gallons All Wet*, Earthsave, http://earthsave.org/newsletters/water.htm; Alberta Beef Industry (n.d.), *When Is a Cow Not a Cow? Just the Facts*, http://www.agt.net/public/jross/JustFaqs.htm

10   Monbiot, G. (2006), *Heat*, Penguin Allen Lane, p 95.

cream empire and spokesman for *Diet for a New America*. His opening sentence explains that he had been 'asked recently whether the figures given in *Diet for A New America* for how much water it takes to produce a pound of meat today are still accurate.' He went on:

> The figure of 2,500 US gallons to produce a pound of meat (roughly 20,000 litres per kilo) comes from a statement by the renowned scientist Georg Borgstrom at the 1981 annual meeting of the American Association for the Advancement of Science.[11]

Borgstrom was now inconveniently dead and unable to explain how he arrived at the figure. But happily the figure had been confirmed in an 'extraordinarily thorough' 162 page report entitled *Water Inputs in California Food Production* – which unsurprisingly only examined beef reared in California. Robbins goes on to quote Marc Reisner:

> In California, the biggest single consumer of water is irrigated pasture: grass grown in a near desert climate for cows. In 1986, irrigated pasture used about 5.3 million acre-feet of water – as much as all 27 million people in the state consumed, including for swimming pools and lawns ... Is California atypical? Only in the sense that agriculture in California, despite all the desert grass and irrigated rice, accounts for proportionately less water use than in most of the other western states. In Colorado, for example, alfalfa to feed cows consumes nearly 30 per cent of all the state's water.

In other words this figure of 20,000 litres per kilo of meat – a mere fifth of Pimentel's estimate – reflects performance in the corner of the United States which Marc Reisner memorably named Cadillac Desert, where extractive irrigation is carried out on a scale that cannot be afforded by any other country in the world.

Robbins continues:

> It is not only Dr Borgstrom that has come to similar conclusions. In their landmark book *Population Resources Environment*, Stanford Professors Paul R. and Anne H. Ehrlich stated that the amount of water used to produce one pound of meat ranges from 2,500 to as much as 6,000 gallons [20,000 to 50,000 litres per kilo]. Dr. Borgstrom, Drs Ehrlich and I all used the word 'meat' to refer specifically to beef.

How very scientific of them! By the same token, no doubt, they feel free to use the word 'beef' when they are referring specifically to feedlot steers fed on irrigated crops grown in Cadillac Desert. It is, frankly, not very encouraging to discover that the cream of the United States' environmental establishment can become exercised in a debate about the volumes of water required to produce a kilo of meat, without bothering to define what species they are talking about, what region they are talking about, or what management system is being employed.

Some wild figures are also available for the amount of water used by slaughterhouses. According to a report cited by Mark Gold, Brazil's biggest pig slaughterhouse at Concordia Santa Catarina uses 10,000 cubic metres of water every working day.[12] That is a flow of over one cubic metre every second –

---

11   Robbins, J, *op cit.*
12   Gold, Mark (2004), *The Global Benefits of Eating Less Meat*, CIWF.

enough water in a day to cover an area of a hectare one metre deep. If it is not completely awash, it must be a very big abattoir indeed.

I checked up with Snells of Chard, where Bramley was slaughtered, and they estimate that washing down a 300 kilo beef carcase would take a maximum of 180 litres. In a day in which they might slaughter 40 cattle, they use in total about 4,500 litres (1,000 gallons). Staceys of Somerton also gave a figure of about 4,500 litres for a day in which they might slaughter 40 pigs and 60 sheep.

Unless there is something which I have missed, all the commonly cited figures for water consumption are wildly inaccurate in respect of grass-fed free range beef, killed in small-scale abattoirs. Yet these figures are routinely cited by commentators without any regard for whether they are meaningful or correct. No doubt, beef fed on forage grown on irrigated land in semi-desert conditions consumes huge amounts of precious water; but this represents only a small fraction of the world's beef supply, and anyway most is produced to feed North Americans, who on average consume huge amounts of almost everything.

As for the concept of 'embodied water' – the idea that by importing corn or beef we are somehow importing somebody else's water – I'm not sure that isn't just hocus pocus. One writer explains 'the concept is very similar to embedded carbon', but actually it isn't.[13] The fossil fuel carbon that goes into making an ingot of aluminium is lost once it has been expended. The water that goes into animals, or into plants, isn't spent, it goes back into the soil or the atmosphere. Certainly, when irrigation is employed, water takes a different, engineered route, and that may lead to problems. But most beef is rainfed. When Bramley died he didn't embody any water other than the 200 kilos or so that made up his bodyweight. All the rest of the water that he ingested throughout his life, he either excreted or transpired, and it went back into the environment to be used by some other living creature.

---

13   IGD (2007), *Embedded Water in Food Production*, carbonhttp://www.igd.com/index.asp?id=1&f id=1&sid=5&tid=48&cid=326

8

THE GOLDEN HOOF
AND GREEN MANURE

Three leaved grass soon yields a threefold profit.
Three volumes may be writ in praise of it.
Andrew Yarraton, 1663

Until now, I haven't mentioned organic farming. The feed conversion ratios, and the ensuing land-take ratios given in preceding chapters, are all based upon chemical agriculture. The product per hectare of the livestock farmer has so far been assessed against yields of wheat in the order of eight tonnes per hectare, which were completely unachievable 30 years ago. Such yields are dependent upon pumping the land year in year out with artificial nitrogen fertilizers whose manufacture requires heavy use of fossil fuels.

Every farming system stands somewhere along two separate axes, one of which runs between chemical farming at one pole and organic at the other, while the second runs between livestock-based and stockfree (or vegan) farming. This means we can identify four separate types of farming method: chemical livestock, organic livestock, chemical stockfree and organic stockfree. There is plenty of room in the middle for various kinds of mixed farming, but because the issue of food production is now so highly charged, many farmers willingly position themselves at a polar extremity: indeed to label your food as 'organic' or 'stock free' you are obliged to.

Any attempt to assess the full nature of livestock's contribution to the agricultural economy has to compare each of these four systems, and it is by making such a comparison that an assessment can best be arrived at. They are distinguished in particular by the different methods they employ to acquire and store fertility, and so before making this comparison, it seems prudent to provide an overview of the history of nitrogen, which starts with the history of manure.

A Serendipitous Revolution

Organic stock farmers, defending meat production, often cite the necessity for manure to fertilize organic arable production. Without it, they claim, organic farming couldn't survive, and hence stockrearing is necessary, and meat is integral to the whole system. Here is John Seymour, participating in a debate on meat-eating published by *The Ecologist* in 1976:

> The great pioneers of high farming in England, who eventually taught
> good farming to the whole world, base their whole practice on the
> beneficial interactions between plants and animals. Coke of Norfolk's
> motto was 'a full bullock yard makes a full stackyard'. He didn't worry

about how much protein a bullock takes to convert into a kilogram of beef, he was chiefly interested in the great tonnages of beautifully composted wheat straw that those bullocks made under their feet. He raised the fertility of his fifty thousand acres of light land in Norfolk so that it produced two tons of wheat to the acre, where before such a figure would have been unbelievable. That was all done by the manure of bullocks and the treading and dunging of folded sheep.[1]

I don't know where Seymour got the figure of two tons an acre from – possibly, since he was in live debate, off the top of his head. Coke in 1816 said he got yields of between 42 and 48 bushels, which (at 60lbs per bushel) is between 1,125 kilos and 1,285 kilos.[2] Even so this was impressive compared with yields in previous centuries, which were as low as 12 bushels per acre – particularly since he started with 'a thin sandy soil [which] produced but a scanty yield of rye.'[3] It wasn't until 1998 that the average US yield of wheat broke the 40 bushel barrier.[4] Modern day East Anglian grain barons are now getting up to 150 bushels an acre, but the dryland-farmers of the US and Canadian prairies – whose viability depends upon the vast amounts of land they have at their disposal rather than what they put unto it – have yet to catch up with Coke.

During the medieval period most land was cultivated under a two or three year rotation, during which half or a third of the land remained fallow for one year, and fertility was enhanced, for example, by folding animals which had been grazing on common land during the day and on the arable land at night. The problem during the later middle ages was that as population increased the amount of corn (ie grain) being grown was increasing, while the amount of pasture and hence of manure was declining: 'The frontier between grass and corn was moving away from grass towards corn all over the areas of mixed farms throughout the Middle Ages,' wrote M M Postan. 'In the course of the 13th century, and perhaps even earlier, the frontier not only approached, but in many places crossed its limits of safety ... By the beginning of the following century, in corn growing parts of the country taken as a whole, pasture and the animal population had been reduced to a level incompatible with the conduct of mixed farming.'[5] Corn yields declined and as they declined there was a need to cultivate even greater areas.

Temporary relief from this spiral of decline occurred in the middle of the 14th century with the timely intervention of the Black Death, which reduced the population by about a third. Permanent escape was provided by the rather more tardy introduction, over the 17th to 19th centuries, of turnips and nitrogen-bearing legumes, such as clover and sainfoin, into the farming rotation. These cultivated crops could feed considerably more animals than could be pastured on a strip of fallow. The Norfolk four-course rotation (pioneered by Coke), of wheat, turnips, barley and clover, never left the land idle and fed everything not consumed by humans to livestock.

The combined effect of legumes and turnips was to reduce the fallow while increasing the output of animal feed. Farmers were able to keep

---

1   The Ecologist (1976), 'Must an Ecological Society be a Vegetarian One', *The Ecologist*, Vol 6:10, Dceember 1976.

2   Ernle, Lord (1912), *English Farming Past and Present*, Longmans.

3   *Ibid.*

4   USDA (1998), 'Record US Wheat Yields, Large Stocks Pressure Prices', *Agricultural Outlook*, Economic Research Service, USDA, August 1998, http://www.ers.usda.gov/publications/agoutlook/aug1998/ao253b.pdf

5   Postan, M (1975), *The Medieval Economy and Society*, Pelican, p 65.

more livestock, and this in turn increased the supply of animal manure. Animal products were the main fertilizer used on arable land, so with more manure available soil fertility improved and yields per acre rose.[6]

This view, from J V Beckett in 1990, echoes that of the agricultural historian, Lord Ernle, 80 years previously :

The field cultivation of roots, clover and artificial grasses ... enabled farmers to carry more numerous, bigger and heavier stock; more stock gave more manure; more manure raised larger crops; larger crops supported still larger flocks and herds.[7]

This was the classic view which I once accepted, but now consider needs adjusting. Although animals were key to the advances of the agricultural revolution, manure played a smaller role than it occupies in the account propagated with such enthusiasm by Lord Ernle, John Seymour and others. The extra nitrogen came from the growing of legume crops for animals, but not necessarily via the digestive tracts of animals that ate them. This might seem a nice distinction but it is crucial to the debate between organic livestock farmers and organic stockless farmers that is the focus of much of this chapter.

Nitrogen is an essential nutrient for growing crops, and in many situations, as in medieval England, a shortage of nitrogen is the factor limiting an increase in production.[8] Fortunately there is plenty of nitrogen in the atmosphere, and from there nature dispenses it in regular and fairly generous doses into the soil. Some of it falls as rain, and some of it is fixed in the soil by plants, algae, bacteria or by other mechanisms. A number of wild or semi-wild ecosystems, notably grassland, accumulate and store it, much as they also store carbon. A hectare of long established pasture, in England, typically holds about 7,275 kilos of stored nitrogen.

When the descendants of Cain grub up grassland, that stored nitrogen becomes accessible to them for their crops. Initially there are high yields. However, nature is prudent as well as generous, and she doesn't hand out her blessings all at once. Plants can only absorb nitrogen when it is in a 'mineralized' or 'free' state, and in a recently ploughed up pasture there will be a decent quota of this in the soil. However most of the nitrogen stored in the field is 'organic' – ie immobilized in organic matter, inaccessible to growing crops, and only becomes mineralized nitrogen, little by little, at a rate of about 1.5 per cent of the total per year.[9] A farmer who does not add extra nitrogen fertilizers can draw only on this free nitrogen for his crops. If he uses from this source more nitrogen than is added to it from the atmosphere every year, then his store of inaccessible organic nitrogen – his capital, so to speak – is depreciating, year by year, and so is his 1.5 per cent interest, and so therefore are his crop yields. When the proportion of this declining 1.5 per cent that is being removed from the field falls to a point

6   Beckett, J V (1990), *The Agricultural Revolution*, Blackwell. A similar phenomenon may have taken place in the early years of the 20th century in the United States, when corn harvests reliant upon the inherited fertility of virgin grasslands began to decline – the Oklahoma dustbowl being the most graphic example. According to Colin Tudge 'US cereal production waned somewhat in the 1920s and 30s until more livestock were introduced'. Tudge, C (2003), *So Shall We Reap*, Penguin Allen Lane.

7   Ernle (1912), *op cit*.

8   It has been argued that in some situations phosphorus was the limiting factor, a matter dealt with shortly, see Newman *op cit*. 13.

9   Allen, R (2008), 'The Nitrogen Hypothesis and the English Agricultural Revolution: A Biological Analysis', *The Journal of Economic History*, Vol 68, No 1, 2008. Allen says 1 to 3 per cent, but for convenience here I call it 2 per cent.

equal to all the nitrogen being deposited annually on the land, the system enters into equilibrium: the farmer is taking the same amount of nitrogen off the land as is arriving from the atmosphere every year, and crop yields are static. The pattern can be seen in the unfertilized crops of wheat that have been planted continuously at Rothampsted since 1850 – for the first 50 years yields declined steadily and for the last 100 years they have been stable.

This state of affairs is indefinitely sustainable: but it is not productive, and if population increases there will eventually be more people than the land can support. By 1350, yields from the three course system, with its two years of grain and one of fallow, after gradually declining over several centuries, had bottomed out, at about 800 kilos per hectare – just ten per cent of a good yield today – with an equilibrium of 3,000 kilos of nitrogen stored in the soil. According to models developed by Robert Allen, farmers at this period were deriving 45 kg per hectare of their nitrogen from the bank of immobilized nitrogen on their fields, the equivalent of what was absorbed from the atmosphere, plus a pathetic four kilos from manure.[10] Given that a modern cow excretes about 80 kilos of nitrogen a year, this last figure is very low, and reflects a state where there was just one hectare of meadow left for every four hectares of arable. These meadows, like the cropland, had been depleted by the continual removal of hay crops to the point that they were only giving their minimum yield of two tonnes of hay per hectare. There was a shortage of meadows because so much land had been brought into cultivation and mined to the point of minimum yield.

There were, according to most historians, three routes out of this predicament – grass, beans and clover. First, as we have seen, the Black Death, in 1350, conveniently removed a third of the population, resulting in fewer mouths to feed, and spare land which could be put over to pasture. This may have increased the amount of manure to a degree, but not enough to result in anything other than a small increase in yields. Much more importantly land managed as pasture (ie which didn't have its hay crop removed every year) absorbed a surplus of nitrogen every year, gradually building up its bank of organic nitrogen until, after decades or even centuries, it reached the upper equilibrium state of 7,275 kilos, where the proportion of the 1.5 per cent of organic lost every year through the less extractive process of pasturing equals the amount of fresh nitrogen absorbed.

The result was that when, after a century had passed and the population started to approach former levels, there was a landbank of meadow and pasture which had accumulated a large store of nitrogen and which, if ploughed up for arable, at first produced yields that were double those of worn out arable land. The value of grassland was reflected in the saying 'to break a pasture makes a man: to make a pasture breaks a man.' By the 1600s the practice of ploughing up grassland, growing corn for some years, and then putting it back to grass had become a long term rotational system, known as convertible or up-and-down husbandry; and it is still reflected today in the common practice on mixed organic farms, of alternating four or five years clover and ryegrass leys with two or three years arable.

But the rise of livestock-rearing after the Black Death, according to Allen, had another effect. In the 15th and 16th centuries increasing numbers of arable farmers replaced spring corn in their rotation with peas or beans, which contained

10   Allen (2008), *ibid*. See also Chorley, G P H (1981), 'The Agricultural Revolution in Northern Europe, 1750-1880: Nitrogen, Legumes and Crop Productivity', *Economic History Review*, Vol 34 pt 1 pp 71-93; and Clark, G, 'The Economics of Exhaustion: The Postan Thesis and the Agricultural Revolution', *Journal of Economic History*, Vol 52 Pt 1, 1992.

higher levels of protein, and whose 'appeal lay as a source of fodder to support the growing number of animals now being stocked'. This would have benefited the fertility of the soil insofar as the manure from these animals was returned to the land, providing a small amount of readily available nitrogen. But Allen's point is that the continual cultivation of nitrogen-fixing peas and beans and the incorporation of their residues into the soil gradually built up the reserves of organic nitrogen, resulting in a slow steady rise in the fertility of arable soils, and in the value of the 1.5 per cent of the stored nitrogen released every year.

The introduction of clover with the Norfolk rotation in the 18th century, famously brought immediate and obvious improvements in fertility from the additional livestock manure and (unlike peas and beans) from a certain amount of free nitrogen released from ploughed in residues. But Allen estimates that 60 per cent of the nitrogen was ploughed in and 'increased the stock of organic nitrogen and contributed to the long run rise in yields'. It is, he claims, the slow rise in yields from the much earlier introduction of peas and beans that has perplexed historians who:

> have been trying to pin the revolution down to a half century
> some time between 1500 and 1800. But which? Was it during the
> parliamentary enclosures? the last half of the 17th century? 1590-
> 1660? The important point of the simulations reported here is that the
> Agricultural Revolution took several centuries. 150-200 years is the
> time frame for the stock of soil nitrogen to move from one equilibrium
> to another.

It is not only historians who were misled, but also the farmers themselves with all their proverbs about the 'hooves that turned sand to gold' and 'muck is the mother of money'. 'Low medieval yields', concludes Allen, 'were not due to a deficiency of manure nor were high 18th century yields due to its abundance. Livestock however were crucial to the yield increases. It was not their dung that mattered, but what they ate, for the legumes fed to them increased nitrogen stocks that ultimately benefited the corn.' By championing the virtues of muck, generations of farmers, without realising what they were doing, improved the condition of English farmland by applying green manures.

This is a point of view that lends considerable historical support to the case for stockfree agriculture. Allen is suggesting that the agricultural revolution, insofar as it was dependent upon nitrogen fixation, could have been carried out just as successfully, in the long run, by using legumes as a green manure rather than as animal feed. In the light of recent research into stockfree organic rotations entirely reliant upon green manures, it looks as though this could be correct.

The Geography of Muck

Unfortunately no sooner had one problem been solved by the so called agricultural revolution, than another was caused by the industrial revolution. As land was progressively enclosed for 'improvement' by capitalist farmers like Coke (even though many of the improvements just described took place some 200 years earlier on open fields) peasants were squeezed out of their communities and impelled into towns, in a process similar to one now taking place on a far larger scale in developing countries. The distance separating the rising population of townspeople from the countryside where their food was grown made it increasingly difficult to return many important nutrients

– including those found in human dung, food waste, horse manure and urban pig and cow manure – back to the land whence they came. It was not impossible to truck or ship or channel them back to the countryside, and sometimes they were; but the financial (and aesthetic) incentive to gather the fruits of the soil and cart them into towns was far greater than the incentive to gather the contents of urban middens and return them to the land. Increasingly they were dumped into rivers or the sea, and the invention of flush toilets and the development of water-borne sewage sytems facilitated the process.

The matter was observed by two people in particular, who offered very different solutions. In *Das Kapital*, Marx observed:

> Large landed property reduces the agricultural population to an ever
> decreasing minimum, and confronts it with an ever growing industrial
> population, crammed together in large towns; in this way it produces
> conditions that provoke an irreparable rift in the interdependent
> process of social metabolism, metabolism prescribed by the natural
> laws of life itself. The result of this is a squandering of the vitality of
> the soil, which is carried by trade far beyond the bounds of a single
> country.[11]

Despite Marx's choice of the word 'irreparable', *The Communist Manifesto* advocated the 'combination of agriculture with manufacturing industries; gradual abolition of all the distinction between town and country by a more equable distribution of the populace over the country,' – a project outlined in more detail by the anarchist Peter Kropotkin in *Fields Factories and Workshops* fifty years later. But, as usual, the social critics were ignored and the solution was instead provided by a technical fix – indeed the invention which could reasonably lay claim to the title 'mother of all technical fixes.'

The originator of chemical fertilizers was Justus von Liebig, an astute German chemist who was invited to London by a government facing a rising sewage disposal problem in its capital city. In a letter to prime minister Sir Robert Peel, von Liebig wrote:

> The cause of the exhaustion of the soil, is sought in the customs and
> habits of the townspeople, ie. in the construction of water closets,
> which do not admit of a collection and preservation of the liquid and
> solid excrement. They do not return in Britain to the fields, but are
> carried by the rivers into the sea. The equilibrium in the fertility of the
> soil is destroyed by this incessant removal of phosphates and can only
> be restored by an equivalent supply.[12]

Liebig was more immediately concerned by phosphates than by nitrogen for a good reason. Phosphorus is less volatile than nitrogen and cannot be extracted from the atmosphere; therefore it is more important to return *the same* phosphorus that is extracted from land back to the land. In a subsistence economy, such as a tribe living off shifting agriculture, this is quite easy. The people cultivate, harvest and feed themselves and their animals, who defecate, die and are buried in the fallow, which one day is cropped again: so the cycle continues, and most of the phophate follows suit. The problem arises when crops are exported, because the phosphates can be returned to the land, but not necessarily in the right

---

11   Marx, K (1894), *Das Kapital*, Volume 3, 'The Transformation of Profit into Ground Rent'.
12   Cited in Girardet, H (2000), *Cities, People, Planet*, Liverpool Schumacher lectures, April 2000. See also Foster, J B and Magdoff, D (1998), 'Liebig, Marx and the Depletion of Soil Ferility', *Monthly Review*, 1 July 1998.

proportions in the right places; and it is even more of a problem when wastes are flushed out to sea, because (aside from those which are recuperated through fish and seaweed) the phosphates are lost forever.

The role of phosphorus in farming has been lucidly explained by the biologist E I Newman in work carried out with the help of historian P D A Harvey.[13] In a typical modern crop of wheat, around 96 per cent of the phosphorus removed from land in a year is in the crop, most of it in the grain, some in the straw. The remaining four per cent is lost through leaching and soil erosion. There is some natural input from weathering of rocks, but this not much more than one per cent. Thus if all of the crop finds its way back to the land, in the form of crop residues, seed sown, manure, sewage, bones, corpses etc, about 97 per cent of the phosphorus is being returned to the land, and it is only necessary to bring in another three per cent from somewhere else to make the farming system sustainable in phosphorus. On the other hand, if all the crop is being exported, and is not returned to the land (ie. it is either flushed out to sea or dumped in the wrong place) then over time there will be a shortfall of phosphorus on the land in question.

A subsistence economy can make up its small shortfall in phosphorus relatively easily, for example by feeding animals on outlying land which is not being cultivated and has a small natural income of phosphorus. The advantage of dovecotes, whose birds foraged far and wide, becomes clear once one understands the phosphorus cycle. But an agricultural economy which exports has to locate substantial sources of nutrients to replenish its phosphorus. In his comparison of various agricultural systems, Newman concludes that Egypt has always been able to do this thanks to the silt deposited by the flooding of the Nile. But the records of a manor farm in Cuxham, Oxfordshire, show that even in the 14th century far more phosphorus was being exported off the farm than was delivered onto it, and though it had relatively high yields for the period, the farm was drawing upon its capital, its store of phosphorus in the soil.

Liebig realized that with the simultaneous development of industrial capitalism and flush toilets, land was being depleted of its phosphorus much faster than before, and somehow the process needed to be reversed. As is so often the case, the Government ignored the sound but unwelcome advice of its hired consultant, and opted to build two massive pipelines to flush the sewage into the Thames estuary rather than recycle it. Liebig responded by starting work on developing artificial fertilizers, which he concluded appeared to be the only way of replacing the nutrients that were so wantonly wasted.

But while the government refused to set an example in London, other municipalities and individuals set about taking advantage of the resource. Graham Harvey describes one of the earliest of these schemes, at Craigentinny Meadows on the sandy land between Edinburgh and the sea:

> Sewage was collected in ponds and allowed to settle for a short time
> before being passed down a wide open drain known as the Foul Burn
> with an outfall in the Forth estuary. Farmers with land alongside the
> burn drew off agreed amounts of the slurry to irrigate their grasslands.
> Most of the fertilized grass was cut for hay and fed to dairy cows
> housed in sheds in the nearby towns of Leith, Musselburgh and
> Portobello as well as Edinburgh itself. Old grassland produced about 40
> tons an acre [100 tonnes per hectare] on the contents of Scottish water

---

13   Newman, E J (1997), 'Phosphorus Balance of Contrasting Farming Systems, Past and Present. Can Food be Sustainable', *Journal of Applied Ecology*, 34, pp 1334-7, 1997.

closets, ryegrass pastures as much as 60 tons (150 tonnes per ha).[14]

150 tonnes per hectare, equivalent to over 25 tonnes of hay, is a phenomenal yield, even by today's standard. The Edinburgh system was engineered to run down a dedicated channel, the Foul Burn, but other schemes simply took advantage of nutrients that were being poured into the rivers. The Duke of Portland constructed 400 acres of water meadows along the banks of the river Maun, into which the growing settlement of Mansfield obligingly dumped all its sewage. The irrigated meadows were described by one observer as 'the pride of Nottinghamshire'[15], and another wrote:

> The eye, after wandering through the glades of the forest, and resting
> on the brown carpeting of fern and heather with which it is clothed, is
> amazed on coming suddenly in view of the rich green of the meadows,
> extended for miles before it, laid in gentle slopes and artificial terraces,
> and preserved in perpetual verdure by supplies of water continually
> thrown over their surface.[16]

A century later, André Voisin described a similar system in the valley of the Elorn, in Brittany:

> Each peasant in this valley owns a small acreage of grass which he
> irrigates regularly with water from the River Elorn; this is charged
> with the waste from the tanneries and the rich sewage from the town
> of Landivisiau [population in 2007 8,000], which dominates that part of
> the valley. The grass is mown to be carried in and fed to the animals in
> the stall. In this way they succeed in obtaining eight or nine rotations.
> In winter they are aided by the fact that the water of the Elorn is
> higher than that of neighbouring rivers. According to the information
> collected (roughly estimated as it was) the peasants of the Elorn valley
> harvest per annum more than 120 tonnes of grass per hectare.[17]

There was an obvious health advantage in applying the effluent to grass, which was not immediately consumed by humans, in contrast to Chinese night soil which was directly applied to crops, and was probably responsible for high death rates from dysentery, typhoid and cholera.[18] Also, the rapid growth of the grass and succession of crops probably helped minimize leaching of nitrogen – though the method of transmitting the effluent from the town was far from ideal.

However, during the second half of the 19th century most of these schemes began to fall into disuse as the shortfall in soil nutrients was plugged, not initially by chemical fertilizers but by the importation of vast amounts of guano – the accumulated droppings of generations of seabirds on distant islands. Phosphate was quarried abroad and also became available as a byproduct of the steel industry (basic slag). Fertility (like most other commodities during the 19th century) was being mined in the four quarters of the world and imported in massive concentrations to the island that was the fount of industrial civilization, and that included industrial agriculture.

14   Harvey, Graham (2002), *The Forgiveness of Nature*, Vintage, p 179.
15   Caird, James (1967) *English Agriculture in 1850-51*, 2nd edition, Kelley, NY, p 206; cited in Harvey, *ibid*.
16   Denison, John (1840), 'On the Duke of Portland's Watermeadows at Clipstone Park', *Journal of the Royal Agricultural Society of England*, Vol 1, p 359, cited in Harvey, *ibid*.
17   Voisin, A. (1959), *Grass Productivity*, Philosophical Library; Island Press 1988, pp 24-25. Quotation is shortened.
18   Buck, J L (1982), *Land Utilization in China*, Paragon 1968 and *Land Utilization in China: Statistics*, Garland NY, cited in Newman, *op cit* 13.

But while the phosphate quarries were still producing, the supply of guano could not last forever, and in 1909 an economic process for manufacturing nitrogen fertilizers from the atmosphere was invented in 1909 by another German, Fritz Haber – a man who might have become as great a hero in the annals of science as Liebig if he hadn't gone on to invent the gas used to kill Jews in Hitler's concentration camps.[19] The Haber/Bosch process, although it creates fertilizer from thin air, is not without its costs, as it requires energy, quite a lot in fact. It takes roughly a tonne of coal to produce one tonne of sulphate of ammonia fertilizer (213 kg N – sufficient for about two hectares of wheat). The European Fertilizer Manufacturer association states that 'of all the energy used to produce wheat ' including other fertilizers and all tractor use, 'approximately 50 per cent is needed to produce, transport and apply nitrogen fertilizers.' The precise figure they give is 52 per cent.[20] Roughly one per cent of the world's manufactured energy is devoted to nitrogen fertilizers which are used in the production of about half the world's food.[21]

The introduction of chemical fertilizers put an end to the requirement to maintain the balance between animals and crops – Postan's 'shifting frontier between grass and grain' – and hence to mixed farming. In theory the application of supplementary quantities of chemical fertilizers in situations where a good balance between animals, legumes and grain could not easily be achieved might have been beneficial, and energy well spent. But, of course, that is not how things happen in a capitalist society. Muck needs shifting and that requires labour. Labour is expensive, while chemical fertilizers are easier to apply, and like most fossil-fuel based products cheap in relationship to the energy they consume. An arable farmer who moved over to chemical fertilizers could forget about livestock completely, cut out the complicated, labour-intensive routine of managing animals and their muck, and concentrate on a relatively simple annual cycle of ploughing, scattering and harvesting.

The result was the decline of the mixed farm, which was replaced, particularly in the US, by monocultural grain farms, and their corollary, the feedlot. Since livestock was no longer an integral part of arable farming, the two activities could be sited hundreds of miles apart and grain from the prairies shipped to huge factories where livestock could fattened and processed in their thousands. Manure became redundant, and muck instead of being an essential component of the farming cycle, became a waste disposal problem. The larger and more industrialized the farm or feedlot, the bigger the problem.

The problem has recently spread to developing countries, particularly China. In the 17th century, 34 per cent of China's manure came from the human residents and 47 per cent from pigs.[22] These remained the main sources of fertilizer up until 1965 when the amount of synthetic fertilizer applied to crops was negligible. Since then the use of synthetic nitrogen has increased spectacularly, as has per capita meat consumption and the size of cities. By the mid 1990s livestock factory farms on the edge of cities, non-existent before 1979, were supplying 15 per cent of China's pork, 25 per cent of its eggs and 40 per cent of its broiler hens.

19   Smil, Vaclav (2004), *Enriching the Earth: Fritz Haber, Carl Bosch and the Transformation of World Food Production*, MIT Press.

20   EFMA (n.d.), *Harvesting Energy with Fertilizers*, European Fertilizer Manufacturers' Association, OECD, http://webdomino1.oecd.org/comnet/agr/BiomassAg.nsf/viewHtml/index/$FILE/EFMAenergyPaper.pdf#search=%22energy%20to%20produce%20nitrogen%20fertilizer%22

21   Smith, B E (2002), 'Nitrogenase Reveals Its Inner Secrets', *Science*, 297, pp 1654-5, cited in Steinfeld, H, *Livestock's Long Shadow*, FAO p 86; and Smil (2004), *op cit 19*, p 159.

22   Zhang Luziang (1956), *Shen's Agricultural Book*, cited in Netting, R (1993), *Smallholders Householders: Farm Families and the Ecology of Intensive, Sustainable Agriculture*, Stanford, p 139.

At a conference in Hangzou held in 2006, Chinese academics warned that these farms were producing 'huge animal wastes, which, mostly discharged untreated, have caused serious pollution to water and air'. Wu Weixiang, a professor at Zhejiang University, reported that in Zhejiang province on the east coast the livestock industry produced 26 million tonnes of manure from some 900 large farms, of which only 6.2 per cent was applied to cropland. 'Many of these substances were discharged directly, without any treatment', he observed. 'Animal manure has been a major contributor to surface water contamination in Zhejiang.'[23]

The more hysterical brand of vegan delights in drawing attention to these muck heaps and slurry tips as another stick to beat the livestock farmer with. Here is the comment of Bruce Friedrich of PETA, speaking in *The Ecologist*'s 1976 debate:

> (Farm) animals produce far more excrement than humans do (130 times as much in the US for example) … and this excrement is swimming with bacteria, antibiotics, hormones, pesticides and other filth. It pollutes water and destroys topsoil. There really can be no such thing as a meat-eating environmentalist.

The same statistic resurfaced twenty years later in a 1997 report. This is from a web article entitled 'Why Go Veg?':

> In December 1997 the Senate Agricultural Committee released a report stating that animals raised for food produce 130 times as much excrement as the entire human population, roughly 68,000 pounds a second. A Scripps Howard synopsis of the report stated: 'It's untreated and insanitary, bubbling with chemicals and disease-bearing organisms … It goes into the soil and into the water that many people will, ultimately, bathe in and wash their clothes with and drink. It is poisoning rivers and killing fish and sickening people … Catastrophic cases of pollution, sickness and death are occurring in areas where livestock operations are concentrated … Every place where the animal factories have located, neighbours have complained of falling sick.' This excrement is also generally believed to be responsible for the 'cell from hell', *Pfiesteria*, a deadly microbe, the discovery of which is detailed in Rodney Barker's *And the Waters Turned to Blood*.[24]

The issue of whether or not the 'cell from hell' is even toxic, let alone deadly, is (unsurprisingly) a matter of dispute.[25] A few years later, Steve Boyan, formerly a political science professor at the University of Maryland, came up with the same figure, and the same language:

> Livestock now produces 130 times the amount of waste that people do. The waste is untreated and unsanitary. It bubbles with chemicals and disease-bearing organisms. It overpowers nature's ability to clean it up. It's poisoning rivers, killing fish and getting into human drinking water. 65 per cent of California's population is threatened by pollution in water just from dairy cow manure.[26]

23 Lei Xiong (2006), *China Needs a New Type of Livestock Revolution*, Worldwatch, Dec 12 2006, www.worldwatch.org/node/4772/print
24 Dropsoul (n.d.) *Why Go Veg?* www.dropsoul.com/why-veg.php, no date.
25 http://en.wikipedia.org/wiki/Pfiesteria_piscicida, 9/10/06
26 Boyan, Steve (n.d.), *How Our Food Choices can Help Save the Environment*, EarthSave, http://www.earthsave.org/environment/foodchoices.htm

All these authors talk about the 130 fold difference in volume between the performance of mostly bovine and human bowels as though there was something wrong with the cow. But for the organic farmer or the Third World peasant, 130 times as much manure is a reason to sing her praises: if anything it is the human digestive system which is inadequate. Anybody who has tried composting their own manure for use in their garden will have been disappointed by the pathetic couple of bucketfuls of dry compost which they end up with – less in a year than a cow produces in three days.

Farmyard manure – FYM as the textbooks call it – becomes a major problem if you have too much of it in one place and you manage it badly; and that happens because pointless overuse of synthetic nitrogen means that manure is both undervalued, and concentrated in the wrong place. Farmers who value manure mix it with straw, stack it and compost it, or they may even generate heat or electricity from it. Robert Netting recounts how the wealth of farmers in the Swiss community of Torbel could be gauged by the size of the muck heap in the street outside their houses, and if it's clean enough for the Swiss it must be clean enough for anyone. Cow manure only 'bubbles with organisms' and 'swims with bacteria' when it is mixed with water. Industrial farmers, dairy farmers in particular, in order to save labour costs, swoosh the muck away with high powered hoses turning it into a 'slurry' which they store in pits euphemistically called 'lagoons', which do literally bubble with bacterially generated methane. Then they spray the stuff on their land with pumps and pipes that easily clog up. The problem is identical to that identified by von Liebig with WCs: mixing faeces with water gets the stuff out of sight quickly, but it creates difficulties further down the line and turns a resource into a disposal problem.

Much of the difficulty with manure management could be addressed, not by outlawing chemical farming, but simply by reversing priorities – by making organic farming the standard procedure, and chemical farming the certifiable exception. At the moment it is the other way round: organic farmers have to prove they are organic, a process which involves considerable bureaucracy and expense, while chemical farmers can just get on the phone and order a tonne of NPK or a drum of weedkiller. Penalizing good practice with bureaucracy and certification fees is a bizarre way to encourage it. We don't, for example, make bicyclists and pedestrians prove that they don't drive, and then award them a certificate – we make motorists buy a licence.

If farmers had to apply for a licence to use chemical fertilizers and pesticides, and if food in supermarkets was assumed to be organic unless it had a label on it saying 'produced with the aid of artificial fertilizers and pesticides', then the tables would quickly be turned. Without in any way restricting the public's right to choose, organically produced food would become the norm, once again, and farmers would be keener to manage manure and nutrients efficiently, and achieve a balance between livestock and arable on their farm.

Stockless Rotations

Thankfully most vegans, especially those who get their hands dirty, are not so fastidious about manure and have a more rational approach to the matter. Many vegan farmers are as opposed to the use of chemical fertilizers as organic livestock farmers – so they have to find other ways of providing the nutrients supplied by manure. Instead of passing legumes, grass or other feeds circuitously through the stomachs of animals, it is more efficient, they argue, to apply them directly as mulch or green manure:

Veganic farmers do not deny that animal faeces and slaughterhouse residues will indeed fertilize the soil and yield fine crops. However the fertility does not originate with these residues, but rather with the grass and other plants which the animals eat. The animal destroys the greatest part of that food energy by its digestion, metabolism and other life processes. Even by the most scrupulous husbanding, only a small portion of that life-force is stored in the animal or passed with its manure ... Since plants are the initial converters of solar energy, they should be our first choice for manurial materials.[27]

Logic – at least reductionist logic – would appear to be on the side of the vegans. There are indeed inefficiencies in passing nutrients through an animal, though in respect of nitrogen, these do not comprise 'the greatest part': an adult beef cow excretes over 90 per cent of the N, P and K it consumes, and a dairy cow about 75 per cent.[28] Even those nutrients which do not end up as meat, milk or manure are not so much destroyed as converted into something else which may or may not be useful, a matter I shall consider in Chapter 12.

Over the last two decades there have been experiments in stockless rotations of organic cereal crops and a rise in the number of stockfree farmers and growers. Experimental research at Elm Farm Research Station and elsewhere suggests that 'all-arable organic farming is economically viable and technically feasible' and that yields can be maintained, at least over 15 years.[29] Some vegetable growers appear to operate successfully using stockless methods, among the most well known being Iain Tolhurst in the UK and Eliot Coleman in the US. The vegans are right to insist that animals are not an essential element in the organic fertility cycle.

However, the plant nutrients, just like those which have been passed through the gut of an animal, still have to come from somewhere, and so stockless organic agriculture usually requires a larger land base than the area of land used for cultivation. If the imported nutrients come from land which could not be cultivated – for instance hay from poor pasture or leaf mould from woodland, or more commonly nowadays, municipal compost – then that would not add to the area of arable land required to produce a stockless crop. But high quality arable land often comes all in one lump, rather than in arable strips conveniently interspersed with patches of poor land, and gathering and transporting these nutrients even a short distance is expensive (except in the case of municipal compost, because it is subsidised to the tune of between two and five times its value).[30] This is precisely the advantage of animals in cultures which have no access to fossil fuels or municipal subsidies: they mop up nutrients from distant pastures and sparse hillsides and deposit them on arable land, with relatively little human energy expenditure. Without stock or cheap fossil fuel, the most convenient way to provide green manure for vegan production is to grow areas of clover, or similar legumes, on part of the area cultivated and disc them in on a rotating basis – effectively a fallow, but a more productive one than would

27   The Khadigar Community (n.d.), 'Ethical Farming in Action', *Vohan News International*, 1.

28   Lory, J (2006), *Calculating Fertilizer Value of Supplemental Feed on Pasture*, University of Missouri Extension, http://extension.missouri.edu/explore/agguides/ansci/g02083.htm; Farm ASyst (2001), *Grazing Livestock and Water Quality*, North Carolina Co-operative Extension Service, http://www.soil.ncsu.edu/assist/grazing/

29   Dave from Darlington (n.d.), '*The Current State of Stockless Farming Research in England*', *Growing Green International* 8.

30   Hogg, D and Hubbel, J (2002), *The Legislative Driven Economic Framework Promoting MSW in the UK*, Eunomia UK, May 2002. http://www.nrwf.org.uk/documents/FULL_economic_framework_MSW_recycling_report_000.pdf

have been possible in the days before the introduction of legumes. A 'stockless rotation' of the kind carried out by Elm Farm, precisely because it is a rotation, has to occur on arable land.

The question here – since we are exploring the relative efficiency of vegan and animal agriculture – is how much extra land is required for this green manure. The unhelpful answer, seems to be anything from as much again to none at all. The panels below give a number of examples, not all of them successful.

STOCKLESS ROTATIONS

1:1   Trout Run, PA: Vegetable farm with short growing season, with 50 per cent under two consecutive crops of clover and vetch green manure.[1]

1:1   Eliot Coleman states that he requires an acre of hay to make sufficient compost to fertilize an acre of food crops.[2]

2:1   Oxbrow cites a stockless system in Norfolk in the 1920s and 1930s where two acres of wheat were grown in rotation with one acre of beans grown as green manure.[3]

2:1   Ballbrado, Ireland: This farm started in 1990 with a stockless rotation at 3:1, but abandoned it, and now does organic grain and beef. The farmer, Richard Auler, commented: 'A stockless organic farm is a dead end, don't even try it. The most important asset on this farm is the suckler herd.'[4]

2:1   Snider Farm, Alberta: Legume and cereal green manure, rye, spring oats and peas intercropped, legume and cereal green manure, wheat, peas and barley intercropped.[5]

2:1   Iain Tolhurst, a successful and well known UK grower, is variously reported as having had 35 per cent and 30 per cent of his field crop land out of production at any one time.[6] He also uses undersown cover crops. On his intensive vegetable plots, he adds 30 tonnes per acre of compost per year; this suggests the use of considerably more than an acre's worth of compost.[7]

4:2   Denmark: K Thorup Kristensen: six course rotation with five years vegetables and barley, and one year clover and undersown cover crops.[8] However a year of peas provided a low return, and the entire crop was ploughed in so it can really only be viewed as a 2:1 ratio. Successfully operated for eight years, but funding withdrawn.

---

1   Rodale Institute (2000), *Crop Rotation Basics: How to Zap Pests, Build Soil with Cover Crops in Strategic Crop Rotations*, p 3, http://www.rodaleinstitute.org/20021001/crop_rotate
2   Abrams-McHenry, M (n.d.), 'Fertility and Land Utilization', *Vohan News International*, 2.
3   Personal communication from S Oxbrow.
4   Irish Farmer's Journal (1999), *Producing Organic Grain and Beef*, http://www.farmersjournal.ie/1999/0821/environment/news.html
5   Canadian Organic Growers (n.d.), *A Farmer's Profile: Organic Crop Rotations*, www.cog.ca/gainingground_FarmerProfile.htm
6   'Tolhurst Organic Produce: A Step into the Future' (n.d.), *Growing Green International*, No 7.
7   Jenny Hall (n.d.), 'To Till or Not to Till', *Growing Green International*, No 7.
8   Thorup-Kristensen, K (2005), *Use of Green Manure, Catch Crops and Deep Rooted Crops in an Organic Vegetable Rotation*, Danish Institute of Agricultural Sciences, Dept of Horticulture.

One of these experiments stands out from all the rest. The Rodale Farming Institute's three and five course rotations have produced a crop every year, and derive all their nitrogen from catch crops of green manure – either red clover crops undersown with the main crop, and left in the ground over winter, or hairy vetch planted in the autumn. Crop yields are comparable with respectable yields achieved on the same site through chemical agriculture, and through a closed organic livestock mixed farming system. The results are impressive. The trouble

---

5:2   Co-operative Wholesale Society: Seven year rotation of wheat, oats and beans with two years of red clover. Had higher yields of wheat than a parallel manure based system, but returns on other crops not available. Phosphorus added to both systems. Mycorrhizal activity higher in manured plots.[9]

3:1   Pre-industrial China: *Shen's Agricultural Book*, a 17th-century Chinese manual describes a system where one third of a hectare of milk-vetch provides 7500 kg of green manure for a hectare of the following irrigated rice crop – this level of only 25 per cent green manure may be achievable because irrigated rice usually obtains some of its nutrients through the application of large amounts of water.[10]

3:1   Elm Farm: three four year rotations with one year's red clover: clover, winter wheat, winter wheat spring oats; clover, potatoes, winter wheat, winter oats; and clover, winter wheat, winter beans, winter wheat. This experiment has been continued for 15 years, and average crop yields have been maintained with no apparent nutrient loss. However phosphate has been added and a number of problems have been experienced including a decline in organic matter and mycorrhizal activity in the soil and a build up of perennial weeds.

4:1   Norway: oats, red clover, oats, spring wheat, barley. This rotation did not accumulate enough N to compensate for the N removed in cereals.[11]

4:1   Terrington: 5 year rotation of clover, potatoes, winter wheat, beans and spring wheat. On grade 1 land, high yields for grain. It started out with two courses of red clover, so over the first 15 years there were three years of clover. Phosphorus added. Problems with establishing clover.[12]

5:0   Rodale, PA USA: A five year five crop rotation of corn, soybeans, oats, corn oats. After 1991 this was changed to a three year rotation of corn, soybean and wheat.[13]

---

9   CWS Farm Groups (2002), *Organic Farming: Experiments 1989-1997 – A Summary of Key Findings*, http://www.pmac.net/focus_on_practice.html

10   Zhang Lu-ziang (1956), *op cit.* in Netting, *op cit.*, p 137.

11   Løes, A K *et al* (2007), *N Supply in Stockless Organic Cereal Production under Northern Temperature Conditions. Undersown Legumes, or Whole Season Green Manure?*, 3rd QLIF Conference, Hohenheim, Germany 20 March 2007, www:orgprints.org/9823/

12   DEFRA, 2002, *Testing the Sustainability of Stockless Arable Organic Farming on a Fertile Soil*, *OF0301*, Final Report 2002.

13   Pimentel, D *et al* (2005), *Organic and Conventional Farming Systems: Environmental and Economic Issues*, Cornell University. http://dspace.library.cornell.edu/bitstream/1813/2101/1/pimentel_report_05-1.pdf

is that this performance has not been replicated elsewhere and others who have tried similar rotations have failed. David Pimentel's 2005 study of the Rodale rotations observes: 'Other researchers have been less successful in maintaining and improving soil fertility levels in organic systems. Rodale's results could also be influenced by geographical soil characteristics and not be universally applicable.'[31] This is a great pity, because if Rodale's results were universally applicable, we would not need to resort to agrochemicals to feed everyone, we would have enough spare land to allow ourselves the luxury of meat, and the world's troubles would be over.

However, the reliable average seems to be about 2:1 – one acre of green manure for every two in crops. Dennis Avery reports: 'The average US organic farm has about one third of its land in green manure crops and fallow to make up for the lack of chemical nitrogen that mainstream farmers take from the air.'[32] Stockless farmer, Bill Cormack, states that 'on very light soils stockless rotations would only work with a very high percentage of fertility crops – as much as 50 per cent of the land would be directly non-productive'.[33] The vegan writer Jenny Hall, in her article 'Stockfree Britain', states:

> The green manuring proportion of a farm rotation would be 25 to
> 40 per cent, but on sandy soils more consideration would need to be
> given to fertility. Therefore in our calculations we have opted for 40 per
> cent to err on the side of caution but would expect this to lower as a
> percentage as stockfree fertility farming techniques improve.[34]

These figures mirror the two and three course rotations used in the medieval open fields, before the development of the Norfolk four course system. Fifty per cent green manure is the equivalent of a two course rotation in which half the fields at any one time are fallow, and 33 per cent green manure is the equivalent of a three course rotation.

If we accept a 2:1 ratio, this means that for every hectare of arable stockfree cultivation, another half hectare of arable land is 'fallow'. This effectively reduces the land use efficiency of stockfree crops by a third, which in turn, affects the conversion ratios we have been examining above, in respect of organic husbandry. So for example if a crop of grain is fed to beef cows at a feed to meat conversion rate of 10:1, the livestock system will be using 6.66 times as much land as an equivalent stockfree system. If the crop is fed to dairy animals at a conversion rate of 3:1, then the livestock system will be using twice as much land as a stockfree system.

These figures assume that in the livestock based system there are enough animals, and hence manure, to provide fertility for grain crops without recourse to additional areas of green manure. In other words, growing grain in a self-sufficient organic mixed farming system requires the correct balance of livestock to arable to ensure adequate fertilization. The standard textbook on organic farming in England by Nicolas Lampkin states that typical whole farm rotations on mixed livestock farms might require five hectares of clover ley or lucerne for two hectares of cash crop (2:5); or on a more intensive farm, six hectares of clover

31   Pimentel, D et al (2005), *Organic and Conventional Farming Systems: Environmental and Economic Issues*, Cornell University. http://dspace.library.cornell.edu/bitstream/1813/2101/1/pimentel_report_05-1.pdf, p 25

32   Avery, D (1999), 'Intensive Farming and Biotechnology: Saving People and Wildlife in the 21st Century', in Tansey, G and D'Silva, J., *The Meat Business: Devouring a Hungry Planet*, Earthscan, 1999.

33   Cited in Dave from Darlington (n.d.), The Current State of Stockless Farming Research in England, *Growing Green International* 8.

34   Hall, Jenny (2008), 'Stockfree Britain', *The Land* 5.

to four hectares of cash crop (4:6). In effect, organic mixed farm rotations, which have roughly two thirds of their land given over to nitrogen fixing plants, are more or less a mirror image of stockfree rotations which have roughly a third given over to nitrogen fixing plants – the difference being that a proportion of the livestock farm's ley is fed to animals before finding its way onto the crops. The 5:2 Co-operative Wholesale Society experiment listed in page 81 is an exact mirror image of the 2:5 rotation described by Lampkin.

A comparison of the performances of these whole farm systems shows that the difference between them is fairly slender. Take for example an imaginary seven hectare farm. A stock free system with two hectares of leguminous green manure for every five hectares of grain crops would produce five hectares worth of crops. An organic system feeding dairy cows would produce two hectares of grain crops, plus 5 hectares of ley for dairy produce, including a little beef. Assuming the nutritive value of the five acres of ley is equal to the nutritive value of the crops, at a conversion ratio of three to one the dairy produce will be equivalent to the crops which could be grown on a third of those five hectares, namely 1.6 ha. So on a seven hectare plot the dairy system would be producing the equivalent of 3.6 ha of crops, against five hectares for the stockfree farm – or a ratio of 1.39:1 in favour of the stockless system. If the livestock were beef fed at a 10:1 feed conversion ration, the ratio would be 2:1, ie the vegan farm would be producing exactly twice as much as the beef farm.

The above example compares a fairly extensive livestock system with a relatively intensive stockfree system. If we compare Lampkin's 4:6 livestock rotation with the more typical 2:1 stockfree rotation, we find that a dairy farm produces four hectares' worth of crops, plus another two hectares' worth from the ley, or six hectares' worth altogether, whereas a stockless farm harvests 6.6 ha of crops – a ratio of 1.1:1 in favour of the vegan system. If the livestock were beef, the ratio would be 1.43:1 in favour of the stockfree farm.

As a matter of interest, a Norfolk rotation, producing wheat and barley for beer and bread for human consumption on two hectares and animal feed (turnips and clover) on two hectares, is a very intensive 2:2 system. If all this animal feed were fed to dairy cows at a conversion ratio of 3:1, it would produce the equivalent in milk products of 0.66 ha of wheat, giving a total of 2.66 ha worth of crops for the entire four hectares. This is exactly the same as the yield from a stockless system with 33 per cent of the land devoted to green manure. However the 2:2 system is probably not a fair comparison since Coke and other practitioners of this rotation were probably also bringing in fertility from outlying land with sheep. Nonetheless, if the outlying land is uncultivable heath, inaccessible without livestock, the four course livestock farmer might reasonably claim that his system, with its extra mutton, is more productive than the stockfree system

Be that as it may, it would be foolish to pretend that feeding nutrients to animals could ever be as efficient as eating the nutrients ourselves. Moreover the performance of any farming system is reliant upon the skill of the farmer to reach a high level of efficiency – and the mixed farmer, dealing with both livestock and crops, needs to be at least twice as skilled as the arable farmer. What these whole farm assessments show is that the difference in productivity between organic livestock farming and organic stockfree farming is nowhere near as great as is often made out. Bandying around figures such as a 10:1 ratio is therefore, in cases where good organic husbandry is practised, a wild exaggeration.

To some extent, the stance which the thoughtful person takes in respect of animals in agriculture must be related to the stance they take on the use of artificial fertilizers. In a world where chemical fertilizers were banned or unobtainable, the advantage of a vegan diet would be slim indeed. If arable crop and meat production took place on organic mixed farms where the whole rotation operated at less than 1.5:1 conversion rate, then the produce of those farms, added to the meat provided by beef and pork fed entirely on pastureland and crop residues would probably produce at least as much food as could be obtained from universal organic vegan production.

Put another way, the superior efficiency of a universal vegan diet can perhaps only result in a significant increase of food production if it is 'subsidized' by petrochemical inputs which a livestock-based system can do without.

Can Organic Farming Feed the World?

The matter also carries worrying implications for advocates of organic farming. If the world went over to an organic farming system many of the potential gains which might be made from reducing meat consumption in wealthy countries, or even eliminating it completely, could not be obtained because much of the land upon which the livestock had been supported would be required either to supply some grain to replace the meat foregone, or else to provide fertility.

Placed under stress, a global organic farming system might find itself in a parallel situation to that experienced in Europe in the 13th and 14th centuries where, in order to feed increasing numbers of people, meat consumption goes down, the proportion of arable crops increase, and the 'grain frontier' makes increasing progress into grass territory – resulting in less meat in people's diet and a decline in yields because of a shortage of animal and green manure. The short term solution to this problem, we should not forget, was the elimination of about a third of the population in the Black Death – an event that the modern historian might consider to be not so much fortuitous as 'Gaian'.

Opponents of organic agriculture have not been slow to point this out. There is a camp of 800 scientists and pundits, including Norman Borlaug (the architect of the green revolution), James Lovelock (of Gaia fame), Dennis Avery (of the Hudson Institute) and Matt Ridley (the UK's best known contrarian and former chairman of Northern Rock) who, under the aegis of the Center for Global Food Issues, have signed a declaration 'In Support of Protecting Nature with High Yielding Farming and Forestry'.[35] The gist of this declaration, laid out most explicitly in supporting information written by Dennis Avery, is that to provide sufficient nitrogen to feed the future population of 8.5 billion people which industrialization will spawn, we will have to resort not only to chemical fertilizers, but also to genetic manipulation. Any attempt to secure nitrogen and other nutrients through natural organic means would require undue encroachment upon natural habitats – if not their total destruction. If we want to feed the world and preserve biodiversity, then we'd better continue with industrial agriculture. Rather than share agricultural land with nature, we should spare land elsewhere.[36] To protect nature we have to farm unnaturally.

Frankly, I dislike these mouthpieces for the agrochemical industry, and make no mistake about it, that is what they are: no less than 21 representatives of Monsanto and seven of Syngenta signed their declaration. I particularly dislike

35   Center for Global Issues (n.d.), *Declaration in Support of Protecting Nature With High-yield Farming and Forestry*, www.highyieldconservation.org/declaration.html
36   The concept of a choice between 'sharing land' and 'sparing land' has been developed by Tim Benson of the University of Leeds, from Rhys Green *et al* (2005), 'Farming and the Fate of Wild Nature', *Science*, 307, pp 550-555, 28 Jan 2005.

the Malthusian complacency with which they assert that the only way forward from the mess that the industrialization of agriculture has got us into is to put ourselves in their hands and pursue yet more of the same. I label them under the acronymically satisfying heading Global Opponents of Organic Farming. The worrying thing is that the GOOFs might be right.

The case they make is put forward most powerfully in a book on the history and legacy of the Haber/Bosch method of producing nitrogen fertilizer by the US academic Vaclav Smil, entitled *Enriching the Earth* – a book which I would advise anyone who campaigns on behalf of organic farming to read. Smil is not banging so loud an ideological drum as the Center for Global Food Issues, and he does not shrink from cataloguing the problems that chemical agriculture has caused. The move to synthetic fertilizers, he states:

> has many undesirable consequences for the soil quality, above all greater soil compaction, resulting in worsened tilth, easier erodibility, lowered water-holding capacity, and weakened ability to support diverse soil biota and to buffer acid deposition ... Higher applications of nitrogen fertilizers have also increased the opportunities for losing the nutrient from fields and transferring it to fresh and coastal waters, to soils and into the atmosphere. Such enrichment must have a variety of consequences for aquatic and terrestrial biota that are now subjected to this steady, and often increasing eutrophication ... Haber-Bosch synthesis has made it possible to sever the traditionally tight link between cropping and animal husbandry, and to move increasing amounts of fixed nutrients not only within individual countries but also among nations and continents. This has given rise to disjointed, one might say dysfunctional, nitrogen cycling. Individual farms, even whole agricultural regions, have ceased to be functional units within which the bulk of crop nutrients used to keep cycling during centuries or even millennia of traditional farming.[37]

This pretty well sums up the main environmental arguments against the use of chemical fertilizers. Nor does Smil hide the fact that Fritz Haber, Carl Bosch and Carl Krauch, the men who around 1910 developed the process which a century later is still used for extracting nitrogen from the atmosphere, were all highly compromised characters.

Haber, who invented the process, was one of 93 scientists, including three future Nobel prize-winners, who in 1914 signed Ludwig Fulda's pro-war manifesto *To The Civilized World*, which claimed that 'were it not for German militarism, German civilization would long since have been extirpated'. (Albert Einstein's pacifist proclamation attracted just four signatures.) Haber's support for the war effort was not altogether surprising given that synthetic nitrogen was more immediately useful for the production of explosives than for the culture of wheat. Haber went on to develop chlorine gas weaponry – although it was clearly outlawed by the Hague Conventions of 1899 and 1907 – and was directly responsible for the fatal gassing of 5,000 French troops at Ypres on 22 April 1915. Two weeks later Haber was preparing to leave Berlin to launch more gas attacks on the Eastern front when his wife Clara, also a scientist, shot herself through the heart with his revolver. No one knows why, but one can hazard a guess. After her suicide, Haber left for the Russian front, and later became head of the Chemical Warfare Service. His downfall came when the Nazis discovered he was

---

37   Smil (2004), *op cit 19*, pp 204-206. Quote slightly shortened.

a Jew: he was removed from his post, and died in 1934, too early to see the gas he developed used against his own people in the concentration camps.

Carl Bosch, the engineer who put Haber's discovery into practice, was less overtly militaristic, and was actively opposed to the rise of fascism. Nonetheless, as chairman of the chemical conglomerate I G Farben, he assisted the German war effort by developing a process for producing a petrol substitute from Germany's inferior coal – which after 1944 provided the only source of German fuel. Bosch died in 1940, depressed and on a diet of painkillers and alcohol. His protégé, Carl Krauch, who had assisted in the development of synthetic nitrogen, was now chairman of I G Farben. Reversing an earlier company decision, Krauch sanctioned the building of a synthetic rubber plant, 'not only impelling the selection of a site near Auschwitz, but also instigating the use of concentration camp inmates as construction workers'.

As Smil acknowledges, synthetic nitrogen was a child of the military industrial complex – as was the hydroelectric manufacture of aluminium in the 1940s ('making Flying Fortresses for Uncle Sam' in Woody Guthrie's words), and a few years later the development of nuclear energy. In the 1930s and during the war, IG Farben also developed the use of organophosphate nerve gases such as Sarin, and the technology was adopted by US companies after the war to make pesticides. This is not a very encouraging pedigree, but it doesn't prevent Smil from concluding that the industrial synthesis of nitrogen is the 'single most important change affecting the world's population' in the 20th century.

The reasoning behind this accolade is best expressed in one particular graph, on page 147 of Smil's book. In the early 1960s China's population was about 660 million, its consumption of synthetic fertilizer was negligible, and almost all its nitrogen was derived from organic sources. By 1996 the population was close to 1.2 billion, applications of synthetic nitrogen had increased more than 50 fold, and 75 per cent of all nitrogen applied to crops was synthetic, produced by the Haber/Bosch process.

In the same period average per capita food consumption increased from about 2,000 calories, to 2,700; and meat consumption increased from a reported figure of 1.4 kg per year in 1961, to a widely accepted figure of around 47 kg per year in the late 1990s.[38] Meanwhile, very little new agricultural land has been opened up, and in recent years the amount of arable land has shrunk, owing to urban development and desertification.

Assuming these figures are broadly correct, and not skewed by anti-Maoist bias, this a phenomenal achievement, even allowing that it has been accomplished with the aid of a magic potion. The feat of feeding twice as many people off the same amount of land, with a diet that was both more ample and contained a great deal more animal protein, Smil argues, could not have been achieved using the traditional organic techniques of the 1950s which were already stretched to their limits. A similar, though less accentuated process has taken place throughout the rest of the world, particularly in Asia. 'Intensive farming,' Smil concludes, 'whose increasing harvests have depended almost entirely on additional nitrogen from synthetic fertilizers, now produces basic food needs for about 2.3 billion more people than the traditional, fertilizer-free practices did … almost exactly 40 per cent of the world's population.' Or more bluntly, without synthetic nitrogen fertilizers, 40 per cent of the people in the world would starve.

---

38   Shurtleff, W and Akiko Aoyagi (n.d.), *History of Soybeans and Soyfoods in China*, unpublished, http://www.soyinfocenter.com/HSS/history.php; World Resource Institute, *Country Profiles*, http://earthtrends.wri.org/pdf_library/country_profiles/ene_cou_156.pdf

I would like to believe that Smil is wrong, for a number of reasons. Quite apart from the environmental problems which Smil mentions, the social effects of the so-called green revolution have been equally disastrous. Chemical fertilizers, and the other inputs designed to achieve yield increases such as pesticides and irrigation, are too expensive for most peasants. Wealthy farmers – those who can achieve the economies of scale necessary to justify the investment – prosper at the expense of the poor, who are eased off their land, through debt or by force. The autonomy and the integrity of a rural community lies in its access to its own resources, the main one (except in maritime communities) being its soil. Imported chemical fertility cannot be had for nothing, and it requires the export of equivalent amounts of biomass (to the city or possibly abroad), making the rural community beholden to the industrial economy for the functioning of its primary resource. The progressive industrialization of the rural economy results in the majority of people being forced into cities, where the rewards of living close to the land are supplanted by the pressure to compete and consume.

It is possible that Smil could be wrong. What would have happened if the Haber/Bosch process and all the rest of the agrochemical armoury had never been developed, either because such developments were technically impossible, or else because, in the context of a pattern of civilization that one might usefully call Sino-Luddite, they were banned? Would the world's population have expanded as it did in Europe in the 14th century? Would it have encroached upon grazing land and suffered declining yields, until humanity succumbed to a global demographic catastrophe analogous to the Black Death?

Perhaps; but it is not beyond the realms of possibility that the world might have pursued an entirely different course, through which population growth was relatively restrained (as it was for example in the 17th century in England, or the 18th century in China) while scientists and innovators concentrated their research upon breeding better varieties, improving the performance of nitrogen fixing crops, and finding better ways of making complex agricultural ecosystems function productively.

In this context, the main charge levelled at the agrochemical establishment by the organic movement is that virtually all funding for research has been pumped into fossil fuel powered agriculture, and that is the main reason why organic yields have so often lagged behind the yields from chemical farming. The watermill wasn't developed until 1,000 years after it was invented, because slaves could do the job cheaper.[39] How many potential agricultural improvements are there which have remained ignored because they can't compete with fossil fuels?

A notable example is the huge discrepancy between conventional UK wheat yields of around eight tonnes per hectare, which are dependent upon high applications of synthetic nitrogen, and average yields of organic wheat which are little more than half as much. Martin Wolfe, of Elm Farm Research Station, has drawn attention to the curious fact that the same does not hold for other grains:

> In 2000, six modern wheat varieties yielded, on average, 10 tonnes per hectare across national trials under standard non-organic conditions. When these varieties were grown organically the yield fell to less than 4 tonnes per hectare. Oat and triticale varieties under the non-organic conditions yielded, respectively, 8.2 and 6.5 tonnes per hectare. However the same varieties grown organically yielded, respectively, 7.1 and 6.7 tonnes per hectare.

---

39   Marc Bloch (1967), 'The Advent and Triumph of the Watermill' in *Land and Work in Medieval Europe*, Routledge and Kegan Paul.

Why should there be such a discrepancy between wheat on the one hand, and oats and triticale on the other? The main reason is that wheats bred for non-organic production are short-strawed with an open canopy, so that they compete less well with weeds than the taller, denser oats and triticales. There is a similar contrast in disease resistance. Modern wheat varieties, adapted to utilizing synthetic fertilizer inputs, may also have lost some ability to interact with soil for their required nutrition, relative to older varieties. There is an urgent need, therefore, to breed organic wheat.[40]

Over recent years commentators such as Jules Pretty, Mae-Wan Ho, Lim Li Ching and others have reported impressive improvements in yields in Third World agriculture through the introduction (or reintroduction) of organic techniques such as intercropping, green manuring, animal composting, introducing ducks into paddy fields to keep down weeds and pests and provide fertilizer, or interplanting crops with prolific legumes such as the velvet bean tree, *Mucuna pruriens*.

One such innovation which has attracted considerable attention is SRI rice, a system developed in Madagascar by Fr Henri de Laulanie, which involves transplanting seedlings further apart so that the plants tiller outwards and develop more vigorous roots and fuller, healthier shoots. The system is not inherently organic, but is usually practised using organic fertilizers. In Madagascar yields improved from around 3.5 tonnes per hectare to eight tonnes, and the system has since been employed with success in about 20 countries, including Laos, Sri Lanka, the Philippines and the Gambia. In China yields of 12 to 16 tonnes have been reported, both by the China National Hybrid Rice Research and Development centre, and by the private sector Meishan Seed Company.[41]

However, when the International Rice Research Institute (IRRI) (the organization founded by the Ford and Rockefeller Foundations which pioneered the green revolution with its IR8 rice hybrid) tested the SRI method they found that there were no significant differences between SRI and conventional cropping systems.[42] The results were reported in an article in the scientific journal *Nature*, entitled 'Feast or Famine', in which the authors suggested that SRI was a diversion from 'more promising approaches' such as genetic engineering, and that 'SRI has no major role in improving rice production generally'.[43] This spat seems to be another example of the Monbiot Maxim – that yield is a function of one's political position.

In his book, Smil draws attention to one promising organic method of fixing nitrogen from the atmosphere: 'Since the late 1950s monsoonal Asia has added an intriguing form of green manuring by cultivating and distributing floating *Azolla* ferns [also called duckweed fern], which harbour N-fixing cyanobacterium *Anabaena*. Fern stocks are preserved in nurseries for distribution to flooded fields

40   Martin Wolfe (2001), *Recognizing and Realizing the Potential of Organic Agriculture*, presentation at Global Ag 2020 conference, John Innes Centre, 19 April 2001, www.biotech-info.net/organic_potential.html

41   Rabenandrasana, Justin (n.d.), *Revolution in Rice Intensification in Madagascar*, ILEIA Newsletter, www.farmingsolutions.org/successtories/stories.asp?id=9; Mae-Wan Ho (2004) *Rice Wars*, Institute of Science in Society, 2004 http://www.i-sis.org.uk/RiceWars.php; *Organic System Doubles Yield*, Foodmarket Exchange news release, 20 Dec 2002, www.foodmarketexchange.com; Gallarde, Juancho (2006) 'Rice-Duck farming in Negor', *Daily Star*, Saturday 28 October 2006, http://www.visayandailystar.com/2006/October/28/negor2.htm; Australian Government Overseas Aid, *Rice Revolution*, 2004, http://www.ausaid.gov.au/closeup/rice_revolution.cfm.

42   Mae-Wan Ho (2004), *ibid*.

43   Surridge, C (2004), 'Feast of Famine', *Nature* 428, pp 360-36, 25 March 2004.

where they could double their mass in just a few days, producing sometimes more than one tonne per hectare of phytomass per day. After the plantlets die and settle to the paddy bottom, their mineralized nitrogen helps to raise rice yields in paddies.'[44] In 1980, China was reported to have 3.2 million acres of paddy fertilized with *Azolla*.[45]

A more recent report on the use of *Azolla* as a nitrogen fertilizer comes from the *Journal of the North Eastern Council*, in Shillong, India, where, we are informed, 'farmers have apathy in using chemical fertilizers'. The report states that 10 to 12 tonnes of *Azolla* manure per hectare reduces nitrogen fertilizer requirement by 30 to 35 kg, suppresses weeds and prevents water evaporation. Azolla is also a good feed for fish, and when dried can be fed to poultry.[46]

The problem with Smil's analysis is that, although he allows that synthetic nitrogen has its downside, he doesn't examine whether things could have been done differently. The acknowledgements at the front of *Enriching the Earth* give credit to corporations such as BASF, and industry bodies such as the International Fertilizer Development Centre – but there is no indication that he has listened to what people on the other side of the fence are saying.

Nonetheless, it takes a big leap of faith to conclude from the success of small-scale ventures such as SRI that China could feed all its population to the current level of nutrition solely through organic farming. This would require all of the 75 per cent of China's nitrogen which Smil estimates is currently extracted from the atmosphere by fossil fuel-powered factories, to be derived through nitrogen-fixing plants such as lucerne, *Azolla*, or *Mucuna pruriens*. This may be theoretically possible, but it would also require extra land, extra water, and a supply of other key nutrients, notably phosphates. I have to agree with Smil that this looks like a pretty tall order.

The challenge is almost as daunting in some other low income countries, such as Egypt, Indonesia and Bangladesh, although it is noticeable that India is considerably more economic with nitrogen fertilizer than China (India has 1.6 times as much arable land per person as China, but China uses 2.6 times as much fertilizer as India).[47]

The situation is very different in many of the developed countries, particularly those in the New World. Smil calculates that the USA, even though it is the world's second largest consumer of nitrogen fertilizers after China, 'in the 1990s could have supplied a healthy diet for 250 million people without using any synthetic nitrogen compounds'. This could be achieved by reducing food exports, the amount of grain-fed meat, and most significantly the 45 per cent of food which is wasted.[48] No doubt countries such as Canada and New Zealand could feed themselves just as easily, and I suspect that France, and even Britain could feed themselves through organic home production, though we British might have to eat more porridge and potatoes.

Britain is fortunate in enjoying a temperate climate, but its famous fertility is not inherent, it has been built up over the last 500 years through a series of historical advantages. The agricultural revolution, by slowly building up soil

44   Smil (2004), *op cit. 19*, p 30
45   Armstrong, W P (1998), *Marriage Between A Fern & Cyanobacterium*, November 1998, http://waynesword.palomar.edu/plnov98.htm
46   NESAC (2006), *Package of Practice for Azolla: A Potential Bio-fertilizer*, http://megapib.nic.in/azolla.htm, 12 November 2006
47   Smil (2004), *op cit.19*, p 174. Smil's figures are for 1996 figures. India's total fertilizer use stagnated between 1998 and 2005; Murray, D (2005), 'Oil and Food Security Linked, *USA Today*, Society for the Advancement of Education, July 2005.
48   Smil (2004), *op cit.19*, p 166.

fertility and introducing legumes and root crops, enabled more stock to be kept, more manure to be produced, more nitrogen to be applied to the land and higher yields to be obtained. The colonization of the New World allowed surplus peasants, who had they remained at home would have increased grain production at the expense of grass, to be exported abroad where they could grow a surplus of crops on virgin land. And fertility has been further enhanced by the importation of guano, rock phosphates, and untold quantities of biomass in the form of wheat, beef, animal feed and other crops (much of which, but not all, we have flushed out to sea).

None of these options were available to China and other South East Asian nations. When China finally emerged in the 1980s from a centuries-long struggle at the brink of carrying capacity, it was not thanks to an agricultural revolution which increased nitrogen levels through the agency of legumes and livestock, but on the wave of a green revolution fuelled by chemical nitrogen and superphosphate. If Smil is right, the option to go organic is a neo-colonial luxury which cannot be afforded by the majority of the inhabitants of the third world, because they do not have enough land, water or biomass to go around. The only immediate prospect of redressing such inequity is to promote large-scale migration of Chinese and South Asian peasants to the less congested farmlands of North America, Australasia and Europe, where they could take up the niche currently occupied by a superfluity of livestock – a scenario which advocates of globalization and the free market ought to be totally in favour of.

A more tempered approach will be to increase research and experimentation into organic agriculture in both the north and in the south, in the hope that the third world, in contrast to Europe, will undergo its biological 'agricultural revolution' after having undergone its chemical 'green revolution', since it failed to do so before. The world is waiting for an organic answer to Justus von Liebig, Fritz Haber and Carl Bosch.

FOOD SECURITY

9

CAN BRITAIN FEED ITSELF?

It is generally agreed that the day of cheap food is at an end.
Kenneth Mellanby 1975

In 1975, the Scottish ecologist Kenneth Mellanby wrote a short book called *Can Britain Feed Itself?*[1] His answer was yes, if we eat less meat. The way in which he worked it out was simple, almost a back of the envelope job, but it provides a useful template for making similar calculations. In this chapter I have adapted and embellished Mellanby's 'basic diet' to show how much land modern UK agriculture might require to produce the food we need under six different agricultural regimes – chemical, organic, and permacultural, each with or without livestock.

There were two main reasons why I decided to repeat Mellanby's analysis. Firstly, although food is now even cheaper than it was when Mellanby so confidently predicted a rise in price, it is nonetheless possible that the UK may one day have to become more self reliant than it is now. Secondly, I am interested to see how organic agriculture in particular performs, because the most convincing argument advanced against organic farming by its opponents is that it takes up too much land. This is of most concern in poor, highly populated countries such as Bangladesh, but Britain cannot afford to be complacent: it is more densely populated than China, Pakistan, Vietnam or any African country except Rwanda.

There are limitations in this kind of statistical exercise; and I do not claim to have carried it out with either the expertise or the thoroughness that it merits. This is, at best, a back of an A4 envelope job. The results should not be seen as anything other than a rough guide, and a useful framework for thinking about such matters.[2]

---

1   Mellanby, K (1975), *Can Britain Feed Itself?*, Merlin Press.
2   This chapter is an edited version of an article first published in *The Land* 4, in 2008. Some material has been cut, to avoid repetition in this book; and the calculations for horse power have been amended. Figures for 1975 are from Mellanby, K (1975), *ibid*. Figures for 2007 are derived, whenever possible, from: Nix, J (2007), *Farm Management Pocketbook*, Imperial College London, Wye Campus, The Anderson Centre; Lampkin, N, Measures, M and Padel, S (2007), *Organic Farm Management Handbook*, Organic Farming Research Unit, University of Wales Aberystwyth and Elm Farm Research Centre; Office for National Statistics (2006), *Annual Abstract of Statistics*, Chapter 21; DEFRA (2005), *Agriculture in the UK*, Chapter 5, DEFRA. Where crop yield figures are not available in these publications I have deduced them from information available on the internet, using at least two sources for each item.
    The nitrogen cycle is the most complicated aspect to assess, and the figures I have used are broad-brush. For manure use I have used a number of sources, in particular Lampkin, N (1990), *Organic Farming*, Farming Press; ASAE (2003), *Manure Production and Characteristics*, American Society of Engineers; and Chorley, G (1981), 'The Agricultural Revolution in N Europe, 1750-1880: Nitrogen, Legumes, and Crop Productivity', *Economic History Review*. The 6:4 ley:crop rotation is also taken from Lampkin (1990); a 5:2 or 7:3 rotation might be easier to achieve and this would mean that the

Mellanby's Basic Diet

Mellanby took as his starting point the UK's total figure for grain production. In 1975, Britain grew 15 million tonnes of cereals on less than 3.6 million hectares at a yield of about four tonnes per hectare. This was the equivalent of 283 kilos per person a year, which is about 2700 calories a day – comfortable enough for every man, woman, child and elderly person in the country. The total population was 53 million.

Working from this figure of 15 million tonnes of grain, Mellanby built up a somewhat more varied diet, subtracting grain from the total as he introduced other foodstuffs. Table A shows us his 'basic rations' of cereal, potatoes, sugar, milk and meat. Every person gets the equivalent of a pint of milk and a pound of potatoes a day, which is what they were actually consuming in 1975: but Mellanby gives them less meat.

The 2400 hectares assigned for dairy are mainly leys – temporary pastures which are rotated with cropland to provide fertility. Another 2400 hectares of permanent grazing are for raising beef. As for sheep, Mellanby retains the 28 million of them that there were in 1975, without bothering to work out how much land they take up or how much meat or calories they provide – in fact they do not contribute much more than one per cent of the total diet.

The three items most obviously missing from Mellanby's basic diet are beer, fat and vegetables. Beer, since it is made of barley and has a calorific value of 100 to 150 calories a pint, is included within the grain figure. Fat is a more serious omission, involving substantial amounts of land, and Mellanby could usefully have included it in his calculations. He may have been deterred by the fact that edible rape oil had barely been invented in 1975, so his self-sufficient Britain would have been dependent for its fat supply on lard. As for vegetables and fruit, Mellanby is content simply to point out that these can be provided in allotments and gardens.

These omissions don't undermine his main point, since there are millions of hectares left over, which could be put over to pigs, more cows (for butter), vegetables, poultry or whatever anybody felt like. There is, in fact, no shortage of land whatsoever.

Mellanby's calculations are for so-called 'conventional' agriculture using nitrogen fertilizers and other chemicals, which makes his task much easier; but he does mention the potential of organic agriculture and concludes that, although less productive than conventional agriculture, it could still probably feed the country using an extra 33 per cent of the land.

---

organic livestock option would require more land for green manure. The figure for human sewage represents half the total nitrogen available from the population's excrement – and twice the amount per person that Wessex Water is currently obtaining from 2.5 million customers. This may be optimistic.

Forestry figures from Forestry Commissions Statistics, http://www.forestry.gov.uk/statistics; and E Agate (2003) *Woodlands: A Practical Handbook*, BTCV.

The performance of grass-only organic dairy cows in Table F, after discussion with farmers and agronomists, I have assessed at 3700 kilos, allowing 500 kilos for the calf. Stocking rate is one productive cow to the hectare, with each replacement also requiring a hectare. This is generous compared to Mellanby's cows which produce 3.666 kilos of milk each from chemically fertilized grassland at an average rate of 1.8 cows-plus-followers per hectare.

The biofuel and horse acreage is based on a figure of 15 per cent of arable land, derived from the studies and models in Jansén, J (2000), *Agriculture, Energy and Sustainability: Case Studies of a Loval Farming Community in Sweden*, Swedish University of Agricultural Science, Uppsala.

The figures in Table H are derived from Centre for Alternative Technology, 2007, *Zero Carbon Britain*, CAT, with some data (e.g. yield of sheep) harmonized with Tables F and G.

The Mellanby Diet Today

Since 1975 a number of factors have changed: the population has risen from 53 million people to 60.6 million, but crop yields have risen much faster. In 2004 Britain grew nearly 22 million tonnes of grains on 3.1 million hectares at a yield averaging over seven tonnes per acre.

Table 4 shows how, as a result, land use has changed in the last 30 years. The total agricultural area has declined only slightly, but there has been a large shift away from temporary grass ley, reflecting the decline of the dairy herd, as well as a smaller drop in arable land and the arrival of set-aside. The amount of land under permanent pasture and forestry has increased correspondingly.

UK LAND USE 1975 AND 2005

	1975	2005
Tillage	4800	4583
Leys	2400	1193
Set aside	-	559
Total arable	**7200**	**6335**
Permanent grass	4800	5711
Rough grazing	6800	6462
Total agricultural land	**18800**	**18508**
Other farm land incl woodland	?	872
Forestry land	2175	2825

K. Mellanby and Office for National Statistics. Early forestry figure is for 1980.

Table 4

Table B is Mellanby's 1975 table updated to 2005, to take account of the rise in population and increases in crop yield. The same diet for 14 per cent more people can now be provided on 86 per cent of the 1975 arable land area. However, beef production nowadays is less efficient than in the 1975 model. There is a reason for this, which I shall explain later on.

I have made one addition to Mellanby's table: some extra hectares to account for vegetables and fruit, which require more land than corn does to produce a given number of calories. About 160,000 hectares are devoted to horticulture in the UK at the moment, but we import about 60 per cent of all our fruit and veg, so we consume over 400,000 hectares worth. This is a substantial amount of land; but I can understand why Mellanby left it out, because calculating the area involved and the number of calories for such a variety of different crops is tricky.

Mellanby Goes Vegan

Mellanby could feed his population quite comfortably by reducing the amount of meat, so what would happen if it went vegan? In order to make a comparison with stockless agriculture providing a non-animal diet, in Table C I have substituted the meat and milk in Mellanby's ration with an equivalent ration of protein (peas) and fat (rape oil). The meat-eaters get their fat from milk (about 24 grams per day) and meat, but both diets are stingy on fat for anyone wanting to lead a physically active or an indulgent lifestyle.

Table C shows that chemical stockless agriculture is by far the most economical in terms of land use and can grow the entire ration on less arable land than that required by chemical livestock agriculture to provide its non-meat component. This is the ideal farming system for any society wishing to reduce the number of its farmers to a minimum, or to grow wide areas of biofuels, or to support large urban populations – all main objectives of modern social policy. With industrial processing of pea, bean and grain protein into artificial meat and milk, a semblance of an animal-based diet could be provided for about 200 million people.

TABLE A: MELLANBY'S BASIC DIET 1975

- 5.3 million hectares available
- 5.7 million hectares of pasture
- 7.8 million spare hectares

Population 53 million. Agricultural land 18.8 million ha.

	Consumption gms/person/day	Calories in diet kcal/person/day	UK production million tons/year	Yield tons/ha	Arable land 1000 ha	Perm: pasture 1000 ha	Rough pasture 1000ha
Cereals for human food	530	1850	10.25	4.166	2460		
Potatoes	453	300	8.76	32	275		
Sugar	32	100	0.625	5	125		
Milk	568	330	11	4.58	2400		
Beef (grass reared)	56	150	1.08	0.45		2400	
Sheep	14	37	276	0.084			3290
Total calories per day		2767					
Land available (excl woodland)					7200	4800	6800
Spare land					1940	2400	3510
Total land use					5260	2400	3290

- **One hectare of arable plus one of pasture feeds 10 people**

TABLE B: CHEMICAL WITH LIVESTOCK

- 4.4 million hectares arable
- 6.4 million hectares of pasture
- 7.6 million spare hectares

Population 60.6 million. Agricultural land 18.5 million ha. Forestry etc 3.69 ha.

	Consumption gms/person/day	Calories in diet kcal/person/day	UK production million tons/year	Yield tons/ha	Arable land 1000 ha	Perm: pasture 1000 ha	Rough pasture 1000ha
Cereals for human food	500	1700	11.06	7.3	1515		
Potatoes	453	300	10	44	227		
Sugar	32	100	0.707	9	78		
Vegetables and fruit	500	150			400		
Milk (inc butter cheese etc)	568	330	12.5	7.0/cow	1252		
Beef	56	150	1.24	0.43		2758	
Cereals for animal feed			6.69	7.1	917		
Sheep	14	37	0.31	0.084			3690
Land Available [Total Calories]		[2767]			6335	5711	6462
Spare land					1946	2953	2772
LAND USED					4389	2758	3690

- **One hectare of arable plus 1.5 hectare of pasture feeds 14 people**

TABLE C: CHEMICAL VEGAN

- 3 million hectares arable
- 15.6 million spare hectares

	Consumption gms/person/day	Calories in diet kcal/person/day	UK production million tons/year	Yield tons/ha	Arable land 1000 ha	Perm: pasture 1000 ha	Rough pasture 1000ha
Cereals for human food	500	1700	11.06	7.3	1515		
Potatoes	453	300	10	44	227		
Sugar	32	100	0.707	9	78		
Rape Oil	35	310	0.774	1.2	645		
Dried Peas	80	207	1.77	3.75	471		
Vegetables	500	150			400		
Land Available [Total Calories]		[2767]			6335	5711	6462
Spare Land					3336	5711	6462
LAND USED					2999	0	0

- **One hectare of arable feeds 20 people**

TABLE D: ORGANIC VEGAN

- 7.3 million hectares arable
- 11.2 million spare hectares

Population 60.6 million. Agricultural land 18.5 million ha. Forestry etc 3.69 million ha.

	Consumption	Calories in diet	UK production	Yield	Arable land	Perm: pasture	Rough pasture
	gms/person/day	kcal/person/day	million tons/year	tons/ha	1000 ha	1000 ha	1000ha
Cereals for human food	500	1700	11.06	4.3	2572		
Potatoes	453	300	10	25	400		
Sugar	32	100	0.707	7.5	94		
Rape Oil	35	310	0.774	0.8	968		
Dried Peas	80	207	1.77	3	590		
Vegetables	500	150			450		
Green manure					2242		
Land Available [Total Calories]		[2767]			6335	5711	6462
Spare Land					-981	4730	6462
LAND USED					7316		

- **One hectare of arable feeds 8 people**

Vegan Organic: Reliance on Green Manure

In Tables D and E I have again updated Mellanby's diet to 2005, but this time for organic husbandry. Both these organic diets, vegan and livestock, take more land than their chemical counterparts. This is partly because average grain yields obtained by organic agriculture today in Britain are less than 60 per cent of those obtained by chemical farmers; in fact organic wheat yields today are similar to those of chemical agriculture in 1975.

But lower yields are only half the problem. To obtain yields above a bare minimum of around 750 kg of grain per hectare, land has to be fed with extra nitrogen. Organic systems by definition do not use synthetic fertilizer, so nitrogen is either imported from other land where it is not required, usually in the form of animal manure; or obtained by inserting into the rotation a crop of leguminous plants such as beans, clover or lucerne, which extract nitrogen from the atmosphere. A dedicated crop of legumes, which is not fed to humans or animals, but ploughed in to provide fertility, is called a green manure. Green manures which occupy the ground for a whole season lower the yield from each hectare still further.

In Table D I have assumed that one hectare out of every three arable hectares is used for green manure – except for the pea crop which fixes its own nitrogen from the atmosphere. This adds an extra 2.2 million hectares to the vegan organic land-take, with the result that it requires 14 per cent more arable land than is in use today (including set-aside). This is not a problem, since it can be taken from the pasture, for which the vegan diet has no use. I have used the 33 per cent green manure ratio because that is the figure given in my main source, the *Organic Farm Management Handbook*. Clearly, other rotations, as listed in <u>Box 2</u> in Chapter 8 would give a different result.

Organic Livestock: The High Yield Paradox

In organic mixed farming systems, nitrogen is provided by manure, and by leys – temporary pasture including clover or other nitrogen fixing plants which after perhaps four or five years is ploughed up for two or three years cropping, and then put back to ley. Essentially these are green manures in which part of the

TABLE E: ORGANIC WITH LIVESTOCK

- 8.1 million hectares arable and ley
- 7.8 million hectares of pasture
- 2.6 million spare hectares

Population 60.6 million. Agricultural land 18.50 million ha. Forestry etc 3.69 million ha.

	Consumption gms/person/day	Calories in diet kcal/person/day	UK production million tons/year	Yield tons/ha	Arable land 1000 ha	Perm: pasture 1000 ha	Rough pasture 1000ha
Cereals for human food	500	1700	11.06	4.3	2572		
Potatoes	453	300	10	25	400		
Sugar	32	100	0.707	7.5	94		
Vegetables and fruit	500	150			450		
Green manure					1696		
Milk (inc butter cheese etc)	568	330	12.5	per cow 5.8	1898		
Beef	56	150	1.24	0.29		4100	
Cereal for dairy cows			2.936	4	657		
Cereals for beef cows			1.656	4	328		
Sheep	14	37	0.31	0.084			3690
Land Available [Total Calories]		[2767]			6335	5711	6462
Spare Land					-1760	-149	2623
LAND USED					8095	4100	3690

- **One hectare of arable plus one of pasture feeds 7.5 people**

nutrients pass through grazing animals before finding their way back to the cropland – though a proportion are creamed off to provide milk, meat, leather etc.

In Table E, three hectares of ley is assumed to fertilize two hectares of cash crops. At this rate the 1.9 million hectares of ley for dairy pasture, plus a small amount of manure from the beef, does not provide enough fertility for all the crops grown and so the organic livestock model also has to rely on 1.7 million hectares of green manure. If there were pigs or chickens, they would provide more manure – but not enough to grow the corn necessary to feed them.

The organic livestock model is worrying because it is very expensive on land. It would require ploughing up 1.76 million hectares of our existing pasture land to provide cropland and leys. This in turn means that there is not quite enough permanent pasture left for the beef – there is a shortage of 149,000 hectares, which has to be taken out of the rough grazing. Only 2.623 million hectares are left for other uses such as wildlife parks or biomass production.

There are two main reasons for the heavy land requirement of organic farming. The first is that average yields of organic wheat and potatoes are only 60 per cent of those achieved with the use of chemicals. With most other crops the difference between organic and chemical is less pronounced, but wheat and potatoes are the staples. The other problem is the cows, particularly the beef cows which take up an enormous amount of land for very little return. There appears to be too much beef, which is strange because it is the same amount per person as in Mellanby's 1975 scenario, and it didn't cause any problem then. Admittedly there are now seven million more mouths to feed, and also Mellanby puts fertilizer on his leys. But even so the beef sector seems to have expanded disproportionately. In 1975 the beef herd occupied the same amount of land as the dairy herd; but now the area devoted to beef (including grain for cattle feed) is almost twice as large.

Here, paradoxically, it is high yields that are causing the problem. The figures in Table E are derived from the *Organic Farm Management Handbook* 2007, and they reflect a more modern management approach in an era of cheap subsidised corn. Whereas Mellanby's cows yielded just 3,600 litres of milk a year, organic

TABLE F: LIVESTOCK PERMACULTURE

Including pigs, poultry, textiles, tractor or horse power and timber

- 7.9 million hectares arable and ley
- 5.9 million hectares of pasture
- 6 million hectares of woodland
- 2.4 million spare hectares

Population 60.6 million. Total agriculture and forestry land 22.205 million ha.

	Consumption gms/person/day	Calories in diet kcal/person/day	UK production million tons/year	Yield tons/ha	Arable land 1000 ha	Perm: pasture 1000 ha	Other kand 1000ha
Cereals for human food	448	1526	9.9	4.3	2302		
Potatoes	453	300	10	25	400		
Sugar	32	100	0.707	7.5	94		
Vegetables and fruit	500	150			100	50 (100)	
Hemp and flax	5 kg/year		0.303	3	100		
Horse/biofuel					876		
Green Manure					430		
Milk (incl butter, cheese)	568	330	12.5	3.7 (3.26 net)	2825	1765	
Beef (grass reared)	33	86	0.735	0.4		1740	
Cereals for pigs	bacon 36	180	1.2	4.3	279		
Cereals for hens/eggs	(egg/chicken)30	50	2	4.3	465		
Sheep	9	24	0.2	0.084			2372
Leather and sheepskin	1.46 kg/year						
Wool	750 kg/ year						
Fish	11	11	0.243				
Timber, firewood				3			6000
SPARE LAND (wild meat)	5	10	0.11	0.031			2407
LAND USED [total calories]		[2767]			7871	3555	8372

- **One hectare of arable plus 0.8 ha of pasture supplies 7.5 people**

cows today average 5,800 litres, only 1,200 litres less than non-organic cows. The trouble is that to achieve this they need fairly large amounts of grain – over a tonne a year each – whereas Mellanby's cows are grass-fed.

The need for grain is not the only problem caused by the high milk yield of these cows. The size of Mellanby's beef herd was dictated by the number of calves that his low yielding cows produced. But now, because there are fewer cows producing the same amount of milk, there are fewer calves. This means that in order to produce the same amount of beef as in Mellanby's diet, we have to run a dedicated beef suckler herd – nearly two million cows which produce nothing except one beef calf a year, whereas Mellanby's calves were all the by-product of cows supplying milk. This cancels out much of the advantage of high-yielding cows and is the main reason why land occupied by beef has swelled from 2400 hectares in 1975 to 4100 in Table E.

I therefore decided to see what would happen if I reduced the beef herd to a size commensurate with the dairy herd and moved back to Mellanby's system of running a larger number of low yielding dairy cows which can subsist entirely on grass. This is akin to what has been happening in New Zealand since they abolished farm subsidies, because it is more competitive – which is why New Zealand butter is advertised as coming from free range cows. And it is what I have done in Table F. If you examine just the cattle figures in it, you will see that the milk yield has been reduced from 5,800 to 3,700 litres a cow, and the total amount of beef produced has been reduced from 1.24 million tonnes to 735,000 tonnes; but the number of dairy cows is increased so that the total amount of milk produced, 12.5 million tonnes, remains the same as in Table E.

TABLE G: VEGAN PERMACULTURE

Including extra veg, textiles, tractor power and timber

- 7.7 million hectares arable
- 6 million hectares arable
- 8.4 million spare hectares

Population 60.6 million. Total agriculture and forestry land 22.205 million ha.

	Consumption	Calories in diet	UK production	Yield	Arable land	Orchard	Other land
	gms/person/day	kcal/person/day	million tons/year	tons/ha	1000 ha	1000 ha	1000ha
Cereals for human food	491	1670	10.9	4.3	2534		
Potatoes	453	300	10	25	400		
Sugar	32	100	0.707	5	94		
Rape oil	35	310	0.774	0.8	968		
Dried peas	80	207	1.77	3	590		
Hemp and flax	7 kg per year		423	3	146		
Vegetables, fruit, nuts	666	180			150	150	
Biofuel					1152		
Green manure					1646		
Timber, firewood			18	3			6000
SPARE LAND (wildlife)							8375
LAND USED [total calories]		[2767]			7680	150	6000

- **One hectare of arable supplies 8 people**

In terms of land-take, the lower yielding cows produce food almost as efficiently as the high yielding cows. The ratio of hectares of land to calories of beef and milk from the corn-fed cows in Table E is 6,983:480 or 14.5:1 whereas from the grass-fed cows in Table F it is 6,330:416 – or 15.2:1. In other words, a 63 per cent increase in milk yield results in a mere five per cent increase in land productivity.

But there is another big difference between the two. The grass-fed cattle in Table F provide over 2.8 million hectares of ley that can be used in rotation to help fertilize over a million hectares of crops – whereas in Table E the 1.9 million hectares of ley that the corn fed cattle bring with them isn't enough to fertilize the million hectares of grain they eat. The low yielding cows are nitrogen providers whereas the high yielding cows are nitrogen takers.

There is one other matter of interest. Since 2004, the net organic yield for wheat has risen from 3.8 tonnes a hectare, to 4.3 tonnes in 2007 and 4.8 tonnes in 2009; but the milk yield for organic cows dropped, from 6,000 litres to 5,800 litres in 2007, when grain prices were high, and has since risen again.[3]

A Permaculture Approach

My main purpose in Tables F and G is to go another step further and see whether the UK could become more self reliant, not only in food, fodder and fertility, but also in fibre and fuel. Our environmental footprint currently stretches across untold ghost acres around the world; if suddenly we had to shoehorn it into the 22 million hectares of non-urban land we have in this country, how would

---

3   The figures for 2009 show a 0.5 tonne increase in the organic yield of winter wheat over 2007, to 4.8 tonnes milled. Mark Measures, one of the authors of the *Organic Farm Management Handbook*, tells me that not too much store should be set by this rise, which represents the better farms. A farm he knows on chalk downland would be content to get four tonnes gross (3.8 tonnes milled). Pers com 2 December 2009.

we cope? Could this be done organically, whilst keeping a reasonable amount of meat in our diet for those who wanted it, and ensuring that a reasonable proportion of the country is reserved for wildlife?

Tables F and G reflect a more permacultural approach, by which I mean permaculture on the macro-scale, involving increased integration of lifestyle with natural and renewable cycles, rather than just mulching, intercropping and herb spirals. Some of the measures taken require a change in our land management systems, and also in human settlement patterns. This is a society in a state of energy descent, with increasing dependence upon renewable resources, more waste cycling and (consequently) a localized economy which is more integrated with natural processes. The approach towards livestock is a default strategy in respect of the cows, insofar as they are grass fed and contribute to the fertility building of the arable rotation; but less so as regards the pigs and chickens since a certain amount of grain is grown to feed them (though in the next chapter I argue that a certain amount of livestock grain is necessary for food security reasons). Here is a list of the main features I have introduced:

Meat and Dairy   The amount of beef in the diet has been reduced both by no longer running a suckler herd, and by reducing the average age at which beef cows are slaughtered. There are 83 grams of red meat per person per day. For a family of four, this is the equivalent of a 5lb Sunday joint, which could probably be spun out till Tuesday or Wednesday. Together with a smidgeon of chicken and fish it comes to 38 kilos of meat per year, which is about half the amount people eat now. The volume of milk consumed is the same as it is at present, and everyone also has a couple of eggs a week. Farm animals provide 670 calories of the daily ration of 2,767, whereas in Mellanby's basic diet they only provided 517.

Pigs   To compensate for the reduced amount of beef in the diet, I have introduced pigs. Although partially fed on grain, these are efficient because their diet consists of two thirds crop residues, whey and food waste. This ought to be possible since in the early 1990s even commercial pig feed consisted of 50 per cent food waste, and on top of that there is all the domestic food waste which currently goes into landfill. The figure of 2767 calories per person (including children and old people) allows for around 700 calories of food waste, which in theory is enough to provide our pigs with all their food. I have kept this margin tight because selling feed to small-scale pig units on mixed farms is an economical way of ensuring that the nutrients in food processing waste cascade back to the land. The pigs also bring fat into the diet, and produce it on less land than rape oil.[4]

Chickens   I have also introduced chickens, which in this model are fed on grain. They take up more land for less calories than pigs, but this is only because the pigs are getting all the food waste. It is possible to feed much of the waste to hens, and they convert it into protein more efficiently than pigs. But the advantage of pigs in a northern country is that they produce fat, when little else does. If resources became scarce, I would expect commercial chickens to be among the first to rise in price (a boiling fowl was a luxury to be had only on special occasions in the 1950s) but there would still be plenty of opportunity for backyard hens fed on household scraps.

Fish   I have allowed the carnivores a small amount of fish, equivalent to about half of current consumption levels. If European countries reverted to local control

---

4   Cf. Chapter 6, reference 18.

of fishing grounds, then management of UK stocks would improve and catches eventually rise. There are some wonderful permacultural systems in Vietnam and China where fish farming is part of the cycle, but I don't know enough about their potential in the UK to include them here.

Sheep I have reduced the number of sheep from 27 million in Mellanby's scenario to 18 million, because they don't produce much food and there is a widespread perception that they have too much of a monopoly of our uplands at the moment. But we might think twice about this because, in the absence of plastic fleeces shipped in from China, we may need more wool than 18 million sheep can produce. Sheep would be bred for heavier fleeces.

Wild Meat I could find no figures for the volume of meat available from wild herbivores, but it is probably minimal. The figure given is roughly the same as the estimated quantity of wild rabbit meat eaten in 1953.

Fruit and Veg In the localized economy envisaged here, a large proportion of fruit and vegetables could be grown more intensively on allotments, in gardens and on urban land. Much top fruit would be grown not on arable land, which needs weeding, but in orchards which could be grazed, or in the vegan case mown. I have reduced the area of arable put down to horticulture in Table F to 100,000 hectares.

Wheat High yields come through breeding for seed production at the expense of stem production. The lower wheat yields associated with organic production can be partially offset by producing thatching straw – another form of biomass that will be in demand if we enter a path of energy descent.

Textiles I have been unable to establish what current UK consumption rates of textiles are, and since so much of it is frivolous there is not much point. Textile fibres do not take up an enormous amount of land. Except for fashion models, most of us eat more than we wear. In Table F I have allocated 7.25 kilos per person per year (a domestic washing-machine load), provided by hemp and flax, wool and leather.

Nutrient cycles Additional nitrogen for crops comes from three main sources. Enough nitrogen to fertilize one million hectares of crops can be obtained from recycling human sewage, preferably on crops for animal rather than human consumption. This requires a society which does not pollute its human waste with heavy metals, through contamination with liquid run-off and effluent. Just over a million hectares can be supplied with nutrients through ley farming. And a further 750,000 hectares could be fertilized with a proportion of the available animal manure. How much can be recuperated depends upon how livestock are managed. In the case of sheep, this might involve bringing them in at night, to shit in the farmyard, as is normal practice in many places on the continent. Any shortfall would have to be met by green manure, at a rate of one hectare for every two cultivated.

In the absence of supplies of imported rock phosphate, phosphorus rather than nitrogen might become the main constraint upon crop yields, in which case we would have to ensure rigorous recycling of animal manures, human sewage, slaughterhouse wastes etc – a further reason for dispersing population around the countryside. A vegan system in particular might have problems maintaining phosphorus levels, since animal feed provides an incentive for recycling nutrients back to the land, and no phosphorus would be brought onto the arable from animals grazing the hinterland.

Biomass   I have not allowed for much intensive biomass energy production, mainly because it takes up arable land that could be better used for food. In non-arable areas, I prefer natural woodland to short rotation coppice, because of its amenity and wildlife value; the prospect of vast acreages of the countryside curtained in eight foot high willow coppice monoculture is not very appealing. However, there is a good case for arable biomass production on farms to provide fuel for tractors. I have allocated ten per cent of the arable land either for biomass to run machinery, or else to grow feed for draught animals.

The Livestock Permaculture land economy outlined in Table F produces all its food, a substantial proportion of its textiles, and the energy for cultivating its fields on 13.4 million hectares, a little over half the entire country. The more orthodox organic system in Table E requires nearly 16 million hectares, it doesn't produce any fuel, it is low on fat, and it produces less meat: only 187 calories in the daily ration, compared with 272 in the permaculture model. The improvement comes through using animals for what they are best at, recycling nutrients and waste – and avoiding feeding them grains.

Woodland or Wildland

We are left in Table F with about nine million hectares, of which 3.7 million hectares are currently classed as woodland or else 'other land on agricultural holdings including woodland', and the rest are rough grazing – including 1.5 million hectares of grouse moor. There are therefore nearly five million hectares of mostly poor quality land spare, for which the most obvious uses are either to 'rewild' it, or else to put it over to woodland.

In the livestock permaculture scenario I have opted to leave slightly over half of this area for wildlife and to convert the other half to woodland. This gives us about six million hectares of woodland, around a quarter of the entire country. This is still a lower proportion than in France (27 per cent), the EU (40 per cent) or the world (29 per cent). Six million hectares of biodiverse woodland, coppice and plantation could produce 36 million cubic metres of timber and pulp – three quarters of what we currently consume (most of which is imported). A saner society, without all the junk mail, newspaper supplements no one reads, tacky throwaway furniture and so on could make do with a lot less.

On the other hand, six million hectares of woodland could also produce enough firewood to heat six million well insulated family homes (at three tonnes per hectare and per home). This is not incompatible with timber production. All pulp and timber, when it comes to the end of its economic life, is firewood.

This leaves three million hectares for wildlife: an eighth of the country, not as much as some people would wish to see. This land, since it is specifically not woodland, would have to be grazed by edible, semi-wild herbivores such as deer, primitive types of ox, or Konik ponies.

The wild area could be increased by reducing the sheep flock still further, at the expense of a small amount of meat and some rather valuable wool; by producing more 'pink veal' (from young grass-fed cattle) and less mature beef; or by reducing the number of dairy cows and the amount of milk consumed. In each case, to compensate, a smaller area of land would have to be converted to crop production and green manure.

Vegan Permaculture

Table G outlines, as far as I am able, a vegan permacultural vision, based on the same data. I have introduced more flax and hemp to make up for the lack of wool and leather; and since the meat-eaters have been allowed pork and eggs, I have increased the variety in the vegan diet by allocating an additional 100,000 hectares for fruit and vegetables, most of which is grown on non-agricultural or orchard land, and fertilized with municipal compost. Perhaps I should allow them more. Nuts are an obvious choice, but reliable information about yields is difficult to find. The vegan system uses human sewage for fertilizer like the livestock system, though there would be more of a problem avoiding applying it to human edible crops.

The obvious, and some would say overwhelming advantage of the vegan system is that it uses less land. However, it is the grazing land that the vegans economize on. They require almost as much arable land as the meat-eaters, mainly because of the lack of manure, and the expense of providing fat or oil. In fact the area of land under annual cultivation in the vegan system in any one year (7.2 million hectares) is considerably greater than in the livestock system where more than a third of the arable land consists of grass leys, and only 4.6 million hectares hold annual crops.

The vegans could perhaps reduce the area of green manure by more efficient use of cover crops, or by importing hay or leaf mould for mulch. There are also the residues from rapeseed oil, biofuel, and products such as oat milk and pea milk which could be used as fertilizer – though vegans might be tempted to trade these with pig keepers.

The disadvantage of the vegan model, from the peasant perspective, is that it results in a lop-sided land economy, with almost all the activity concentrated in the arable area; and overall it appears to provide less employment on the land than the livestock system. The less arable areas of Britain would become agriculturally redundant. All that empty space in the grassland area gives the relatively small growing area a rather compacted urban feel, and I worry that the spare land might get filled up with monoculture energy crops. But that depends upon what the vegans decide to do with it, and that is not really for me to say. When these scenarios were originally published I invited vegans to outline their own vision of what could done with the large areas of UK land that would be liberated or abandoned, depending on one's viewpoint.[5] Jenny Hall responded with a four page article in a subsequent issue of *The Land*, where she suggested that 10 million hectares would be dedicated to woodland, one million to berries and nuts, and 3.8 million to wildlife conservation and 'wildland'.

That is not to say that the people's choice has to be either one thing or the other. Vegan and livestock land use systems can coexist well enough side by side, as long as boundaries are drawn and fences maintained. Instead of being strictly vegan or enthusiastically carnivore, it is entirely possible to have a level of compromise between the two approaches outlined in Tables F and G, and indeed that is more likely.

---

5   The ideas expressed in Mark Fisher's website www.self-willed-land.org.uk are highly compatible with a vegan land economy. Jenny Hall wrote an article outlining a vision of a stockfree rural economy in Britain, in response to the publication of 'Can Britain Feed Itself?', Hall, J (2008), 'Stockfree Britain', *The Land* 5, Summer 2008.

TABLE H: LOW MEAT & BIOENERGY

Including extra veg, textiles, tractor power and timber

- 10 million hectares arable
- 3 million hectares of pasture
- 7 million hectares of woodland
- 2.2 million spare hectares

Population 60.6 million. Total agriculture and forestry land 22.205 million ha.

	Consumption gms/person/day	Calories in diet kcal/person/day	UK production million tons/year	Yield tons/ha	Arable land 1000 ha	Perm: pasture 1000 ha	Other land 1000ha	Woodland
Cereals potqtoes	568	1971	12.6	5.04	2500			
Beans	150	391	3.33	4	833			
Vegetables	1130	401	25	30	833			
Rape for biofuel			1.16 (meal 1.74)	3.5	833			
Miscanthus for bioenergy					2500			
TOTAL ARABLE CROPS					7500			
GREEN MANURE					2500			
Milk	171	108	3.8	2.2		500	1238	
Beef	8	23	0.177	0.177		500	500	
Sheep	1	3	0.022	0.84			262	
SR Coppice								1000
SR Forest								3000
Woodland								3000
Wildland							2200	
LAND USED [total calories]		2897			10000	1000	4200	7000

- **One hectare of arable, plus 0.3 hectares of pasture supplies 6.6 people**

Minimal Livestock/Bioenergy

After the foregoing projections were published in *The Land* magazine in 2008, the Centre for Alternative Energy (CAT) at Machynlleth published a first draft of Zero Carbon Britain, an alternative energy strategy for the country, in which they offer a projection for a relocalized rural economy with more emphasis on biofuel crops and greatly reduced numbers of livestock. According to the report's main author, Peter Harper, this was a first stab at constructing a model, and contains some inconsistencies, but I believe it begins to paint a picture of what a fairly localized biofuel economy might look like. I have taken the liberty of transcribing it into the same matrix as the other scenarios, after making some minor adaptions.[6]

CAT's rural economy is based around a 12 course rotation, in which (typically) the land would be under green manure during three years, and sown to *Miscanthus* for bioenergy over another three years, with grain, rape for biofuel and root crops taken over the other six years. The authors warn:

> There is much justified anxiety about biofuels. Sometimes presented as a means of maintaining post peak oil mobility, a few simple calculations show that biofuels come nowhere near to providing the World's (or even Britain's) current transport demands. A key reason why they are not a panacea is that under typical field or forest conditions, photosynthesis is simply not very efficient. For a given

---

6   I have had to guess how the specified amount of milk would be produced, being given little option but to place it in the rough grazing (which would be both unproductive and inconvenient). However the CAT strategy only requires 20 million hectares because its designers consider that 2 million hectares of agricultural land are needed for 'roads and buildings'. This is a massive overestimate: the entire area of non-agricultural development land in the UK, including the whole of London and all our towns, cities and villages only comes to around 2.5 million hectare, while the category of 'other farmland' which includes buildings, ponds, tracks, farm woodland etc is 872,000 hectares.

area, windmills produce 20 times more energy, photovoltaic panels 100 times more … Despite these drawbacks this strategy recognizes that bio-energy can play a useful role in a sustainable energy strategy.

The requirement for biofuel bumps the arable area up to ten million hectares, significantly more than in any of the other projections. In total there are six million hectares of woodland (the same amount as in Tables F and G) and 4.3 million hectares are devoted to bio-energy, of which 3.3 million are arable and one million short rotation coppice (SRC). In total this is about two thirds of the entire area given over to non-edible biomass. This is achieved at the expense of two things: first there is very little meat and dairy, scarcely a fifth of the amount provided in Table F; secondly there are are only 5.2 million hectares of permanent grass and wildland, compared to 8.3 million in the Livestock Permaculture model, and 8.4 million hectares of spare land in the Vegan Permaculture model. Ecologists would have something to say about the loss of biodiversity, and it is also possible that conversion of grassland to arable would result in a loss of soil carbon that would negate some of the advantage of planting biofuels.

The diet in Table H provides about 10 per cent more calories than the other models. However there is very little fat in the diet, and I suspect that much of the rape oil would be eaten rather than used for biofuel.

The CAT model provides a picture of how regimented and lacking in biodiversity an agricultural economy reliant on bio-energy might be. Every summer a quarter of the arable land – ten per cent of the entire country – would be a curtain of eight foot high *Miscanthus* grass. Given that wind turbines are more efficient in respect of land, and are reliant on grazing animals to keep the area around them clear, I suspect that meat-loving members of the public who currently object to their installation might, in a zero carbon Britain, come to see them as the lesser of two evils.

Conclusion

The main conclusion to be drawn from this exercise is that organic livestock-based agriculture, practised by orthodox methods and without supplementary measures, has the most difficulty sustaining the full UK population on the land available, while other management systems can do so with a more or less comfortable margin.

However, organic livestock agriculture becomes more efficient and sustainable when it is carried out in conjunction with other traditional and permacultural management practices which are integral to a natural fertility cycle. These include: feeding livestock upon food wastes and residues; returning human sewage to productive land; dispersal of animals on mixed farms and smallholdings rather than concentration in large farms; local slaughter and food distribution; managing animals to ensure optimum recuperation of manure; and selecting and managing livestock, especially dairy cows, to be nitrogen conveyors rather than nitrogen stealers.

These measures demand more human labour, and more even dispersal of both livestock and humans around the country than the chemical or vegan options. Effective pursuit of livestock-based organic agriculture of this kind requires a localized economy, and some degree of agrarian resettlement. Other management systems based on synthetic fertilizers or vegan principles lend themselves more easily to the levels of urbanization currently favoured by the dominant (and mostly urban) policy makers.

10

ON GRANARIES

Menenius: For the dearth, the Gods, not the Patricians make it …
and you slander the helms o' th' state; who care for you like fathers.

First Citizen: Care for us? True indeed, they ne'er car'd for us
yet. Suffer us to famish and their storehouses cramm'd with grain.

Coriolanus Act I Scene 1, William Shakespeare

Vegan propagandists sometimes argue that meat causes famine, or at least is implicated in it:

> The livestock industries of the world scour the globe in search of cheap sources of protein … At the height of the 1984 famine which inspired the historic Band Aid concert, Ethiopia exported feed crops to the UK. Similarly, in 1997, during times of extreme food shortages, North Korea exported 1,000 tonnes of maize to Japan for poultry feed.[1]

And according to another vegan writer, quoting from a Scandinavian report on the meat industry:

> During the years of famine in the early 80s, most Sahel countries reported an increased export of agricultural products. This human famine was not a natural disaster but clearly of political and economic origin. In other words, the starving countries of sub-Saharan Africa were exporting feed for the west's animals while its own people were dying. It is difficult to imagine anything more obscene. This scenario is being repeated in all the poorest regions of the world, but the devastation is usually presented as an act of God, of war, or of incompetence. The fingerprints of the true culprits are never identified.

The true culprits are identified, a few lines further on, as 'multinationals, who are devouring the world's vegetation like a plague of locusts so they can profit from livestock production.'[2]

The export of food under famine conditions is indeed an obscenity, but the accent on livestock is misplaced. It is largely because of demand for animal feed that any surplus food is grown in the first place. Under famine conditions the animal feed should have been turned over to human use, because that is partly what animals are for: to ensure that there is always a surplus of grain. The reason why it wasn't (as the Scandinavian report correctly identified) was because the people who were starving didn't have control over it.

---

1   Gold, Mark (2004), *The Global Benefits of Eating Less Meat*, Compassion in World Farming (CIWF).
2   Wardle, Tony (n.d.), 'Hype, Hypocrisy and Hope', *Growing Green International* 9.

It is now widely recognized that although famines may involve an absolute food shortage in the affected locality, there is very often no absolute food shortage in the country whose government is responsible for agricultural policy and food security. The export of food from famine prone areas is a regular occurrence, and it doesn't seem to make much difference whether the food is originally destined for humans or for animals. Throughout the Irish famine of 1845 wheat was being exported from the east of the country to England, and wheat tends to be grown for human consumption as much as for animals.[3] During the catastrophic famine in India in 1876, which killed about ten million people, Londoners were eating rice and wheat exported from India. 'It seems an anomaly', wrote Cornelius Walford at the time, 'that with her famines at hand, India is able to supply food for other parts of the world'.[4]

But the anomaly continued: 'Between 1875 and 1900, years that included the worst famines in Indian history, annual grain exports increased from three million tonnes to 10 million tonnes … By the turn of the century, India was supplying nearly one fifth of Britain's wheat consumption'.[5] During the 1899 famine in the Indian province of Berar, when 143,000 Berari died, 747,000 bushels of grain and tens of thousands of bales of cotton were sent out of the province.[6]

Nor does it really make much difference whether the produce is exported abroad, or is hoarded by local merchants and élites. An English medical officer, interviewed after the 1873 Bengal famine, testified:

A: In Bengal they died in front of bulging food shops.
Q: Bulging with grain?
A: Yes, they died in the streets in front of shops bulging with grain.
Q: Because they could not buy?
A: Yes.[7]

Crop failure, or dearth, in recent times, has usually only turned into famine when a group of people has no money or access to limited but still available supplies of food. As Amartya Sen put it in his influential study *Poverty and Famine*, 20th century famines have occurred when 'a moderate short fall in production' is 'translated into an exceptional shortfall in market release'. To understand how easily that sort of thing can happen we have only to point to the current housing situation in the UK. There is no real shortage of buildings, or indeed of houses (for many lie empty or are second homes), and the building industry until 2007 was booming – yet, many people cannot afford a home and the number of people in emergency accommodation is higher than it has ever been, because they can't buy into the market. In the poor world, the same thing happens with food.

If the causes of famine are usually political and economic, what about the causes of the 'moderate shortfalls in production' which trigger famine? How much are these shortages attributable to animal husbandry, and how much to agriculture? Should we pin all the blame upon Abel? It might be that Cain has better PR?

The great famine of 1876-8 in India affected two areas most severely, Bengal and the Deccan plateau in southern India. The Deccan suffered badly because

---

3   Thompson, E P (1991), *Customs in Common*, Penguin, p 264.
4   The Corner House (2002), *The Origins of the Third Word: Market, States and Climate*, Corner House.
5   *Ibid.*
6   *Ibid.*
7   Thompson (1991), *op cit*.3.

policies initiated by the British had undermined the agricultural economy:

> After the 1857 Mutiny, the British pursued a relentless policy, especially in the Deccan, against nomads and shifting cultivators whom they labelled as 'criminal tribes'. Although the agro-ecology of the Deccan for centuries had been dependent upon the symbiosis of peasant and nomad, upon valley agriculture and hill-slope pastoralism, the colonial state's voracious appetite for new revenue generated irresistible pressure on the *ryots* (peasants) to convert 'waste' into taxable agriculture. Punitive grazing taxes drove pastoralists off the land, while cultivators were lured into the pastoral margins with special leases.
>
> The traditional Deccan practices of extensive crop rotation and long fallow, which required large farm acreages and plentiful manuring became less numerous. Between 1843 and 1873, cattle numbers in the Deccan fell by almost five million. The 1876-8 drought killed off several million more, with cattle populations plummeting by nearly 60 per cent.[8]

It is only slightly reassuring to note, in passing, that the proportion of people who died in the famine – about 25 per cent in many regions of Mysore and Madras states – was less than the number of cattle which died. Some farmers clearly had enough food reserves or money to ensure not only their own survival, but their stock's as well. But the main observation to be made here is that a low impact grazing regime was replaced by an ecologically inappropriate agricultural regime. (We have seen that happen in the UK, with less loss of life, where large areas of downland pasture have been ploughed up for subsidized wheat.) We are not told in this account where the pastoralists were shifted to, but it stands to reason that they increased grazing pressure wherever they went.

This is a common story, repeated up until modern times. For example, Vandana Shiva reports that: 'Under Ethiopia's Third Five Year Plan, 60 per cent of the lands brought under cultivation in the fertile Awash valley were devoted to cotton production. The local Afar pastoralists were evicted from their traditional pastures and pushed into fragile uplands, contributing to the deforestation that has been partly responsible for Ethiopia's ecological decline.'[9]

The 1985-9 famine in the Darfur and Kordofan regions of Sudan was exacerbated by increased ecological pressure resulting from a doubling in the number of cattle over the previous 30 years. Once again, much of this pressure was caused by the conversion of traditional tribal grazing lands to industrial farming producing sorghum and other crops for export. The state-run Mechanized Farming Corporation acted as a conduit for a loan from the World Bank to clear and cultivate around five million acres of land by 1984. As the soil, inappropriate for intensive agriculture, became exhausted, the farms spread further into the nomadic herders' territory.[10]

Two main ethnic groups the Baggara of Arab origin and the Dinka, of African origin, found themselves competing for diminishing grazing land, and the Islamic Government in Khartoum saw an opportunity to foster a conflict which would undermine the strength of rebel resistance forces based in the south of the country. Allegedly with the collusion of Chevron Oil who were prospecting in the region, the government armed Bagarra and other Arab militias to carry out raids on Dinka, stealing cattle, burning granaries and killing people. By 1989 an estimated 30 per cent of the population of the South had been uprooted,

---

8   The Corner House (2002), *op cit*.4.
9   The Ecologist (1993), *Whose Common Future*, Earthscan, 1993.
10   Keen, David (1995), 'The Benefits of Famine', *The Ecologist*, 25:6 Nov/Dec 1995.

the economy of the Dinka was severely disrupted with the loss of hundreds of thousands of their revered cattle, and the rate of people dying of starvation well surpassed that in neighbouring Ethiopia. The conflict in the Darfur, as we know, continues to this day.

The causes of famine are many and complex; but one has to hunt high and low to find any example of a famine being caused through 'inefficient' animal husbandry displacing 'efficient' arable agriculture.[11] In all the above cases it is the reverse, and the Irish potato famine in particular offers a classic example of the dangers of overefficient vegetable farming. Subsistence farmers in the south and west of Ireland were forced by extractive rents to live on the most efficient form of production available, the crop which provided most nutrients from a given bit of land, namely potatoes, and very little else. They even concentrated all their efforts on the 'lumper', the most efficient varieties of potato. When the potato harvest collapsed, because of blight, they had little in the way of animal or alternative vegetable produce to turn to – while the grain that was grown in the east of the Ireland was exported to England. Total reliance on the most efficient form of production took several million Irish people to the brink of starvation and then pushed them over the edge.

But the distinction between livestock and agriculture is a red herring. The conflict is not between animal and vegetable, or between peasant farmer and nomad herdsman, who often, as in the example of the Deccan given above, are in 'symbiosis'. The conflict is more often between a locally rooted and proven tradition of land-use which invariably has its animal element, and a superimposed 'efficient' agricultural improvement, often a monoculture, designed to extract and deliver resources for the international market. Whether the commodities so delivered are to provide consumers with meat that they don't need, cotton tee-shirts they don't need, or palm oil they don't need, is almost immaterial. I say 'almost', because if the commodity is animal feed it can at least be used to feed humans in an emergency, given the political will.

The Rumbim Tree

Throughout human history, meat consumption has not often threatened food security; on the contrary a main role of animal husbandry has been to *provide* food security: 'The purpose of domestication was to secure animal protein reserves and to have animals serve as living food conserves'.[12] Animals, fattened on rich summer grasses, and lured into domestication by titbits, could be kept ticking over on hay as store animals until the back end of winter, when there was nothing else left. Nomads' herds of cattle or sheep are walking rat-proof larders, full of high value protein and carbohydrates, which serve as a buffer against famine when the rains fail or times get hard.

In the early 1960s anthropologist Roy Rapaport examined a particularly sophisticated use of animals as a buffer against starvation amongst the Maring of New Guinea (though a similar pattern has been observed in a number of

11   The enclosure of English common fields for sheep in the 14th to 17th centuries displaced agricultural peasants to make way for an inefficient animal based industry, but it can hardly be said to have caused a famine. The great famine of 1888-1892, which affected virtually all of Ethiopia but was most severe in the north, was largely caused by the exceptionally rapid spread of the disease Rinderpest amongst cows. It was the loss of oxen for ploughing that triggered the famine and added to the very slow recovery. Spiess, H (1994), *Report on Draught Animals under Drought Conditions in Central, Eastern and Southern Zones of Region 1 (Tigray)*, United Nations Development Programme, Emergencies Unit for Ethiopia, http://www.africa.upenn.edu/eue_web/Oxen94.htm
12   Dando, W A (1980), *The Geography of Famine*, Arnold.

other New Guinea populations).[13] In the following section I examine this rather unusual strategy in some detail, partly because it shows that humans are capable of taking a very different approach towards meat-eating from our own, and partly because it offers a parable for the dilemma which humanity now faces. In no way am I advocating that we should adopt their system.

Within this tribal group, each clan of about 200 members lives off slash and burn – clearing areas in the forest, farming them for a year or two, and then moving further afield. Any surplus food they grow is fed to pigs who also derive a portion of their food by roaming the forest floor. The young pigs are raised in the household by the women who treat them as pets, scratching and cossetting them like puppies. They are given names: Jokai, Kikia, Kombom or Prim. When they are tiny they are carried out to the gardens where the women work, then walked out on a leash tied to their front leg, and soon they learn to trot along behind, just like a dog. When they are a few months old, they are encouraged to go off into the bush during the day to forage, but in the evening they return of their own accord to the household for an evening meal of manioc and sweet potato.

But instead of maintaining a stable population of pigs by eating them at a constant rate, the Maring refuse to kill them and allow the population to build up, and up, until the pigs are eating them out of house and home – breaking into their gardens and consuming so much of the produce that the soil becomes exhausted. As long as the population of pigs remains low, and a woman has perhaps just one pig to care for, then they can be fairly easily fed on waste and surplus produce from nearby gardens. But adult pigs eat as much as a human. When each household has to support four or five of them it becomes a huge burden upon the economy. As the gardens near to the village become exhausted, new gardens have to be established at a greater distance, and women have to trudge further and further afield to maintain their crops, and fetch food for parasitic pigs which nobody eats.

After 10 years or so of steady increase in the numbers of pigs, the women find themselves encroaching on the territory of another clan whose women are similarly engaged trying to find land to feed their burgeoning pig populations. Arguments break out, pigs that break into gardens are killed by furious gardeners, reprisals are made, the women complain to the men that they can't cope with the work load, and eventually the elders come to the conclusion that there are too many pigs.

At this point two or more conflicting clans each hold a year long series of feasts, called a *kaiko*, where they sacrifice pigs to the ancestors and invite their allies amongst the neighbouring clans to participate in orgies of pig-eating. At the *kaiko* observed by Rapaport, held by a tribal group of some 200 adults and children, about 110 full sized pigs were eaten leaving just 59 mainly young pigs alive. The festivities and dancing provide an opportunity for social intercourse and trading between clans, and for young adults to find a spouse. They are also an opportunity for guests from clans which have not declared *kaiko* (and hence are not allowed to eat their own pigs) to share the pork provided by a host clan. It is mainly women and children who get the pork, as adult males are not allowed to eat pig until their own clan calls a *kaiko* – which is fair enough, given that the women do most of the work keeping the pigs fed.

When the festivities have come to a climax the conflicting clans, by agreement, participate in clearing a patch of the forest for battle and then they declare war. The ensuing hostilities:

---

13   Rapaport, Roy A (1967), *Pigs for the Ancestors: Ritual in the Ecology of a New Guinea People*, Yale University Press.

lead to many casualties and eventual loss or gain of territories. Additional pigs are sacrificed during the fighting, and both the victors and the vanquished soon find themselves entirely bereft of adult pigs with which to curry favour from their respective ancestors. Fighting ceases abruptly, and the belligerents repair to sacred spots to plant small trees known as *rumbim*. Every adult male participates in this ritual by laying hands on the *rumbim* sapling as it is put in the ground … From now on, the thoughts and efforts of the living will be directed toward raising pigs.[14]

The saplings, as they grow, act as a guide to the passage of time (the Maring can't count up to 10), and as long as they remain in the ground, a truce is maintained. When, after 10 or 12 years the pigs have again multiplied beyond their carrying capacity, and *kaiko* is declared once again, the *rumbim* are uprooted, and the cycle once again shifts from green to red.

Why do the Maring indulge in this boom and bust approach to pig husbandry, rather than simply rearing and eating the pigs at a constant and sustainable rate? At first sight this resembles the tribal herdsman's strategy of keeping large numbers of cattle on the hoof as a hedge against drought or famine. Indeed this is an agreed reason why some Melanesian tribes keep large herds of pigs:

A way in which Melanesian populations are able to adjust to such unpredictable variation [in weather and crop yield] is through the practice of trying to plant more crops every year than can be or need to be consumed by the planters in a year not appreciably disturbed by adverse weather. Planting will be more than enough … should the weather be bad … The vegetable surpluses of normal years become available for the feeding of livestock … The practice of feeding vegetal surpluses to the pigs in years of normal and maximal crop years is described as 'banking' in Oliver's account (1955) of a Solomon Island culture, and the term seems appropriate, for the pigs are indeed food reserves on the hoof.[15]

But this is not satisfactory as an explanation for Maring pig-accumulation, since the Maring don't have to deal with drought and famine. The worst that ever happens is that the 'dry season' is wetter than usual, which makes the burning of clearings for new gardens a bit more tiresome. In any case, says Rapaport: 'In view of the shortcomings of pigs as energy storehouses and in view of the high cost of pig maintenance in both energy expenditure and land, it might well be asked if pig husbandry could be regarded as adaptive if the storage of energy for use in periods of crisis were the main advantage gained by those who practice it. The indications are that in many situations people would serve themselves better by giving up pig husbandry, letting surplus tubers rot in good years and in years of shortage themselves consume substandard tubers, which are edible, even if undesirable.' Why feed food into pigs at an inefficient conversion rate of 5:1 (or whatever it may be), when a sufficient reserve of vegetable protein and carbohydrates can be stored with no loss of nutrients?

Rapaport suggests that the Maring prefer to invest surplus food in pork because it is nutritionally superior, and because it tastes good, and no doubt

14   Harris, Marvin (1974), *Cows, Pigs, Wars and Witches*, Vintage Books.
15   Vayda, A P *et al* (1961), 'The Place of Pigs in Melanesian Subsistence', *Proceedings of the 1961 Annual Spring Meeting of the American Ethnological Society*, ed. V E Garfield, pp 69-77; cited in Rapaport (1967), op cit. 13, p 64.

he is right: the natives of Papua New Guinea are renowned for their love of pigmeat. But this doesn't explain why they 'bank' the pigmeat over a period of up to ten years. If they love it so much, the temptation to eat it must be hard to resist over such a long period. If the weather and crop yield is regular, why don't the Maring eat pork regularly?

Marvin Harris contends that the *kaiko* excesses are a corollary of the need to wage interclan warfare because this is the only means of limiting population. Each clan has more to gain by throwing huge pig parties and getting the other clans on their side, and by attending their allies' pig parties, than by eating pork and two veg every other day. Pig accumulation is a sign of military prowess. 'A group that cannot manage to accumulate pigs is not likely to put up a good defence of its territory, and will not attract strong allies.' It sounds rather like US foreign policy.

Harris' interpretation explains why Maring males are motivated to increase their pig herd to the maximum in the face of increasing complaints from their women folk – but this mechanism simply serves the interests of one clan against those of another. It would appear to be counterproductive for the Maring people as a whole. It is true that the public feasts ensure that the pig meat gets spread around the entire population, and the taboos associated with the *kaiko* ensure that the bulk of the meat goes to those who need it most: women and young children. But neither Harris, nor Rapaport provide a clear explanation why saving all the pigs up for ten years and then killing them should be a better strategy for the population as a whole than eating them as and when they reach maturity. After all, it is very wasteful keeping all those pigs alive in an agricultural economy which has rather limited food surpluses.

Yet, paradoxically it is precisely that wastefulness which makes the *kaiko* system an effective ecological strategy. The bulging pig population acts as a cushion between the human population and the limits of the land's carrying capacity, and it is the nutritional and ergonomic inefficiency of the pigs, once there are too many of them to keep alive on surplus food, which makes it so.

The point which Rapaport is at pains to emphasise (and provides numerous tables to support) is that *kaiko* is not called because the clan has reached the limits of its land's carrying capacity. Pressure from the women to start sacrificing the pigs occurs when the amount of work and distances travelled which they occasion becomes insupportable; and when the pigs begin to cause 'international incidents' by breaking into other people's gardens – not when the yields from agriculture begin to decline. The clan runs out of the energy and forbearance necessary to maintain a growing population of pigs, before it runs out of space to create healthy new gardens. It is the inefficiency of maintaining the pigs which prevents the carrying capacity being exceeded. If the produce of those distant gardens were fed directly to humans it would be worth continuing to till them – more children could be fed so there would be more people to help with the labour. But there comes a point where feeding the produce of the gardens into pigs who are converting it into human nourishment at a very inefficient rate becomes an insufferable waste of time.

So what would happen if all the clans got together and signed an agreement that the pigs shouldn't be allowed to multiply unchecked, but should be eaten on a regular basis to keep their number low and stable, and their production efficient? In this case, the humans would multiply (as they do) though more slowly than the pigs did in the *kaiko* cycle, because humans breed slower. More gardens would be needed to sustain them, but there would be no labour problem because every new mouth to be fed would nourish a new pair of hands able to clear and cultivate more gardens, rather than a parasitic pig. Nor would there be so many 'incidents' because humans have greater moral qualms against breaking down fences than pigs do.

Initially the clans would flourish, but as population continued to grow there would be no mechanism to stop them exceeding the carrying capacity of their lands: they would run out of new sites for gardens, their existing sites would become overused and degraded, and there would be increasing pressure to expand into the territory of another clan who were experiencing the same problem. Eventually they would go to war. In this 'sustainable meat' scenario, there would be three main differences from the *kaiko* cycle: firstly it would take longer, perhaps 50 or 60 years rather than ten, because humans breed more slowly than pigs; secondly, perhaps five or six times as many people would have to be killed in the hostilities in order to give the land a similar rest, because there would be no surplus pigs to kill; thirdly, the carrying capacity of the land would have been breached, possibly resulting in permanent damage; and fourthly, there wouldn't be any big feast to invite friendly neighbours to.

Now, imagine that there were voices in the tribe saying: 'Pigs eat garden produce that we could more efficiently eat ourselves, we should get rid of all of them, and just grow enough surplus tubers to see us through a bad year.' If each clan opted for a vegan diet, it would get rid of its pigs, but it would still multiply and eventually run out of space for gardens, and then go to war. In this case there wouldn't be any pigs to kill; and the cycle would again take a long time, though, curiously, slightly less than the 'sustainable meat-eaters' cycle.[16]

In fact, in both the above scenarios the cycle might well not happen at all, because when pressure on land became severe, raids on gardens would be committed by hungry people rather than by mischievous pigs. Reprisals would be likely lead to fatal skirmishes which temporarily reduced population and garden pressure, to the point where a full-scale battle might not be necessary. This state of affairs is potentially self-perpetuating: if no solution were found, neighbouring clans would find themselves permanently on the brink of their carrying capacity, in a condition of chronic petty warfare.

It also seems likely that the alternative 'sustainable meat' and 'vegan' cycles, even if they could be made to work in every other respect, would fail because they take too long. A cycle of ten years is repeated several times over a normal person's lifetime. A fifty year cycle is lived through once, or very rarely twice, in a lifetime: a Maring tribesman who witnessed the planting of the *rumbim* trees at the age of ten would be 60 by the time they were uprooted. It is much harder for a culture to maintain its continuity or its collective memory over such a slow cycle. Like the proverbial frog in a saucepan which doesn't notice that the water is edging imperceptibly towards boiling point, a vegan tribe would be more likely to fail to recognize the slowly advancing symptoms of overpopulation. Besides acting as a buffer, the pigs also accelerate the process of population growth and ecological degradation, so that human individuals can see what is happening several times during a lifetime and respond sensibly – which in this case involves throwing a party ending in a not particularly vicious battle every ten years.

---

16   This is assuming the sustainable meat-eaters start the cycle from a lower human population base than the vegans, because pigs take up some of their land. Since the pig population is stable, the sustainable meat-eaters have the same amount of land to multiply into as the vegans, at the same rate of population growth; and since the sustainable meat-eaters start from a lower population base, this will take more time.

The Feed Buffer

There is a sense in which all of us on planet Earth benefit from the same accelerated vision as the Maring. When George Monbiot states 'the world produces enough food for its people and its livestock, though some 800 million are permanently malnourished' we may struggle to understand the contradictory nature of the statement. Like the Maring with their pigs, our livestock have brought us to the brink of our global carrying capacity, and it is a good job that it is they who have brought us there, otherwise it would be us who would have to pay the price. It is time for us to slaughter some of our pigs and plant our rumbim trees.

In other words, it is a fundamentally reassuring fact that if the world does start to run out of food, this is a relative rather than an absolute state of affairs. As campaigners frequently point out, we can solve the problem by eating less meat:

> If the earth can sustain two billion omnivores, it might in theory support 20 billion vegans.[17]

> Reducing meat production by just ten per cent in the US would free enough grain to feed 60 million people.[18]

The above statements are taken from articles where they have been used as an argument for veganism. But to my mind they are an argument for eating meat, or at the very least for having eaten it. If we had overstretched the limits of our carrying capacity by eating nothing but super-efficient grains, there would be 20 billion vegans in the world, acre upon acre of soybean or wheat or rice monoculture largely dependent upon fossil fuel fertilizers and no easy way out.

Fortunately we haven't, and there is a solution, which is not dissimilar from the Maring's *kaiko*. It is highly likely that people in the United States, Europe, Japan and other excessively carnivorous countries will have to cut down their consumption of meat to accommodate further population increases, and to effect a more equitable distribution of resources. That doesn't, obviously, mean slaughtering untold numbers of cows and pigs in one fell swoop, and dishing them out to the Third World for Christmas. It means reducing the size of the overdeveloped world's herd so that we can be fed more efficiently off a higher percentage of inedible waste products and release a bit of ecological space. Fortunately a little bit of meat restraint releases a lot of nutrients (especially if the meat comes from feedlots which do attain something close to the legendary ten to one ratio of inefficiency).

It is a fundamentally reassuring fact that if population growth outstrips increases in crop yield, and food reserves run short, we can remedy the situation by getting rid of some of our more inefficient animals. However, if we have to resort to that remedy, it should in no way be taken as a reason for abandoning animal husbandry altogether. That would be the height of stupidity for the Maring, and it would be the height of stupidity for the world as a whole. It would mean that in future there would be no buffer left between a growing human population and famine.

Even if things never reached that point, a vegan agricultural economy would have no obvious incentive to 'bank' resources for a poor year. The Maring, we noticed earlier, are lucky enough not to suffer from large variations in crop yield from year to year; but this is not the case for many Melanesian tribes, nor for most agricultural economies around the world; and it is not the case for the

---

17   Vegan Organic Network (n.d.), *Vegan Food: Six Reasons Why*, Vohan News International, 1.
18   Motavalli, Jim (n.d.), *So You're an Environmentalist; Why are You Still Eating Meat?* www.Emagazine.com, cited in *Growing Green International*, 9.

world as a whole. As climate change increases, crop yields are likely to become more erratic and unpredictable.

The overriding advantage of meat is that demand for it is elastic. People don't need it but they like it, and up to a point, however much you produce, they'll keep on buying it. The demand for cereals for human consumption, on the other hand, tends to be inelastic. People need their pound of grain a day, but they don't need much more, and they won't buy any more unless they have sufficient wealth to invest the grain in animals, either to produce higher value food, or else to keep it 'on the hoof', for a rainy day (or a drought).

The existence of meat means that a farmer can sow wheat, barley, oats, beans, maize, and so on with reasonable confidence that, in the event of a good harvest, someone will buy it, because even if everybody has sufficient, it can be fed to animals. This dynamic is not restricted to a money economy. It works exactly the same for Melanesian subsistence farmers who can sow enough sweet potato and manioc to cover a bad year knowing that it is not a waste of effort, because in a good year the surplus can be fed to pigs.

Take the animals, the elastic element, out of the equation and the business of sowing grain suddenly becomes far more risky. In a vegan world, if a farmer sowed wheat or rice and there was a bumper crop resulting in a surplus, where would it be sold? And if there was a poor crop, where would the reserves come from, and why would they have been kept? In order to ensure that the world's farmers sow enough seed to cover a bad year, they must have a market which is elastic enough to buy in a good year.

This elementary matter of the need for a feed buffer fails to feature in most of the literature that is written about meat-eating and vegetarianism, including full length books; the only people who seem to take it into account are the FAO and some free market economists. The FAO observe:

> A remarkable characteristic, important for global food security, is the capacity of the livestock sector to draw on many different types of feed resources and to contract and expand with resources availability and market demand. During the two recent global food crises in 1974/75 and 1982/83, reductions in total cereal supply were almost entirely absorbed by the livestock sector adjusting to higher prices with reduced output, higher productivity and use of alternative non-food feed items.[19]

And Bjorn Lomborg adds in the same vein:

> Increased security is due … also to the so-called feedgrain buffer. A bad harvest means that consumption gets scarcer. But when world market prices increase as a response, less grain will be used to feed livestock, thus partly compensating for the initial scarcity. When the grain supply dropped and prices increased in 1972-4, the reduction in US feed consumption was as large as the total global production shortfall.[20]

Or as the vegan commentator cited a few paragraphs above might have put it : 'Reducing meat production by just ten per cent in the US freed enough grain to feed 60 million people.'[21]

---

19   De Haan. C et al (1997), *Livestock and the Environment: Finding a Balance*, FAO.
20   Lomborg, Bjorn (2001), *The Sceptical Environmentalist*, Cambridge, 2001.
21   Motavalli, *op cit*.18.

Since 1968, the amount of grain produced per person globally has been fairly stable, varying between a low of 292 kilos per person in 1970 and 2002, and a high of 343 in 1984 (with the median being 314) suggesting that a feed buffer of around 50 kilos per person might be desirable, equivalent to perhaps 10 to 15 kilos of meat per person in a good year.[22] This is about 16 per cent of average production, whereas Tristram Stuart suggests that 'agronomists reckon that in order to generate food security, nations should aim to supply around 130 per cent of nutritional requirements.'[23] However not every year is a good year and a certain amount of buffer is provided by food waste which would decline under conditions of duress, so I will stick with 50 kilos of grain per person. If this is regarded as feed for default meat – which seems reasonable since it is making use of a necessary surplus – then it adds substantially to the default livestock budget estimated in Chapter 4. How to ensure that this surplus is distributed equitably raises politically contentious questions too complex to be considered here.

Until recently, there were few other ways that this flexiblility of demand could be achieved, other than by directing surplus grain towards animals. The main alternative was to turn grain into alcohol. In the early years of the 19th century, the United States was not only a 'Hog Eating Confederacy' but was also awash with whisky, thanks to the bumper crops of corn coming off the virgin soils of the mid west. 'Come on then if you love toping,' William Cobbett wrote from America, 'for here you may drink yourself blind at the price of a sixpence.'[24] Nowadays we have a more mundane use for corn alcohol which is to mix it with diesel and use it to power motor cars. The growth of biofuel manufacture over the first decade of this millennium has been so rapid that it now rivals meat production as a means of absorbing surplus grain. Biofuels have already attracted from environmentalists and social justice campaigners the same criticisms that have been levelled against meat – that they divert food from the plates of hungry people into uses that are, by comparison, extravagant. The rise in grain prices of 2008 was partially attributable to demand for biofuels; and it is clear that there is no need for two major world commodities to take on the comparatively modest role of grain buffer.

The main problem with biofuels is that, because they take food out of the food chain, they are even less efficient at using grain than meat animals. A tonne of wheat produces about 350 litres of bioethanol, which is the equivalent of about 230 litres of diesel. However very roughly 60 per cent of this energy is required to manufacture the ethanol, so the net return is around 90 litres per tonne.[25] The 50 kilos of grain that might be a sufficient feed buffer for each person would therefore supply about a gallon of diesel – enough to drive 50 or 60 miles in a

22   Earth Policy Institute, 2006, *EcoEconomy Indicators* (USDA data) http://www.earth-policy.org/Indicators/Grain/2006_data.htm

23   Stuart, Tristram (2009), *Waste: Uncovering the Global food Scandal*, Penguin, 2009.

24   Pollan, Michael (2006), *The Omnivore's Dilemma*, Bloomsbury, p 100, citing W J Rorabaugh, *The Alcohol Republic: An American Tradition*, 1979, which includes the Cobbett quote on p 59.

25   These figures are derived from Murphy, Jerry D (2007), *Biomass, Biofuels, Biogas and Beverages, Presentation to the Institute of Brewing and Distilling*, 23 Feb 2007, and are based partly on results obtained from a plant in Sweden. These figures are broadly in accord with Richards, I R (2000), *Energy Balances in the Growth of Oilseed Rape for Biodiesel and Rape for Bioethanol*, Levington Agricultural Ltd, the British Association of Biofuels; however, Richards has higher energy costs for processing of around 75 per cent. A DTI study gives lower energy costs for processing than Murphy, see Elsayed *et al* (2003), *Carbon and Energy Balance for a Range of Biofuel Options*, Sheffield Hallam for the DTI Sustainable Energy Programme; while David Pimentel and Tad Patzek have concluded that ethanol production from corn requires 129 per cent as much energy as the fuel produced, see Pimentel, D and Patzek, T (2005), 'Ethanol Production Using Corn, Switchgrass, and Wood; Biodiesel Production Using Soybean and Sunflower', *Natural Resources Research*, Vol. 14, No. 1, March 2005.

small car – compared to 10 to 15 kilos of meat. Most people would plump for the meat, although another option that would appeal to some would be eight bottles of whisky.

In fact, this comparison is unfair because the manufacture of ethanol from grain also results in a byproduct known as distillers' grains, which are normally used as animal feed and are equivalent to roughly a third of the feed value of the whole grain before it is distilled. According to a study by Jerry Murphy of University College Cork, the distillers' grains from a given quantity of grain could be anaerobically digested to produce gas equivalent to the amount of ethanol derived from same amount. In other words, with your 50 kilos of grain you could drive not just 60 but 120 miles, in lieu of your 10 to 15 kilos of meat. That's less than two per cent of the distance the average person in the UK travels a year, against around 20 per cent of the meat they eat.[26]

The other matter to bear in mind is that distillation for biofuels is normally carried out in centralized facilities, and although it probably could be operated at a local scale it would require considerable investment. Feeding animals grain, on the other hand, is a task that can be done anywhere, that requires nothing more than a sack and a bucket, and that in a default livestock economy ensures that nutrients cascade back to the land.

The other way in which food security could be ensured in a vegan world is by banking food in state controlled granaries which maintain sufficient surpluses to guide the economy through periods of dearth and glut. This, until the 19th century, was the approach taken by China, which has never placed great reliance upon an animal feed buffer as a defence against famine. Unlike the inhabitants of India, the Chinese do not consume milk (though they are developing a taste for it), and in 1949 animal produce, mainly pork and duck fed to a large degree on waste produce, supplied just five per cent of the total protein consumed.[27] China maintained a population at the brink of the carrying capacity of the land – much as Ireland did in the 19th century – but unlike Ireland, it guaranteed food security through state managed granaries:

> China provides an example of successful bureaucratic management
> of food supplies, during the Qing dynasty in the eighteenth century.
> The Chinese state undertook far-reaching measures to feed the people
> during times of scarcity; these included public granaries, the provision
> of loans, discouragement of hoarders, encouragement of circulation
> by canals and roads … The Chinese peasant did not beg for charity,
> he demanded relief and saw the bureaucracy as bound by its office to
> provide this, and the rich as bound by duty.[28]

The system collapsed in the 19th century, partly because of the expense and insecurity caused by the Opium Wars, and catastrophic famines occurred which were eventually alleviated, to an extent by the reimposition of state control under Mao Tse Tung, and then by the development of a coal powered nitrogen fertilizer industry. Instead of a legume-based agricultural revolution in the 17th and 18th centuries, China underwent a chemical based 'green revolution' in the second half of the 20th century. In the last 50 years China has built up a formidable feed grain buffer, mainly based upon pork and poultry. Per capita consumption of

26   National Statistics On Line: http://www.statistics.gov.uk/cci/nugget.asp?id=24.

27   Quingbin Wang *et al* (1993), *China's Nutrient Availability and Sources*, 1950-9, Working Paper 93-WP113, Centre for Agricultural and Rural Development, Iowa State University, http://www.card. iastate.edu/publications/DBS/PDFFiles/93wp113.pdf

28   Thompson (1991), *op cit.3.*

meat and eggs increased from 5.6 kg per head in 1961 to 75 kg in 2007, which is approximately the same as in the UK (although milk consumption in China is much lower).[29]

The centralized approach to grain management doesn't attract many adherents at the moment – the experience of the USSR didn't do a lot for its reputation – but we should not discount its potential effectiveness, since it apparently worked well in China for a century or two. If the market fails to contain population and consumption, and the world is pushed towards a more vegan diet, then a centralized system of food control may have to come with it, hopefully more humane than Stalin's. In any case, the market has proved woefully inadequate as a means of distributing the existing abundant surplus of grain to the poor, so it would be foolish to expect it to dispense a much smaller food buffer with any degree of equity. But to pursue this line of enquiry any further would lead me to political considerations that are not the theme of this book.

---

29   FAO (2002) *Some Issues Associated with the Livestock Industries of the Asia-Pacific Region, FAO Regional Office for Asia and Pacific*, http://www.fao.org/DOCREP/005/AC448E/ac448e03.htm; Macdonal, Mia (2009) 'China and Industrial Animal Agriculture: Prospects and Defects', *Food First Backgrounder* 13:1, Spring 2009, Institute for Food and Development Policy; Evans, T (2009), *Good News on Global Egg Consumption*, The Poultry Site, http://www.thepoultrysite.com/articles/1575/good-news-on-global-egg-consumption; Annual Abstract of Statistics, 2009, Chapter 21.16.

11

FOOTLOOSE FOOD

The bearded goats,
The toosie stots
An' a' the braxy carcases;
It's a' the Markiss's
The land is a' the Markiss's.

James Mactavish 1901, as adapted by Dougie Maclean.

'The developing world's undernourished millions are now in direct competition with the developed world's livestock,' says the Vegan Society's website. Indeed they are, at least with the developed world's grain-fed livestock; but this is a question of economic equity, rather than an inherent problem with livestock. The world's undernourished are also in competition for land with the developed world's chocolate guzzlers, soap users, cotton wearers, motorists and holiday makers.

To make their case more effectively, proponents of veganism need to be tackling a different question: are the world's undernourished in competition with their own livestock? Would people in poor countries do better if they got rid of their animals, and consumed the grain which they fed to them?

We are encouraged to think of meat as food for the rich, and of meat consumption as an indicator of wealth. Much of the literature on the environmental impact of meat homes in on the fact that meat is a means of directing nutrients towards the stockyards and paunches of the well-to-do. Words like 'pecuniary' (from the Latin *pecus* meaning cattle), 'chattel' and 'capital' (same origin as cattle, from *caput*, the Latin for head) 'fee' (from the same root as *vieh*, German for cattle) and 'stock' remind us of the close association between the keeping of animals and the accumulation of property.

But these words only reflect one kind of property: liquid, mobile, transferable assets.[1] Another set of words, such as 'stake', 'domain' and 'territory', whose origin had nothing to do with animals, evokes an altogether different kind of property: landed property, what North Americans call 'real estate'.

If animals are capital, vegetation is real estate. The owner of land is normally the owner of all the vegetation on it because the vegetable kingdom has roots in that land and cannot move. The landowner can, if he wishes, feed that vegetation to animals, so he has the option to become an owner of livestock – and to accumulate several years of vegetable growth in that livestock.

The landless, almost by definition, can never own vegetation. They can rent it, or the land on which to grow an annual crop; they may have some sort of stake,

---

1   Incidentally the word 'assets' comes from the Latin *ad satis* meaning 'enough', compare French *assez*.

secure or not, in perennial plants growing on common land; but the landless cannot normally acquire perennial ownership over vegetation, other than by becoming landowners. They can, however, own animals, because animals are mobile, they have legs not roots and so are not attached to land. If vegetation is real estate, animals are chattels, and this fact is of immense advantage to the poor. Ownership of a cow, a goat or a pig is a way for the landless to sequester whatever scraps of vegetation may come their way and harvest them, either daily in the form of milk, or at one final reckoning in the form of meat.

For this reason, although the rich between them own more animals than the poor, if you are poor or landless, you are more likely to keep livestock than if you are rich. Sixty per cent of all rural households in poor countries keep livestock, according to the FAO's RIGA database of households in 12 third world countries, plus Bulgaria and Albania. In nine of these countries, the poorest households are more likely to keep livestock than the wealthiest; and in five countries the poorest households actually own more animals on average than the wealthiest. Across all countries and income bands, households with livestock derive on average 16 per cent of their income from their animals, while the average number of animals kept is 0.8 Tropical Livestock Units – the equivalent of 80 chickens, 8 goats or a cow and a calf. Male members of the household tend to be in charge of cows and other large animals, while women more often have control of poultry management.[2]

I have been unable to find out how many rural households in the over-developed nations keep livestock, but it is nowhere near 60 per cent. One survey shows that in the six poorest countries of the EU 74 per cent of rural households carry on some kind of subsistence production involving vegetable production or livestock, whereas in the 12 wealthiest countries in the EU, a mere 16 per cent do so.[3] Those of us who live in countries where social security payments cover the unemployed's supermarket bills may need reminding that while meat is the rich man's luxury, in many parts of the world it is the poor man's necessity.

Not all animals are chattels. Some are wild, and the landless may have access to these as well, though what that usually means is that they have access to the land those animals inhabit. Once ownership is established over land, either by private individuals or by state forestry organizations then ownership over the animals that run on it is usually claimed as well. However some animals may remain wild and unowned because there is little competition to eat them. A study by the US Quartermaster Corps could only come up with 42 societies around the world in which people eat rats – yet there are a great many rats in the world.[4] Animals which are not sought after for food not infrequently become pests, so their consumption by a minority is tolerated or even welcomed – in the UK rabbits are now in this category.

In some cases different species of animal competing for a resource will represent the interests of different groups. The double satisfaction of dining on revenge can be seen in this glimpse of 19th century Africa: 'The Tswana people of South Africa recognized their poor by their reaction to the appearance of locusts. Whereas the rich were appalled lest the locusts ate the grass needed by their cattle, the poor who had no cattle rejoiced because they could themselves eat the locusts.'[5]

    2   FAO (2009), *Livestock in the Balance*, State of Food and Agriculture Report for 2009 (SOFA 2009), unpublished consultation draft, 11 August 2009, p 33.
    3   Pichler, F *et al* (2006), *First European Quality of Life Survey: Urban Rural Differences*, European Foundation for the Improvement of Living and Working Conditions.
    4   Harris, Marvin (1986), *Good to Eat*, Allen and Unwin, p 14.
    5   Renehma, Majid (1992), 'Participation' in W Sachs ed, *The Development Dictionary*, Zed Books.

However it may be the livestock of the wealthy who are raiding the fields of the poor. In medieval England, pigeons flew out daily from the dovecotes of the manor to feast on the peasants' corn, bringing their nutrients back to deposit on the floor of the dovecote – and the rabbits from the Lords' warren's cropped their grass. In the 18th century the deer of the Crown forest estates grazed, and had a right to graze, on the crops of neighbouring small farmers.

But the scales can be tipped back in favour of the poor. 'Most of all did it rejoice the farmer's heart to slay secretly for his own pot one of the legion of privileged birds from the dovecote,' suggests Trevelyan.[6] And an 18th century victim of royal deer incursions in Winkfield, Berkshire, was reported as saying:

> I know how to resent and how to revenge it, which every farmer knows too … and this is the true reason why game in all forests is so very scarce, and why, probably, some resolute people take an insufferable liberty to kill the deer.[7]

For centuries the poor and landless of Britain fought a running battle with landowners over the right to hunt and eat the beasts and fowl that moved back and forth over the wealthy's estates. The battle culminated in the Black Act of 1723 which made 50 offences punishable by execution and which was not repealed until 1827, when it was judged to be of greater benefit to send the most spirited elements of the peasant class to the antipodes rather than hang them. Landowners stocked their woods with 'tame and docile birds, whose gay feathers sparkled among the trees before the eyes of the half-starved labourers breaking stones on the road for half a crown a week.'[8] The war was won by the landowners, who maintained ownership rights over the wild animals and birds found on their land, but, despite the executions and the transportations, it was a Pyrrhic victory. Innumerable peasants improved the diet of their families and neighbours by poaching, and often did so with the open approval of their community. Poaching as a subsistence activity faded out in Britain in the 20th century when food started to become cheap, even for the poorest sector of the rural population. Now pheasant shoots almost give their birds away, and rabbits are imported from China.

Similar struggles took place in virtually every country in Europe over the same period.[9] Ortega y Gasset in his *Meditations on Hunting* writes:

> In all revolutions, the first thing that the 'people' have done was to jump over the fences of the preserves or to tear them down, and in the name of social justice pursue the hare and the partridge. One of the causes of the French Revolution was the irritation the country people felt because they were not allowed to hunt, and consequently one of the first privileges which the nobles were obliged to abandon was this one. And this after the revolutionary newspapers, in their editorials, had for years and years been abusing the aristocrats for being so frivolous as to spend their time hunting.[10]

During the 19th and 20th centuries, the colonial administrators commandeered the lands of indigenous hunters and pastoralists and demarcated them as

---

6   Trevelyan, G M (1944), *A Social History of England*, Longmans, 1944, p 23.
7   This is the voice of Will Waterson, 18th-century schoolmaster, speaking on behalf of his neighbouring poacher/peasant farmers. Thompson, E P (1976), *Whigs and Hunters*, Allen Lane.
8   Hammond, J L and Barbara (1911), *The Village Labourer*, Guild, 1948.
9   See for example the magnificent collection of material at the poaching museum in St Pancraz, Austria.
10   Ortega y Gasset, José (1942), *Meditations on Hunting*, Wildlife Adventures Press, 1995, p 40.

'forest reserves' or 'game reserves'. A member of the Baiga tribe in central India, whose lands were confiscated by the British forestry administration, took much the same attitude as our Berkshire peasant: 'Even if the Government passes a hundred laws we will do it. One of us will keep the official talking; the rest will go out and shoot the deer.'[11] And the same struggle is being played out today in areas of the world where forests are being taken over by Government departments and state approved logging companies.

Nowadays when the rights of indigenous or local people to take wild meat are restricted or withdrawn it is on conservation grounds, rather than by reference to the divine right of landlords; but the language of ecology often masks social struggle. Communities usually evolve ways of managing their natural resources to ensure stability and security, and erosion of these resources is often a result of external or urban forces. This is Jane Goodall on the hunting of wild animals in Tanzania:

> The decline is due in part to habitat destruction as human populations increase and need ever more land for crops, livestock and settlements. But the greatest threat is the bushmeat trade – the *commercial* hunting of wild animals for food. For hundreds of years the indigenous people have lived in harmony with their forest world, killing just enough animals to feed their families and villages. Now things have changed. In the 1980s foreign logging companies moved into the last of the great African rainforests. And even if they practise 'sustainable logging' they open up the forests with roads. It is these roads that are the problem. Hunters ride the logging trucks to the end of a road and shoot everything from elephants and chimpanzees to antelopes, birds and reptiles. The meat is cut up and smoked then transported to town. There, the urban elite will pay more for bushmeat than for chicken or goat. It is their cultural preference. The situation is made worse because indigenous hunters are paid to shoot meat for the logging camps – for maybe 2000 people who were not there before.[12]

Even in Alaska, where population pressures are still low, the tension between indigenous and urban interests colours the conservation debate. There are few subsistence farmers in Alaska, but there are many subsistence hunters or fishermen – 95 per cent of rural households consume wild fish or game in their diet.[13] Sarah Palin's gun-toting is an object of derision for liberals in the Lower 48, but bagging the annual moose is as normal for Alaskans as growing vegetables is for the English. Subsistence hunting is enshrined in the Alaskan constitution, which rules that fish and wildlife are 'reserved to the people for common use' and that 'no exclusive right or special privilege of fishery shall be created or authorized'.

Unfortunately the States' founding fathers, when they drafted the constitution in 1959, failed to anticipate that ten years later the land would be coveted for other uses. Indigenous hunting and fishing rights threatened to hold up construction of the Trans Alaska Oil Pipeline, and in 1971 the federal government bought out these rights through the Alaska Native Claims Settlement Act – promising that the native subsistence lifestyle would nonetheless be upheld. In order to keep this promise, the federal government, with the support of wildlife conservationists

---

11   Gadgil, Madhav and Guha, Ramachandra (1993), *This Fissured Land*, Oxford University Press.
12   Goodall, J (2004), 'When Primates become Bushmeat', *World Watch*, July/August 2004.
13   United Fishermen of Alaska (2007), 'Subsistence Basics', *Subsistence Management Information*, http://www.subsistmgtinfo.org/index.htm

mandated that subsistence hunting and fishing would be prioritized for rural residents – Alaskans living in Anchorage, Fairbanks, Sitka and other cities could no longer claim by right their salmon and their moose. The Alaskan State fought this ruling through the courts for about ten years, with the result that now all Alaskans have subsistence rights on 40 per cent of the territory, and rural Alaskans have priority on the other 60 per cent.[14] City dwellers now consume an average of 10 kilos of subsistence meat and fish per year, whereas rural Alaskans consume an impressive 170 kilos – about a pound a day. This is understandable given that nothing else grows in Alaska, except for monstrous 75 pound cabbages in the Matanuska Valley. Sarah Palin would have created a better impression on the East Coast if she had been photographed holding one of these.

Fish

Outposts like Alaska are a reminder of how bountiful the world was before the rise of the agrarian city state. Alaska calls itself 'the last frontier' but it is only the penultimate, for beyond lies the ocean – indeed 60 per cent of Alaska's subsistence meat is fish. Only a tiny fraction of the world's terrestrial meat is hunted, but almost all of the marine fish that provide five per cent of the world's protein are wild animals, captured in the chase. Wild fish do not respect borders or fences, they belong to no one and so fishing has traditionally offered opportunities for the poor and landless. But that frontier, too, is under threat. As societies and nations have become urbanized and detached from nature, encouraging their populations to multiply beyond the carrying capacity of their hinterland, they have sought access to food from distant sources, commandeering other nations' lands and exploiting the oceans that are still viewed as a common treasury for all.

The collapse of the North Sea herring fishery in the 1960s, of the Peruvian anchovy fishery in the 1970s, and of the Newfoundland cod fishery in the 1980s have slowly alerted decision-makers and the public to the increasing pressures upon world fish stocks. But while consumers have become aware of overfishing, and are now encouraged to buy fish certified by the Marine Stewardship Council, they remain poorly informed about the political question that underlies all debates about marine stewardship – who owns the seas? This dispute, waged on the high seas and in the corridors of bureaucracies such as the World Bank, the FAO and the European Commission, sometimes pits nation against nation – but it more commonly pits large corporate fleets represented by economists against small local fishing fleets represented by anthropologists. It is not hard to determine who has the upper hand, but the battle has not been entirely one-sided.

The economists take their philosophical standpoint from Garret Hardin who, in a much cited article published in *Science* in 1968 coined the term, 'Tragedy of the Commons'. [15] In fact fishery economists such as H Scott Gordon and Francis Christy had arrived at the same conclusion some years before. Hardin's paper described how individual 'rational' graziers sharing a common pasture would inevitably overstock the common with their own cattle in order to derive more

---

14   United Fishermen of Alaska (2007), 'Federal or State', *Subsistence Management Information,* http://www.subsistmgtinfo.org/index.htm

15   Hardin, Garrett (1968), 'The Tragedy of the Commons', *Science*, December 1968. Gordon, H Scott (1954), 'The Economic Theory of a Common Property Resource: The Fishery', *Journal of Political Economy*, 62:2, pp 124-42; and Christy, F and Scott, A (1965), *The Commonwealth in Ocean Fisheries*, Johns Hopkins University, Baltimore.

private gain at public expense. In the same way, the fishery economists argued, the common resource of the oceans was overstocked with fishing boats: there were 'too many vessels chasing too few fish'.[16] Their answer was the same as that advocated by Hardin for pastures (and for everything else): the introduction of private property rights. On land this took the form of enclosure with fences; in fisheries, since fish cannot easily be fenced,[17] it took the form of metaphorical enclosure through licences, or more frequently quotas – preferably tradable quotas, because their accumulation by powerful actors leads to 'fewer boats' (though bigger ones).[18] The economists were, of course, supported by these increasingly corporate vessel owners who stood to profit from making the global fishing fleet leaner, meaner and greedier.

The anthropologists, of whom there were dozens working on this subject in the 1980s and 1990s, reacted angrily (by academic standards) to what they viewed as a slander on the ability of fishing communities and peasant societies generally to manage their own affairs. Hardin's observation that common rights can be exploited for individual gain has long been common knowledge to all who share common resources, and any society that survives has, almost by definition, found a means of mediating unsustainable acquisitiveness. Arthur McEvoy remarks 'Farmers in Garret Hardin's 'Tragedy of the Commons', are as radically alienated from each other as they are from the grass on which they feed their cows … Hardin's commoners don't know how to talk to each other.' Another anthropologist, James McGoodwin, adds: 'What I find most objectionable about the Tragedy of the Commons model, at least when it is applied to the fisheries, is the cynical view of the mentality, character and personality of fishers implied in the explanation of how the tragedy develops.'[19]

Throughout the 1980s and the 1990s in particular, anthropologists working in the field consistently reported that the communities they studied had developed often sophisticated methods for managing resources to ensure that they were not over exploited and were distributed reasonably equitably – though this generosity did not necessarily extend to everybody in the community; in some cases a resource may have been kept common only to an elite or a particular social group. Some of these studies focussed on land-based activities such as hunting or pastoralism, but more than half involved fishing communities, for it was the marine resource base that had best survived the colonial and industrial land grabs of the 19th and 20th centuries. Coastal communities, in both

16   Edward Loayza, cited in *The Ecologist*, Vol No 2/3 March-June 1995, p 42; exactly the same words are used by FAO Director General Jacques Diouf in J Diouf (2001), *FAO Director General: Too Many Vessels Chasing Too Few Fish*, Reykjavik Conference on Responsible Fisheries in the Marine Ecosystem, 2 October 2001, reported at http://www.waddenzee.nl

17   The ultimate form of marine enclosure is fish-farming. In the 'cage system, areas of ocean or inland waters are literally fenced off to provide an area for the private rearing of fish or other sea creatures; in other cases fish are reared in excavated lakes or ponds. Like terrestrial farmers, fish farmers provide their stock with feed, and protect them from disease with chemicals. The farming of fish is often advocated as a sustainable option because their feed conversion ratio can be as low as 2:1. On the other hand farmed carnivorous species are sometimes fed extravagantly on wild fish such as anchovies; and there are issues of disease and pests infecting wild stock. Just as the privatization of land results in landless people, so the privatization of marine resources may result in poor people being made 'sealess'. Regrettably I have not had time or space to cover fish-farming satisfactorily in this book.

18   For a typical formulation of the marine tragedy of the commons argument, see Loayza, E with Sprague, L (1992), *A Strategy for Fisheries Development*, World Bank Discussion Paper, Fisheries series 135.

19   McEvoy, Arthur (1990), 'Towards an Interactive Theory of Nature and Culture', in Worster, Donald (ed), *The Ends of the Earth: Perspectives on Modern Environmental History*, Cambridge, 1990, p226; J. McGoodwin, *Crisis in the World's Fisheries, People, Problems and Politics*, Stanford University Press, 1990, p 196.

industrial and undeveloped countries, have evolved a wide variety of methods for controlling fishing effort, for example: banning or restricting commercial sales, taboos on eating certain foods, limits on the number or size of pots, nets or boats, restriction on certain areas, seasonal restrictions, lotteries for favoured fishing sites, queuing systems for fishing favoured areas in turn etc. [20]

In addition, fishing in most coastal communities is often a part time activity with fishermen having 'one boot in the boat and the other in the field.' In Northern Sweden in the 1920s 'one called oneself "fisherman", but to this had to be added: log driver, stevedore, mason, logger, seal hunter, carpenter, boatbuilder, industrial worker, shoemaker, ropemaker, bicycle repairman ... A fisherman was usually a jack of all trades.'[21] Another researcher in the South Pacific noted that 'almost no one is willing to be a full-time anything'.[22]

The ability to turn to several occupations is characteristic of peasant communities in general, and of many rural communities in more developed countries; it is a way both of adapting to seasonal variations, and of spreading risk, so that a poor harvest or the closure of a local sawmill is not a disaster. When a fishing community spreads its own risks in this way, they are also spread for the fish. Whereas a dedicated commercial fishing fleet, obliged to pay off its capital investment, seeks to maintain catch levels when fish are scarce, maritime or riverine peasants are more likely to turn their hand to other occupations until the fishing picks up.[23] This approach to fishing is an aquatic adaptation of the FAO's 'default land user strategy', harvesting what rises effortlessly to the top of the food chain, rather than pursuing an 'active sea-user strategy' designed to squeeze out of the sea what she only grudgingly yields. The two types of fishery also have contrasting approaches to waste. Modern trawlers discard about half the fish they catch, because they are targeting a single species for an industrial processor, or because they have exceeded their quota. A local coastal fishery, managing its own resources, discards nothing of value, for everything can be consumed by or sold to dealers who can distribute anything edible inland along decentralized local food networks – and the less desirable species are available for the poor.

In a low impact fishing economy of this kind, the lack of technological ability to overfish is rooted in the lack of any need to overfish. Not only is there little reason to get tangled up in profit-seeking fisheries development schemes or to embrace new technologies that would increase production at the expense of the community's food security; there is also every reason to avoid such schemes. When fish stocks are abundant, fishing is easy with inexpensive gear. When fish stocks are limited, technological improvements simply enhance one fisherman's chances at the expense of another's, until, in the words of a Maine fisherman arguing for a ban on a new kind of lobster pot, 'soon everybody will have the damn things'.[24]

But local inshore fishing economies, while they may be well managed and internally stable, are vulnerable to external agents, who either infiltrate levels of technology and capitalization that local fishermen cannot match, or simply arrive with big offshore fleets that mop up all the fish. Olvar Löfgren's essay

20   *The Ecologist*, special issue on fishing, vol 25 2/3, 1995.
21   Loftgren, O (1982), 'From Peasant Fishing to Industrial Trawling', in Maiolo, J and Orbach, M, *Modernization and Marine Fisheries Policy*, Ann Arbor Science, Butterworth, London, p 154-60.
22   Ruddle, K, Hviding, E and Johannes, R (1991), *Customary Marine Tenure: An Option for Small Scale Fisheries Management*, Centre for Development Studies, University of Bergen, 1991.
23   The above paragraph is taken from Fairlie, S, Hagler, M and O'Riordan, B (1995), 'The Politics of Overfishing', *The Ecologist*, Vol 25 No 2/3 March-May 1995.
24   Acheson, J (1982) in Maiolo and Orbach, *op cit*.21.

'From Peasant Fishing to Industrial Trawling' recounts how the transition to a corporate dominated fishery took place in Sweden and elsewhere on the North Atlantic fringe, over the course of the twentieth century, leading to the collapse of herring stocks as a result of overfishing for fish meal in the 1960s. David Matthews' book *Controlling Common Property* describes how local Newfoundland fishermen could foresee the collapse of their cod fishery in the 1980s, yet were powerless to prevent corporate trawling fleets scouring the ocean bed for the last remaining stocks, with the complicity of fishery economists and bureaucrats.

As waters in the northern hemisphere have become overfished, corporate fleets from the EU and South East Asia have targeted the waters around Africa. These boats are stealing the resources of the poorest people on the planet. There was a poignant moment on the BBC World Service late in 2008 when a reporter managed to telephone through to a ship captured by Somali pirates, one of whom explained: 'We've had no government for 18 years. We have no life. Our last resource is the sea, and foreign trawlers are plundering our fish'. Another BBC reporter quoted a resident of Garowe as saying 'Illegal fishing is the root cause of the piracy problem: they call themselves coastguards.'[25]

Almost every inshore fishery around the world has a similar story to tell, and its a wonder more haven't resorted to piracy to make their point. Artisanal fisheries organizations have campaigned long and hard for trawler bans in countries such as India and Indonesia, with partial success. During the 1990s, the influence of the 'anthropologists' persuaded organizations such as the FAO to promote schemes for community management of local fisheries. However, both the Indian and the Indonesian government have been reluctant to sacrifice export earnings derived from licensed trawlers, and ineffective at defending their waters from the incursions of foreigners. In India, local fisheries have been encouraged to build up their fishing capacity with larger boats and outboard motors in order to compete with the offshore trawlers, but the main effect has been to exacerbate competition amongst each other. In Kerala the local fishing fleet doubled in horsepower between 1991 and 1998 yet there was no increase in catch; it was, says a representative of the fishing union, 'entirely the result of internal competition'.[26] This little 'Tragedy of Technology', to my mind, epitomises the human condition more accurately than Hardin's 'rational graziers'.

Fishing communities in India wishing to look into the future might do well to examine the fate of the UK fishing industry. In Britain the number of fishing boats over 15 tonnes declined from 41,723 in 1872 to 2,142 in 1970. Now there are just 1,411 boats over 10 metres long and fewer than 6,000 boats in all. When Britain joined the EU its waters became part of the 'European pond', with much of the quota going to Spain's massive trawler fleet, or to the select elite of huge Scottish pelagic vessels based in the Shetlands. 'Our resource base has been "prostituted" like a maritime Highland Clearance', writes Alastair MacIntosh. 'Some three-dozen millionaires scoop up Scotland's entire catch of herring and mackerel. About 45 ocean-going ships involving some 450 crew members now monopolize a community resource that once supported more than 1,000 boats, 10,000 crew members and an even greater workforce around Scotland's coastline at the end of World War II.'

25   Harper, Mary (2008), *Mummy Can I Phone the Pirates*, BBC News, 29 November 2008; Hunter, Robyn (2008), *Somali Pirates Living the High Life*, BBC News, 28 October 2008.

26   Vivekanandan, V (2003), *Socio-Economic Dimensions of Overcapacity and IUU South Indian Federation of Fishermen Societies*, Asia Pacific Fishery Commission www.apfic.org; Fegan, Brian (2003), 'Plundering the Sea: Regulating Trawling Companies is Difficult when the Navy is in Business with Them', *Inside Indonesia*, http://www.insideindonesia.org/content/view/339/29/

Former Scottish fisherman David Thomson is one of many who have been fighting against the combined influences of free market economists and EU bureaucrats to resurrect the Scottish inshore fishery and the coastal communities it supports. In 1997, he drew up for the FAO the table reproduced as Table 5 below which shows more graphically than any paragraph I could write how ecologically and socially damaging it is to place ownership of the world's marine biomass in the hands of a few powerful and overcapitalized corporations.

Table 5: The World's Two Marine Fishing Industries – How They Compare

	LARGE SCALE	SMALL SCALE
Number of fishermen employed	AROUND 500,000	OVER 12,000,000
Annual catch of marine fish for human consumption	AROUND 29 MILLION TONNES	AROUND 24 MILLION TONNES
Capital cost of each job on fishing vessels	$ 30,000 – $ 300,000	$ 250 – 2,500
Annual catch of marine fish for animal feed and industrial reduction to meal and oil.	AROUND 22 MILLION TONNES	ALMOST NONE
Annual fuel oil consumption	14 – 19 MILLION TONNES	1 – 2.5 MILLION TONNES
Fish caught per tonne of fuel consumed	2 – 5 TONNES	10 – 24 TONNES
Fishermen employed for each $1 million invested in fishing vessels	5 – 30	500 – 4,000
Fish destroyed at sea each year as by-catch in shrimp & trawl fisheries	6 – 16 MILLION TONNES	NONE

Credit: David Thompson/FAO

The one matter that Thomson's chart doesn't show is who eats the fish. One can be reasonably confident that fish caught by the small scale sector are feeding local people (though not necessarily all local people) and contributing to a region's food sovereignty. The global fleet of factory trawlers, on the other hand, is an engine for hoovering up protein in certain parts of the world and ferrying it to consumers thousands of miles away who have the wherewithal to pay for it. And the flow of protein is from the waters of people who need it to the tables of people who don't.

Sacred Cows and Infernal Goats

The term 'waste' is interesting, not so much for its derivation, from the Latin *vastus* meaning desolate, as for the modern meanings that have subsequently been assigned to it. Waste originally signified land that wasn't worth owning – in economists' language, land that is so worthless that it doesn't command any rent: hence anyone can use it. As well as tracts of heathland, bog or seashore, waste can include balks and banks between fields, verges by the side of roads ('the long acre') – and in modern times railway tracks, vacant lots, the detritus of markets, rubbish tips, supermarket skips and discarded reusable containers. It is easy to see how the meaning of the word developed.

It is because waste is by definition economically valueless that it is of great value to those who live at the margins of the monetary economy. Waste provides a crucial security net for the destitute. Its existence relies on the fact that there will inevitably be activities that are worth someone doing for their own benefit, but not worth employing anyone else to do. It is the existence of land of this quality that provides the basis for David Ricardo's theory of rent: waste is land, or resources, for which you don't have to pay any rent.

A second salient feature of waste, is that it is predisposed to the provision of meat and dairy produce. There are some forms of waste which provide food of uniquely vegetable composition: the gleaning of previously harvested crops for example, or the detritus of vegetable markets. But the majority of crop wastes are by-products which are inedible to humans – straw, leaf material, the fibre that is left over after processing, and substandard crops. When the labour costs of feeding these materials to animals outweigh the potential profits, they become waste products rather than byproducts, available to the landless to feed to their own livestock for the cost of carrying them away. When the labour costs of grazing an animal on a tiny wedge of land or an inaccessible low grade pasture outweigh the potential profits, that too becomes waste.

The use by the poor of balks, headlands and the long acre (the side of the road) to feed livestock has long been abandoned in the UK, with one exception: travellers still tether their horses and goats there. The disuse explains why nowadays the most fertile land in Britain, the hedgerows, can be found just five metres away from land that has been stripped of its nutrients by years of intensive silage cropping.

But in many poorer countries the practice of letting livestock glean food that would be inconvenient to gather by hand is still widespread. In towns in Mauritania, for example, poor incomers from rural areas bring their pastoral habits with them:

> Goats are the main livestock. The main products are goat milk and
> fattened sheep. This stock rearing is carried out by low-income
> families and is done by women. Animals wandering in the streets are
> characteristic of this system; they feed on urban waste but always
> receive a high quality complementation of kitchen waste, wheat flour,
> groundnut cake, lucerne etc. and are watered daily.[27]

But it is in India where scavenging by animals is not only a central pillar of the food economy, but has also been elevated to the status of religion. No discussion of India's adoption of the sacred cow can avoid borrowing from

---

27   Soule, Ahmedou Oulde (2003) 'Mauritania', *Country Pasture Profiles*, FAO, http://www.fao.
org/ag/AGP/AGPC/doc/pasture/forage.htm

Marvin Harris' essays on the subject *Mother Cow*, and *The Riddle of the Sacred Cow*.[28] Harris writes:

> Tourists on their way through Delhi, Calcutta, Madras, Bombay and other Indian cities are astonished at the liberties enjoyed by stray cattle. The animals wander through the streets, browse off the stalls, defecate all over the sidewalks and snarl traffic by pausing to chew their cuds in the middle of a busy intersection. In the countryside, the cattle congregate on the shoulders of every highway and spend much of their time taking leisurely walks down the railway tracks … What are those animals doing in the markets, on the lawns, along the highways and the railway tracks and up on the barren hillsides? What are they doing if not eating every morsel of grass, stubble and garbage that cannot be consumed by human beings and converting it into milk and other useful products?

These miscreant cows are tolerated (as the cliché we use to describe something immune from criticism reminds us) because they are sacred. Harris explains the economic forces and the class politics that lie behind the reverence accorded to the cow in India, which allows her not only to mop up waste but also to act in a Robin Hood capacity (rather like the Tswana's locusts mentioned on p 120):

> One reason why cow love is so often misunderstood is that it has different implications for the rich and the poor. Poor farmers use it as a license to scavenge, while wealthy farmers resist it as a rip-off. To the poor farmer, the cow is a holy beggar; to the rich it is a thief. Occasionally the cows invade someone's pastures or planted fields. The landlords complain, but the poor peasants plead ignorance and depend on cow love to get their animals back. If there is competition, it is between man and man or caste and caste, not between man and beast.

The Hindu taboo against harming cows contributes to the adaptive resilience of the human population 'by protecting cattle that fatten in the public domain or at the landlord's expense'. The tolerance accorded to them allows large numbers of cattle to graze in a highly populated country where only 3.9 per cent of the land is pasture. The grass itself has evolved to accommodate them: Allan Savory observes that, in contrast to the western United States, 'in the densely populated brittle environments of India, where there are millions of animals, particularly sacred cows, grasses that can withstand high levels of overgrazing dominate.'[29] Harris, writing in the 1970s, estimated that 'probably less than 20 per cent of what the cattle eat consists of humanly edible substances: most of this is fed to working oxen and water buffalo.'

From this virtually free resource, the poor and landless of India obtain four benefits from the cow: milk, manure, muscle-power and meat. The amount of milk produced by these hardy Zebus is pitiful by industrial standards, but it is protein for the poor and the feed is free. The cow's manure piles up for her owner when she is stalled, and is bonus biomass to the cowless when she is grazing. As soon as a pat hits the street, a child rushes out and sweeps up the little gift as if it were manna from heaven. The dung is also used as fuel, its slow-burning qualities making it particularly suitable for cooking. 'If India's wandering bovines were fed fodder crops rather than straw or grass, their dung would be

28   Harris, Marvin (1974) 'Mother Cow', Cows, Pigs, Wars and Witches, Vintage Books; and Harris, Marvin (1986) 'The Riddle of the Sacred Cow', Good to Eat, Allen and Unwin.
29   Savory, Allan (1999), *Holistic Management*, Island Press, p 198.

too liquid, rendering it useless for fuel'.[30] Male calves are sold to hauliers or farmers who still require fifty million or so of these tough, economical draught animals throughout India.

The matter of meat is more perplexing since the Hindu social system is founded on the belief that cows are holy and should not be killed or eaten. Some cows are slaughtered commercially and the beef sold to Muslims or Christians. But traditionally the meat derived from animals that died a natural death, braxy as it is called in English,[31] was disposed of and eaten by the lowest, poorest caste in Hindu society, the Untouchables. Nowadays the Untouchables, call themselves Dalits (meaning 'downtrodden') and are still employed to carry out the dirty work of civilization, like stripping off and diving into the sewers of Delhi to remove obstructions. But initially they might have received some recompense for their ostracism in the carcases discarded by their more pernickety betters.

The origins of cow worship, and of the Untouchables are only dimly visible in early Hindu and Buddhist texts, but there is an agreeable logic in Marvin Harris' interpretation which accords with earlier Hindu writings – documented at length in a recent book by D N Jha entitled *The Myth of the Sacred Cow* – and provides the entire basis for the chapter on the Holy Cow, in Rifkin's *Beyond Beef*.[32] The Aryan conquerors of Northern India, who first arrived around 1800 BC, were herdsman from the steppes in search of grass, and they were beef eaters. However, unlike their cousins who continued to advance westward, the Aryans had turned into a blind alley, and with the Dravidian people occupying the Southern end of the peninsula, soon found themselves boxed in. As the population grew, the pastoral diet of beef became increasingly unsustainable – a pressure upon land and a burden on the poor. Beef eating became reserved for ritual sacrifices, of which, it can be assumed, the members of the priestly Brahman caste were the main beneficiaries.

Some Indian scriptures corroborate the view that cow sacrifice and worship was a response to ecological restraints. The *Skandha Purana* states that early in the Treta Yuga (the second and most violent of the four world eras in the Hindu cosmology) the universe was afflicted by severe famine:

> The sages told the people in general that they could sacrifice animals
> to the Gods and eat their remains, but only if they did so as a religious
> sacrifice and not merely for personal survival. 'After this allowance
> was put in place' the Purana tells us, 'gods, kings and men performed
> animal sacrifices and ate the meat that resulted from it. Soon, however,
> the famine abated.' [33]

But the pressure of population continued and in the five centuries before Christ, the bovine sacrifices pontificated over by the Brahman priests came to seem increasingly extravagant. Since the produce of multiple acres of land was concentrated into the carcases of the slaughtered cows, the issue of sacrifice was no mere doctrinal spat but a matter of paramount social and economic importance. 'At a time when ordinary people were starving and in need of oxen to plough their fields, the Brahmans went on killing cattle and getting fat from eating them'. This ostentatious overconsumption accounts for the growing

30   Rosen, Steven J (2004), *Holy Cow*, Lantern Books.
31   For an explanation of this word, and a transcription of the poem from which the quote at the opening of this chapter is taken, see Brown, Ivor (1960), *Chosen Words*, Penguin.
32   Jha, D N (2002), *The Myth of the Sacred Cow*, Verso, 2002; Rifkin, J (1992), *Beyond Beef*, Dutton
33   Rosen (2004), *op cit*.29.

popularity of Buddhism in India and its eventual espousal by the Emperor Ashoka in 257 BC. Buddhists preached a gospel of *ahimsa*, non-violence, opposing animal sacrifices, and the killing of animals in general. However their religious tenets did not prohibit the eating of braxy, nor did Ashoka forbid the killing of beef where necessary. To explain what happened next Harris turns to Rajendra Mitra, a 19th century Sanskrit scholar who wrote:

> When the Brahmans had to contend against Buddhism which
> emphatically and so successfully denounced all sacrifices, they found
> the doctrines of respect for animal life too strong and too popular to
> be overcome, and therefore gradually and imperceptibly adopted it in
> such a manner as to make it part of their teaching.

Indeed, to regain control the Brahmans felt they had to outshine the Buddhists in their dietary restraint. The Dalit campaigner B R Ambedkar wrote:

> Without becoming vegetarian the Brahmans could not have recovered
> the ground they had lost to their rival namely Buddhism. What could
> the Brahmans do to recover the lost ground? To go one better than the
> Buddhist Bhikshus ... become vegetarians – which they did.[34]

The result, as Buddhism's influence waned and Hinduism became the dominant religion, was that the holier-than-thou Brahmans became completely vegetarian, while lower classes could eat meat but refrained from beef. However, the cow was not rejected as unclean, as was the pig by Muslims, but instead venerated as sacred. The very plausible reason Harris advances for this is that the cow was necessary to provide oxen to plough the relatively dry lands of Northern India. Taboos against eating draught animals are by no means unusual. In some parts of Europe there were prohibitions against killing oxen for meat, and in the UK there is still a reluctance to eat horse meat. According to Keith Thomas, 'the rise of the cult of the roast beef of England closely paralleled the decline of the ox as a working animal.'[35]

Since, in India, it was necessary to go to the expense of feeding cows to produce draught animals, it followed that milk (rather than pigs as in China) should become the dominant animal protein. Milk became the ritual food and the cow became, in Gandhi's words 'a mother to millions of Indians', supplanting the bull symbolism of the macho Aryan invaders. There remained the issue of the braxy meat and the hides, which together constituted a by no means negligible resource. The answer was to be found in the 'Broken Men', as Ambedkar calls them, taking their name from this passage from Sir Henry Maine's account of early Irish history:

> Much of the common tribeland is not occupied at all, but constitutes,
> to use the English expression, the 'waste' of the tribe. Still this waste is
> constantly brought under tillage or permanent pasture by settlements
> of tribesmen, and upon it cultivators and servile states are permitted
> to squat, particularly towards the border. It is part of the territory over
> which the authority of the chief tends steadily to increase, and here it
> is that he settles his *fuidhir* or stranger-tenants, a very important class
> – the outlaws and 'broken men' from other tribes who come to him for
> protection, and who are only connected with their new tribe by their

---

34   Ambedkar, D N (1948), *The Untouchables: Who Were They and Why They Became Untouchables?* 1948, http://www.ambedkar.org
35   Thomas, Keith (1983), *Man and the Natural World*, Penguin, p 54.

dependence on its chief, and through the responsibility which he incurs for them.[36]

Ambedkar suggests that in early India, the chief function of Broken Men, who were either the remnants of defeated tribes, or outcasts from existing ones, was to live on the edge of the settlement and 'do the work of watch and ward against the raiders belonging to nomadic tribes.' With a shortage of land for cattle there would have been little in the way of wastes (in the old sense of the word), as there were in Ireland, so what better than to entrust the broken men with the wastes (in the new sense of the word) that issued from the refusal to eat beef. The Broken Men were given the responsibility of disposing of carcases and tanning the leather, and fed in part on the proceeds. It is for this reason, Ambedkar argues, that the Broken Men, who previously had simply been outcasts, became Untouchables. Since the cow was sacred and other meats eaten, the Brahmans fixed the notion of uncleanness not on their pigs, but on their knackers.

So it is that, despite a slaughter ban in all states except Bengal and Kerala, some 20 million cows are butchered every year by low ranking castes, the meat gets eaten, and the hides get tanned, providing the basis for a leather export industry worth $3.5 billion. In Gandhi's words, 'Mother cow is useful dead as when alive.'[37] An increasing proportion of this processing is now carried out in regulated slaughterhouses catering to middle-class non-Hindus. But as an informant of Harris's remarks, it is the cow who dies of old age that the poor get to use: 'It is good for the untouchable if a cow dies of starvation in a village, but not if it gets sent to an urban slaughterhouse to be sold to Muslims or Christians'.

The Hindu three-diet system is an ingenious adaptation to the constraints of a default livestock economy, which, though rooted in an oppressive caste system, has an elegant justice about it. The Brahmans at the top are confined to an ascetic vegetarian diet, as are members of the Vaishya mercantile class. The warrior Kshatriya caste and the Shudra labourers in the middle can eat meat, but not beef; and the beef is reserved for the 60 million poorest who might otherwise be most at risk from malnutrition. The people who do the physical work get the bulk of the meat – in sharp contrast, say, to 19th century Britain. India's millennium-long doctrinal dispute over meat-eating and animal sacrifices is perhaps the best documented example we have of a civilization imposing cultural taboos to conform with ecological restraints. It shows it can be done, and our own civilization, faced with the need to reduce meat consumption, has a great deal to learn from it.

But in other respects the system is manifestly unjust. The caste system preceded cow worship, but the slaughter ban, if Ambedkar is to be believed, precipitated untouchability. Moreover the protection afforded to Indian cows has long been criticized on animal welfare grounds by foreign observers, who viewed that the way many of the animals were treated, they would be better off dead. To an extent this may be due to a misplaced comparison of the hardy Zebu with the Hereford and Holstein that live off Britain's fat pastures. But Gandhi was in agreement: 'I do not know that the condition of the cattle in any part of the world is so bad as in unhappy India'. He 'remembered the magnificent specimen of the cattle in England where, while they certainly did eat beef, they bestowed the greatest care on their cattle wealth'.[38]

---

36   Maine, Henry (1875), *Lectures on the Early History of Institutions*, Lecture III.
37   Gandhi, M K quoted in Rosen (2004), *op cit*.29.
38   Gandhi, M K (1954), *How to Serve the Cow*, Navajivan Publishing House, Ahmedabad, pp 7 and 26

As the 20th century progresses, these animal welfare criticisms start to get picked up by the writers of economic reports. A 1959 Ford Foundation study concluded that about half of Indian cows could be regarded as surplus in relation to feed supply. Alan Heston, an economist from the University of Pennsylvania, reported in 1971 that India had 30 million unproductive cows, producing a tenth the amount of milk that US or European cows produced, which he argued were redundant, and ought to be slaughtered. 'Slaughter' said Mahatma Gandhi, 'is a thing that suggests itself easily to Western economists. That is why they cut the Gordian knot, by slaughtering the inferior breed of cows and bulls.'[39]

Marvin Harris sets out to show that the economists are missing the point:

> Many experts assume that man and cow are locked in a deadly competition for land and food crops. This might be true if Indian farmers followed the American agribusiness model and fed their animals on food crops … In his study of cattle in West Bengal, Dr Odend'hal discovered that the major constituent in the cattle's diet is inedible byproducts of human food crops, principally rice straw, wheat bran and rice husks. When the Ford Foundation estimated that half of the cattle were surplus in relation to feed supply, they meant to say that half of the cattle manage to survive, even without access to fodder crops.'[40]

Moreover, to concentrate on their milk yield is to misunderstand the default role of the cow in the Indian economy. You don't expect much milk from an animal that lives on straw, hedgerow and city detritus – though the little that it does produce plays an important role in meeting the nutritional needs of the very poor. But the Zebu doesn't just produce milk. She supplies dung, meat, leather and above all calves. An apparently barren cow, all skin and bones, after a favourable monsoon, may fatten up and give birth once again. Harris considers that the underlying rationale for the slaughter ban is to preserve a sufficient pool of cows to meet the continual requirement for 80 million beasts of burden. His view is that:

> The inevitable effect of substituting costly machines for cheap animals is to reduce the number of people who can earn their living from agriculture and to force a corresponding increase in the size of the average farm … Less than five per cent of US families live on farms, as compared with 60 per cent 100 years ago. If agribusiness were to develop along similar lines in India, jobs and housing would soon have to be found for a quarter of a billion displaced peasants.

And that is indeed what has happened. Urban populations have swelled to previously unimaginable proportions as a result of India's failure to control its population, which has doubled since Harris was writing in 1973, and its green revolution, which benefits commercial farmers able to afford fertilizers and tractors at the expense of small farmers. Meanwhile the foreign exchange pouring into the computer and call centre industries has led to the rise of an Americanized middle class whose allegiance to cow worship, even from those who remain nominally Hindu, is slim. 'Junk food is fashionable. Eating meat is regarded as progressive. Modernization is equated with changing from being vegetarian to non-vegetarian, even while the rest of the world attempts to reverse

---

39   *Ibid.*
40   Harris, Marvin (1974), op cit., 28.

this trend.'[41] Abattoirs meeting modern standards of hygiene are constructed on the edge of the new conurbations. Notwithstanding a tide of opposition against cow slaughter from the BJP, the party that represents the Hindu equivalent of *Daily Mail* readers, the days of the sacred cow appear to be numbered.

In order to meet the demand for dairy products from an extra 600 million consumers, Indian agronomists have introduced blood from higher yielding western cows – with considerable success. In 2000, the Indian dairy industry surpassed that of the United States to become the largest in the world, and in 2006 produced 92 million tonnes – over one and half litres a week for everyone in the country. Even more encouraging, according to an FAO document called *Dairy Giant Walking Barefoot*, the White Revolution was achieved with 'dairy animals largely fed on crop residues', although 'high producing animals are supplemented with mostly home mixed concentrate feeds.'[42] Nor has the dispersed structure of the farming sector been completely undermined. 'The dairy sector in India is predominately smallholder and unorganized in nature' state the FAO, adding that the 'unorganized sector' – I like this expression – 'in India accounts for more than 50 per cent of total production and handles more than 77 per cent of the milk marketed. The total number of households in production is more than 67 million; out of this 11 million can be characterized as farmers, with an average of two to three animals.' A further 17 per cent of milk is delivered to 110,000 farmers' co-operatives which are sufficiently 'organized' to supply the milk to cities through milk processors based in 176 districts.[43] The dairy sector in India employs 18 million people, 5.5 per cent of the entire workforce, of whom 58 per cent are women, and 69 per cent from disadvantaged groups.[44] Whereas the green revolution in India has been widely criticized for forcing peasants off the land, the white revolution has been widely applauded for providing livelihoods to the land-poor.

This success is testimony to the smallholder sector's resilience and ability to deliver, which is mirrored by a similar successes for small scale dairy production in Kenya, and by figures from the Polish dairy industry before the EU laid their hands on it.[45] It is also testimony to the ability of a still partly default livestock system to deliver a substantial quantity of animal protein under conditions of considerable population pressure. It makes a revealing comparison with China where rather more animal protein is provided per capita, through grain fed pigs and chickens. The Chinese have about 70 per cent as much agricultural land per person as the Indians; but they use 2.4 times as much nitrogen fertilizer per person.[46] Meat is increasingly being reared in factories, with the result that between 1985 and 2005 the number of farming households keeping poultry declined from 44 per cent to 14 per cent – while 140 million former peasants are now migrant workers in cities, dependent upon fossil fuels and the buying power of the West for their employment and their wellbeing.[47]

41   Gandhi, Maneka (1999) 'Factory Farming and the Meat Industry in India' in Tansey, G and da Silva, J (eds), *The Meat Business: Devouring a Hungry Planet*, Earthscan.

42   Barbaruah, M I and Joseph, A K (2008), *India: Dairy Giant Walking Barefoot*, FAO; http://www.aphca.org/workshops/Dairy_Workshop/Country%20Sessions/India.doc.; see also Brouwers,Tiny (2006) *India Report Part 1: Dairy Giant Walking Barefoot*, 2006, http://www.milkproduction.com/Library/Articles/India_part_1dairy_giant_walking_barefoot.htm

43   Milk Production.com (n.d.), *The Anand Model Throughout India*, http://www.milkproduction.com/Library/article_series/The_Anand_model_throughout_India.htm

44   FAO (2009), *Livestock in the Balance*, State of Food and Agriculture Report 2009.

45   For Kenya see FAO (2009) *ibid*, p 46; for Poland see Chapter 17, p 298.

46   Smil, Vaclav (2004), *Enriching the Earth*, MIT Press.

47   FAO (2009), *op cit*.43, p 44.

The divergent paths taken by India and China are only partly due to the fact that monogastrics are more susceptible to industrialization than ruminants because of their need for concentrates. In fact a firm called Keggfarms in India has had remarkable success breeding a high yielding 'smallholder chicken', the Kuroiler, a dual purpose bird for eggs and meat which also has magnificent plumage – an important feature for many of the women who look after them. The day old chicks are reared for three weeks in 1,500 'Mother Units' scattered around the country, and then distributed to villages by dealers called *pherriwallahs* who transport them in baskets on bicycles or by public transport.[48]

The differences between India and China probably have more to do with political circumstance than with their respective preferences for milk and white meat. India's peasantry has been represented by a decentralized but influential network of Gandhian self-help organizations and unions, whereas China's was championed by a Maoist government whose authoritarian excesses led to its downfall and replacement by a cadre of urban capitalists. There is a danger that India's dairy industry might yet take the same direction, largely because of aggressive moves being taken by supermarket firms such as Tesco and Carrefour to monopolize and centralize milk distribution.[49]

In any event, the white revolution in milk production has already taken some steps down the road towards industrialization by breeding to improve dairy production at the expense of draught ability. 'Milk production is increasingly becoming a major objective of bovine rearing in India,' note the FAO. 'There is marked replacement of animal power with mechanical power.' An article in the Indian environmental magazine *Down to Earth* warns that most of India's 27 different cattle breeds had been developed for draught use, but are now not being maintained or improved. Nonetheless, with almost no technical support for draught animals, more than 50 per cent of the land is still not tractor ploughed; and the number of oxen in India declined from 80 million in 1973 to a still significant 65 million in 2004.[50] The tractor population is now three million, 60 per cent of the global norm of 18 tractors per 1,000 hectares, whereas, if all of India's farmers had a tractor, the figure would be well over the norm.[51]

The people who have been hit hardest by the white revolution have been those at the very bottom who might once have been able to keep a poor cow, but are priced out by the upbreeding. Kamlabai Gudhe, a Dalit woman from Wardha, Maharashtra, whose husband died in a wave of farmer suicides in 2006, was given a crossbred Jersey cow valued at over £200 – but found that its appetite was insatiable. 'This brute eats more than all of us in this house put together,' she complained. 'And we don't get more than four litres a day from it'. The animal was given her as part of a heavily criticized scheme promoted by Maharashtra Chief Minister Vilasrao Deshmukh to distribute 40,000 'quality' cows to people in need. Vijay Jawandia, a local farming activist, stated that the money would be better spent reviving the growing of *jowar* (a millet) which, before it was phased

48   Ahuja, V *et al* (2008), *Poultry Based Livelihoods of the Rural Poor: Case of Kuroiler in West Bengal, SouthAsia Pro-Poor Livestock Programme*, 2008, http://enrap.org.in/PDFFILES/Doc012-PoultryBasedLRP-Kuroiler-reduced.pdf
49   Wiggerthale, Marita (2009), *To Checkout Please!*, Oxfam Germany, http://www.oxfam.de/download/to_checkout_please.pdf
50   Down to Earth (2004), 'Farm Power Divide: 65 Million Draught Animals', *Down to Earth*, 31 Oct 2004.
51   Indian Institute of Technology (2008), *National Meet on Tractor and Allied Machinery Manufacturers (TAMM-2008) & FIM Workshop*, Agricultural & Food Engineering Department, Indian Institute of Technology, Kharagpur West Bengal http://www.dak.iitkgp.ernet.in/news/showannouncedescr.php?newsid=327

out, occupied 30 per cent of the land in the district and provided a substantial quantity of fodder.[52]

A frequent alternative, over the last 20 years, has been to give the poor and landless goats, through government relief programmes and self-help schemes. To an agricultural economist, the advantage of goats over cows in the Indian context is transparent: there is no widespread religious proscription against killing them, so they can be sold for meat, and there is an export market to the Middle East. Yet they browse the commons just like cows, so they can be given to landless labourers who might otherwise rely on one of those decrepit, unslaughterable, unexportable cows, or even, God forbid, demand land. As an Uttar Pradesh government advice sheet listing the advantages of goat farming puts it: 'No religious taboo against goat slaughter and meat consumption prevalent in the country … Goats make a valuable contribution to the livelihood of economically weaker sections of society. Amongst the livestock owners, goat rearers are the poorest of the lot'.[53]

Former Indian Minister for the Environment, Maneka Gandhi, writing in 1999, explains how goats are taking over the niche traditionally occupied by the cow:

> Our goats are not grown on farms. In fact apart from poultry no
> animals are. People with land grow crops instead. People with no land
> raise animals. They are given pairs of goats as part of government relief
> schemes and self-help schemes. With no land to grow fodder on, these
> goatherds raise and graze their animals on open ground – hillsides,
> parks and forest areas.[54]

However, this is where, in Maneka Gandhi's view, the similarity with cows ends, for she regards the goat as a menace and the sponsored goat programme as an ecological disaster:

> Protected wildlife sanctuaries are the last green areas in India and
> amount to a paltry eight per cent green cover for the whole country.
> But even this figure is dwindling because goat owners need green
> for their goats and this is where they find it. Each goat, according to
> a government survey, consumes ten acres of land before it is killed at
> the age of two. Nor does it simply eat the grass like a cow – it actually
> pulls up the plant by the root, destroys young shoots and leaves the
> earth bare. The topsoil flies off and pretty soon the land is barren. In
> Rajasthan, which is a desert to begin with, the goats destroy whatever
> vegetation there is and encroach into the sanctuary area all the time.
> The strong winds then scatter the sand from the ravaged area and the
> desert creeps up further … Even city planting efforts fail for lack of tree
> guards. Why do we need tree guards? Because goats come and eat the
> plants. So if the municipal authorities do decide to plant, they choose
> poisonous Oleanders or *Astonia calaris*, which is known as the Devil
> Tree because no bird or animal will touch it.

Perhaps Maneka Gandhi is exaggerating. According to her figures, India's 115 million goats must now be 'consuming' a quarter of the country's entire land

---

52   Sainath (2006), 'Till the Cows Come Home', *The Hindu*, 3 November 2006, http://www.thehindu.com/2006/11/23/stories/2006112305660900.htm

53   Government of UP (n.d.), *Commercial Goat Farming*, Planning Dept, Government of Uttar Pradesh, http://planning.up.nic.in/innovations/inno3/ah/goat.htm

54   Gandhi, Manekha (1999), *op cit*. 40.

area every year. Nevertheless anyone who has kept both goats and cows will know how much more destructive goats can be, particularly towards trees, if left to their own devices – and one of the main advantages of the sacred cow is that it can be left to its own devices.[55]

Maneka Gandhi also alleges that the spread of goats is part of the reorientation of the agricultural economy – including the informal economy of the commons – towards export orientated industries which will enhance the GDP of the country. Meat production in India, she claims, between 1976 and 1994 (more or less the interval between Marvin Harris' analysis and Maneka Gandhi's) shows a 17 fold increase in the production of beef and veal to almost 1.3 million tonnes, and a nine-fold increase for buffalo to 12 million tonnes. The value of meat exports multiplied over tenfold from 1980 to 1996 to seven billion rupees. 'India only eats 25 per cent of these animals. The rest is exported primarily to the Middle East ... Every dead goat, for example, fetches approximately R250 (£4). '

However, exports of goat meat from India are miniscule – a mere 82 tonnes in 1999 when Maneka Gandhi's article was published in the UK. They have since risen to 572 tonnes in 2006, but this is still a tiny amount from a national herd over 100 million strong.[56] Most of the goat meat is exported, not from India to the Middle East, but from the countryside to the town. In 2005, an editorial in *The Times of India* made the following observations:

> Though largely unobserved, a dietary shift is taking hold in India. A vast number of vegetarians – 20 per cent of India's population – have begun to try out flesh foods outside the home ... Though urban areas eat three-quarters of India's poultry meat, the consumption of egg, fish and meat has gone up in rural homes ... Chicken sells four times as much as goat meat simply because it is cheaper. Pig meat sells much more than goat meat does thanks to its lower cost ... A persons' wallet decides his meat preferences. A non-vegetarian Muslim Bangladeshi eats less meat per capita than a partly vegetarian Indian because Bangladeshis are poorer.[57]

In other words, members of India's Hindu elite are rejecting its 2,000 year old dietary obligations, and choosing a western diet, just as they are choosing to wear western clothes, listen to western music, speak a western language, and adopt a western name when they work in the call centre. The worry is that the beef that was once reserved for India's poor is being replaced by goat meat for its wealthy.

Although Marvin Harris is observing from a western meat-eating perspective, and Maneka Gandhi, 25 years later, from an Indian vegetarian perspective, their views and conclusions are similar. Gandhi, towards the end of her essay, states:

> I do not think that India can be seen in terms of capitalists or communists, it is instead a cowdung economy. If you take the cow or its cowdung away, we are done for. We will die as a people. I say this knowing full well the scorn and ridicule that our dependence on cattle attracts from the rest of the world ... The Indian reverence for the cow may seem like sentimentality but it is born from solid

---

55   This should not be understood to mean that there is anything inherently wrong with goats. I used to work as a goatherd in Southern France in an area where milking goats could live for the entire winter, without any hay at all, off the climax vegetation of holm oak, without preventing it growing into a sizable tree, which could then be coppiced for charcoal and tannin.

56   FAOSTAT.

57   Kala, Arvind (2005), 'The Flesh-Eater of India', *The Times of India*, 25 October, 2005.

practical considerations. In India all essentials are transported by road. The entire rural economy moves on four legs. A government study estimated that draught animals saved us 780 billion rupees on fuel and transport costs.

Harris rounds off his essay with an assessment of the energy efficiency of the Indian cow-protection system:

In Singur district in West Bengal, Dr Odend'hal discovered that the cattle's gross energetic efficency ... was 17 per cent. This compares with less than four per cent for American beef cattle raised on Western range land. As Odend'hal says, the high efficiency of the Indian cattle complex comes about not because the animals are particularly productive, but because of scrupulous product utilization by humans.

And he concludes:

More calories go up in useless heat and smoke during a single day of traffic jams in the United States than is wasted by all the cows of India during an entire year.

Here are two writers, one American, one Indian: one an enthusiastic carnivore, advancing the interests of the poor; and the other coming from from a Hindu/ Gandhian tradition proclaiming her 'belief in vegetarianism', and advancing the interests of the environment. Yet they are in agreement that India is a 'cow-dung economy' and a cow-transported economy. The meat-eating anthropologist actively supports an ill-enforced vegetarian ideology; the vegetarian minister is careful never to make an anti-carnivore statement. And both agree that a policy which concentrates India's commons into carcases and sells them off to wealthy consumers will be her ruin.

ENERGY AND CARBON

12

ANIMAL FURLONGS AND
VEGETABLE MILES

*The environmental problems created by industrial systems mostly
derive from their geographical location and concentration.*

FAO SOFA 2009

I return to a statement quoted earlier: 'Fertility does not originate with [animal faeces and slaughterhouse residues], but rather with the grass and other plants which the animals ate. The animal destroys the greatest part of this food energy by its digestion, metabolism and other life processes. Even by the most scrupulous husbanding, only a small portion of that life-force is stored in the animal or passed with its manure.'[1]

It is the word 'destroy' that must be questioned here, for even someone like myself, with a limited understanding of the laws of physics, knows that the energy which is not converted into flesh or manure is not entirely wiped off the face of the earth, but is diverted into something else, for example heat, or movement. This energy may sometimes be lost for human purposes, but it can also be useful.

The heat from animals has been used in the past to warm houses in winter, by keeping them on the lower floor of a house, underneath the living quarters. If this arrangement has gone out of fashion, it is because at present we do not need the energy. If energy prices rocket, then perhaps double decker farmhouses could make a reappearance. Animal heat, embedded in manure, can also be used for heating greenhouses, or it can be captured through anaerobic digestion in the form of methane gas. However, most animal heat is unavailable for human purposes. Feeding straw to animals might be a more efficient way of keeping humans warm than burning straw in power stations – but only if we allow them into our living room.

Besides being heaters animals are also automobiles. The movement of animals is most obviously available in the form of animal traction, notably horses, mules, donkeys and oxen in Europe, camels, llamas, yaks etc elsewhere. Animal movement has been employed for a wide variety of useful activities including ploughing and cultivating, mowing, threshing, carrying and carting, pulling barges, logging, hunting, compressing roadstone, street sweeping, etc. However the use of animal traction for farm, forestry or road work is extremely low in Western Europe at the moment, though still prevalent in the Third World, and undergoing something of a revival in the USA.

It is more normal nowadays, in industrialized countries, to employ the mobility of animals to achieve certain land management objectives. Pigs may be

---

1   The Khadigar Community (n.d.), 'Ethical Farming in Action', *Vohan News International*, 1.

used to clear ground, or to stir up forest floors, cattle to trample bracken, sheep to keep grass down in an orchard, or poultry to eat insect pests and clear land. Set against the sum total of human nutritional needs, these benefits may seem minor, and they are more usually a by-product of the meat, but on the farm they can be significant. For example a stockfree holding has to mow its orchards, whereas a mixed holding with sheep can graze them as part of the stock rotation.

But historically the most important automotive function of livestock has been the processing and transport of nutrients. Animals do not create fertility; they process and transport it. It was not just out of delicacy that farmers in Coke's day referred to the Golden Hoof, rather than to that part of the anatomy more usually associated with manure.

Human beings, it has been calculated, consume, either directly or through animals, only about 3.9 per cent of the world's biomass.[2] There are, it seems, ample nutrients out there, the problem is not locating them but harvesting them and getting them to the relatively few places where arable culture can be carried out. Probably the best method of achieving this so far devised has been by transporting them in animals' guts.

The practice of grazing animals on poor, rough and inaccessible pastures during the day and enclosing them on arable land, or in the farmyard, at night has been practised across Europe for centuries. In England sheep were grazed on downs or commons during the day, and folded in fields at night, where they paid their rent in droppings. In a truly elegant system, sheep were also used to retrieve nutrients which had been washed away by heavy rain, carried downstream, diverted by sluices into water meadows and incorporated into lush grass which they grazed and brought back up to high ground, or ate as hay. The practice of folding was used to particular advantage on chalk downland: Cobbett mentions sheep folds a number of times in his *Rural Rides* through Hampshire. Thomas Davis, who was steward at Longleat in 1794, noted that sheep were folded overnight at a rate of 1,000 per acre, and states that 'the first and principal purpose of keeping sheep is undoubtedly the dung of the sheepfold, and the second is the wool'; and a surveyor for Winchester College made the same observation two years later.[3] Daniel Defoe, passing through Wiltshire, saw the practice as an important way of improving land that otherwise would be both too poor and too distant to cultivate:

> Now, so much of these downs are plowed up, as has increased the
> quantity of corn produced in this county, in a prodigious manner, and
> lessened their quantity of wool; all which has been done by folding
> their sheep upon the plow'd lands, removing the fold every night to a
> fresh place, 'till the whole piece of ground has been folded upon; this
> and this alone, has made these lands, which in themselves are poor,
> and where in some places, the earth is not above six inches above the
> solid chalk rock, able to bear as good wheat, as any of the richer lands
> in the vales, though not quite so much: many of these lands lie so
> remote from the farmers' houses, and up such high hills, that it would
> not be worth their while to carry dung from those farmhouses, to those
> remote lands.[4]

2   Vitousek, Peter *et al* (1986), 'Human Appropriation of the Products of Photosynthesis', *BioScience*, 36(6) 368-73.
3   Cited in Betty, Joe (2007), 'The Floated Watermeadows of Wessex: A Triumph of English Agriculture', in Cook, Hadrian and Williamson, Tom, *Watermeadows: History, Ecology and Conservation*, Windgather Press.
4   Defoe, Daniel (1727), *A Tour Through England and Wales*, Everyman, 1928, p 283.

In the 1970s, I worked on sheep farms in the South of France where the flock was grazed during the day on mountain terrain, a good deal rougher, higher and more remote than the hills of Wiltshire, and folded under cover in the *bergerie* at night. The sheep trampled their droppings into a fine golden powder, which for much of the year stayed bone dry, so there was no point in mixing straw with it. This was treasured by vine farmers who could broadcast the manure by hand from a sack as easily as if it were a chemical fertilizer, and who paid a very handsome price for it.

Animals not only perform the work of transporting nutrients, they also digest and process them. Ruminants, in a matter of hours, convert unpromising fibrous biomass into a bioactive fertilizer and compost catalyst (and in the case of cows a useful building material as well).[5] Then, as in Cokes' cattleyards, they deposit it on the floor, piss on it for good measure, and trample it into the straw with their hooves. A flock of sheep or a herd of cows in a well-designed farming system is the most energy-efficient compost-making machine yet devised. It can carry out the functions of harvester, tractor and trailer, chipper/shredder, bioactivator, and muck-spreader; and it usually does this for free, since its operating costs are covered by the value of the meat or the milk. At Polyface Farm, an organic farm in Virginia, pigs are used instead of a tractor frontloader to turn manure heaps. As the winter bedding from the cows builds up, grain is thrown into it which, when pigs are let in, encourages them to root through it, turning and aerating it. The system also avoids the use of a mechanical grain crusher, since the grain ferments and becomes digestible to the pigs.[6]

It may be nutritionally more efficient to make compost directly from dispersed vegetable matter without the aid of animals, but it's hard work, which these days is usually performed by petrol driven machinery when it is done on any scale. The nearest modern equivalent to a herd grazing the commons is perhaps the subsidized municipal compost made from tree cuttings, lawn and hedge clippings and garden waste. The material is strimmed, flailed or mown, stuffed (or in the case of fallen leaves, blown) into bags, driven to the tip in a multitude of cars and vans, transferred to another site, chipped and shredded, piled into windrows, turned, turned again, and then again, and then packed in plastic bags, delivered to retail centres and sold to people who drive it home and put it back in the very gardens where the raw material first came from. I have been unable to find a life cycle analysis for the entire process, but the shredding and turning alone requires 29 kilowatt hours of electricity per tonne – or 167 kilowatt hours for compost made from a mixture including food waste and paper.[7] If and when oil runs out, we may well go back to carrying out this sort of activity 'on the hoof'.

The automobility of animals also means that livestock farming is less dependent upon machinery than arable farming. The fact that vegetables can't move is often a disadvantage: everything has to be done for them (though on the other hand, they do not try to escape). Crops of cabbages cannot make their way to the waterhole in dry weather, or pick themselves up and find their way to the packing shed at the farmer's call like cows to the milking parlour. Cabbages have to be carried to market, while livestock can be driven to market on the hoof, and in low carbon local economies around the world, they still are. Very large herds

---

5   Anybody who has ever tried to wash dried dung off the flanks of a cow will understand why it was valued for its adhesive properties.

6   Polyface Inc, (n.d.), *Product Descriptions*, http://www.polyfacefarms.com/products.aspx

7   Komilis, Dimitris P and Ham, Robert K (2004), 'Life Cycle Inventory of Municipal Solid Waste and Yard Waste Windrow Composting in the United States', *Journal of Environmental Engineering*, November 2004.

of animals can be, and in some places still are, managed without any form of machinery whatsoever – whereas over a certain scale most arable farmers rely on tractors or animal traction. When things go wrong a livestock farmer will likely have his arm up a cow's backside, while an arable farmer will have his head underneath his tractor.

Muscle or Machine?

A Turner prize-winner called Mark Wallinger has designed a 50 metre high white horse to be erected near the Thames estuary, along with 10,000 new homes, shops, offices and an international railway station. While the public are no doubt relieved that a traditional subject for public works has been preferred to the alternative, a 'stack of five open cubes featuring a laser light, by Daniel Buren', the fact that the horse 'will be a faithfully accurate representation of a throroughbred' means that it will be a misrepresentation of the traditional British horse – equivalent to erecting a statue of a giant Ferrari in Cowley.

In the census of horses conducted by the Board of Trade in 1920, there were 19,743 thoroughbreds in Britain, out of a total of 2,192,165 horses. More than two thirds of all these were draught horses of which 774,934 were in active agricultural work, ploughing the bulk of the 4.5 million hectares under cultivation (about the same area as today). By 1939, the number of horses on British farms had declined to less than a million, and during the 1950s virtually all of them disappeared, displaced, some say, not so much by tractors as by tractors with front-loaders. Those of us brought up in the 1950s are the last to remember working Shires, horse-drawn coalmen and rag-and-bone men, not to mention details like the hessian feed bags attached round the muzzles of horses during lunch break so that nothing spilled onto the clean suburban streets.

Britain, being at the forefront of civilization, yielded to petrol hegemony quicker than most nations. Western Europe has followed suit, but even in the 1980s you could still see, for example, a leathery French peasant guiding his yoke of oxen every morning along a certain stretch of the Route Nationale 9, not far from Rodez. I like to think they delayed the construction of the Autoroute 9 until the poor fellow was dead and buried, and in a broader sense they did. In Eastern European countries such as Poland and Romania, horse cultivation is still common practice, though under attack from EU modernizers. In the USA horse cultivation is flourishing in the boondocks, inspired by the commercial success of the Amish communities, and aided by the fact that there is plenty of land to keep them on.[8] In large parts of the Third World, animal traction remains the only economic choice for small farmers.

It is at this stage in the evolution of farming technology that we are discovering that fossil fuels are causing more problems than they solve, and we are faced with need to find alternatives. 'There are three horses in the race to replace petroleum – biofuels, electricity and hydrogen – and at various times you see the fortunes of these various horses ebb and flow,' a motor industry expert called Roland Hwang was quoted as saying in the *New York Times*.[9] But however true that may be of the automobile industry, it will be some time time before we see electric or hydrogen-powered tractors rivalling the use of diesel on the farm. The two main contenders are biofuels and the runner Hwang thought had been retired from the race – namely the horse.

8   For information on the latest development in horse-powered technology, see *Small Farmers' Journal*, published by Lyn Miller, in Sisters, Oregon.
9   Mouwad, J (2008), 'Pumping Hydrogen', *NY Times*, 23 September, 2008.

These two sources of on-farm renewable energy are not incompatible, and there is no reason why they shouldn't be carried out equally satisfactorily on adjacent farms. However, it is useful to compare their performance, and for vegans there is an interest in proving that the non-animal solution is superior. The late Dave from Darlington wrote a stimulating article on the subject in an early edition of the vegan magazine *Growing Green*, but unfortunately much of the information he had at his disposal for horses was so out of date that the comparison is almost valueless.[10] I wrote a short paper on the subject in 2007, but that too I find is flawed and out of date.[11] The main problem is that one is faced with a superabundance of evolving data about the performance of biofuels, and a dearth of information about the performance of horses.

Much of the casual commentary on horse use is coloured by allusions to the amount of land they take up. We are variously told that when horses tilled all our land they required a quarter or a third of it to feed them. Darlington suggests that 40 per cent of farmland is required to feed horses, but this is derived from a secondhand reference to a conference held in 1973 in Alberta, where yields of crops per acre are extremely low. The most extreme example of this approach comes from one of the Global Opponents of Organic Farming, Dennis Avery of the Hudson Institute, who (citing Vaclav Smil) states:

> if American farming were horse-powered today, it would take 250 million horses to match our tractor and harvester engines. We'd need 740 million acres of land to feed the horses – twice as much arable land as the United States has. Instead of exporting food to densely populated Asian countries, the United States would be hard put to feed itself.[12]

Although this statistic is meant to illustrate the inefficiency of horses, it doesn't take too much thought to see that it actually reflects the inefficiency of US agricultural machinery and its extravagant fossil fuel expenditure. 250 million horses is almost one horse per US citizen, enough to plough 2.5 billion acres which is enough land to feed the world. If the USA is employing that much horsepower to produce its crops, it needs a radical rethink.

The most reliable calculation probably comes from Jan Jansen's study of the agricultural economy of a Swedish village in three years during the 20th century. In 1927 there was one working horse for every 8.4 hectares of land, and horses consumed 18 per cent of all the energy harvested in the community.[13] However, there is little to be gained from estimates of the amount of farmland devoted to horses in days gone by because crop yields have increased dramatically, while the area which a horse can cultivate has probably increased somewhat as well, thanks to improved technology. In 1920, one and a half times as much land was put down to oats as to wheat (whereas now the oat crop is negligible) and this was a reflection of the need to feed horses. But in 1920 the average yield of oats was 1.7 tonnes per hectare.[14] Now the UK average is four tonnes for organic oats and 6.5 tonnes for non-organic, and it is even higher in Ireland. Leaving aside the

10   In particular he was comparing horses fed on land in Canada in the 1970s with a very low yield, with biofuel produced on land with a very high yield. Dave from Darlington (n.d.), 'Horse Power: The Use of Draught Animals in Agriculture', *Growing Green 5*.

11   Fairlie, S (2006), 'Biofuel, Horsepower and Hectares', *The Land* 2, Summer 2006.

12   Dennis Avery (2000), *Earthday Celebrate the Gas Powered Tractor*, Hudson Institute; http://www.hudson.org/index.cfm?fuseaction=publication_details&id=569.

13   Jansen, Jan (2000a), 'Energy Analysis of Early, Mid and Late 20th Century Swedish Farming Systems: A Local Case Study', in *Agriculture, Energy and Sustainability: Case Studies of a Local Farming Community in Sweden*, Doctoral thesis, Agraria 253, Swedish University of Agricultural Sciences, Uppsala.

14   Mitchell, B R (1971), *Abstract of British Historical Statistics*, Cambridge Univesrity Press, 1971.

hay and grass that will always be a proportion of a horse's feed, we now need roughly a third as much arable land to feed a horse as we did in the 1920s.

So how much land does a working horse need nowadays, and how does that compare with a biofuel tractor? It is difficult to calculate because it depends how much oats and how much non-arable grass it is getting. Ken Laing advises: 'A starting point would be to allow one tonne per horse per year for a horse worked frequently' with the rest of the feed consisting of pasture or hay. One benchmark for comparison is to take a fairly high yielding field of wheat and determine how much horsepower and biofuel power it could, theoretically, generate. Wheat yields in the UK are around eight tonnes per hectare, against 6.5 for oats (which is what horses prefer); but since horses also eat grass from non-arable areas, and since the feed value of oat straw is more than of wheat straw, the figure of eight tonnes is reasonable. Eight tonnes of grain, grown on a hectare of prime land, will provide 22 kilos of grain every day for a year, which is enough in terms of calories (72,000 per day) to keep two horses working at a moderate pace.[15]

Two horses are normally agreed to be capable of cultivating ten hectares. According to horseman Charlie Pinney, 'when farm horse numbers were at their highest, the generally accepted horse-per-hectare ratio was around one pair per ten hectare, with that number increasing by one horse per each ten hectare increment in farm size. If the average farm unit is 40ha it therefore needs five horses,' – which is 16 hectares per pair.[16] It looks as though, theoretically speaking, it requires the equivalent of one hectare of good quality agricultural land to provide the horse power to cultivate 10 hectares. This figure is confirmed by Jansen who calculates that his Swedish village, were it entirely horse-powered today, would need to devote 10.2 per cent of its arable land to feeding its horse, plus another 0.8 for the energy necessary to manufacture the equipment.[17]

How does this compare with biofuels? This is even trickier to determine, not from a paucity of information, but because there are so many different kinds of biofuel, and so many different analyses of their potential performance. A hectare of land, producing about eight tonnes of grain, according to figures from what was then called the Department of Trade and Industry, which are the most optimistic I can find, will produce bioethanol equivalent to 1,000 litres of diesel.[18] It takes about 100 litres of fuel to cultivate a hectare of wheat,[19] and if we allow another ten litres for the embodied energy cost of the machinery[20] that means a hectare of bioethanol could power a tractor to cultivate that one hectare and a further 9 hectares.[21]

15   Laing, K (n.d.), *Horse Power for Organic Farms*, http://www.ruralheritage.com/horse_paddock/index.htm.

16   Pinney, C (n.d.), *The Case for Returning to Real Live Horse Power*, http://www.feasta.org/documents/wells/contents.html?six/pinney.html.

17   Jansen, Jan (2000b), 'Horses or tractors in a Bio-Energy Powered Future Swedish Agriculture', in *Agriculture, Energy and Sustainability, op cit*.13.

18   Elsayed, M A *et al* (2003), *Carbon and Energy Balances for a Range of Biofuel Options*, Sheffield Hallam, for DTI Sustainable Energy Programme. Most other estimates assign much higher processing costs for bioethanol, which would make the tractor far more greedy for land than the horse. See for example: Richards, I R (2000), *Energy Balances in the Growth of Oilseed Rape for Biodiesel and Wheat for Bioethanol*, Levington Agriculture Ltd, for British Association for Bio Fuels and Oils, 2000.

19   Richards (2000), *ibid*.

20   Richards and several other estimates all hover around 100 litres per hectare. The following studies suggest that 10 per cent is a reasonable estimate: Nagy (1999), *Energy Coefficients for Agriculture Inputs in Western Canada*, Canadian Agricultural Energy End Use Data Analysis Centre; Barber, A and Scarrow, S (2001), *Kiwifruit Energy Audit*, Zespri International. However 10 per cent seems low given the figure of 8 per cent allocated by Jansen for the much lighter and probably longer-lived horse equipment.

21   A hectare of biodiesel from rape oil is much more inefficient and will only provide power to cultivate only a further five hectares. Richards (2000), *op cit*.18.

There is therefore a broad equivalence between the energy performance of a draught horse, and the energy performance of a tractor running on bioethanol from wheat. Circumstantial factors and mitigating circumstances could be argued on both sides. Horse advocates could point out that nearly all studies view bioethanol processing to be even less efficient than the DTI figure I have used,[22] while biofuel advocates might point to studies showing that if the wheat straw were used to provide heat, or distillers grains' anaerobically digested, then the process would be more efficient.[23] Horse advocates might argue that the cultivations necessary to grow ten hectares of grain shouldn't take a pair of horses more than 100 days, leaving another 160 days (plus two rest days a week) on which the horses can do moderately heavy work at no extra fuel cost – whereas any additional work performed by the tractor would require extra fuel.[24] Tractor advocates may riposte (illogically in this context) that you don't have to feed a tractor when it's in the shed.

And so it can go on with, I suspect, the balance in favour of horse cultivation at the moment, simply because it is an existing and proven technology, whereas on-farm biofuel is not. However, vastly more investment and research is being directed into biofuels than into animal technologies, and it is anticipated that within the next few years a 'second generation' of biofuels will outstrip the performance of existing methods. Whereas current biofuels are generated inefficiently from high value crops such as wheat, corn, rape oil and (less inefficiently) sugar cane, the second generation will focus on converting cellulose from fibrous crops such as switch grass and *Miscanthus*, which can be grown on lower quality land. Protagonists claim that *Miscanthus* can 'produce about 2.5 times the amount of ethanol we can produce per acre of corn'.[25] If that is the case then there is a high chance that it will prove to be more efficient than animal power. Jansen examines a range of biofuel options for his Swedish village, which require between 14 per cent and six per cent of the arable land to produce the energy needed for manufacturing and powering the machinery.[26] The low figure of six per cent is achieved by using willow coppice rather than rape oil to power the machinery. The fact that a horse emits about a fifth as much methane as a dairy cow might also weigh against animal traction.[27]

If the intrinsic efficiency of biofuel over animal power is confirmed, then it still remains to be shown that the tapping of this energy does not require an infrastructure so energy intensive that it defeats the object. A significant advantage of animal power is that it requires only the animals themselves, a few steel tools that last for generations, and a bit of leather – and animals reproduce of their own accord. In a study of horse-powered farming systems, Chet Kendell found that on a thirty acre farm in Michigan, over a period of 40 years, a farmer derived a revenue of $21,000 from the sale of his horses' progeny, whereas a fossil-fuel-powered farm of the same size, whose small tractor is traded in for a new one every 10 years, would have incurred costs of $70,000.[28] On-farm processing

22   See Fairlie (2006), *op cit .11*.

23   Murphy, Jerry D, *Biomass, Biofuels, Biogas and Beverages*, Presentation to Institute of Brewers and Distillers, Guinness Store House, 23 February 2007.

24   Laing (n.d.), *op cit.15*.

25   Yates, Diana (2008), *Miscanthus can Meet U.S. Biofuels Goal Using Less Land than Corn or Switchgrass*, University of Illinois News Bureau, 30 July 2008, http://news.illinois.edu/NEWS/08/0730miscanthus.html

26   Jansen (2000b), *op cit*.

27   Crutzen, P J (1986), 'Methane Production by Domestic Animals', *Tellus B Chemical and Physical Meteorology*, Vol 38, p 271.

28   Kendell, C (2005), 'Economics of Farming with Horses – Operational Cost for Horses', *Rural Heritage*, Spring 2005, http://www.ruralheritage.com/back_forty/economics_cost.htm

of biofuels for a tractor would require additional equipment, though the expense of this would be offset by savings on fuel bills. Centralized generation involving the transport of biomass to power stations, and the subsequent redistribution of the energy and other products back to into the wider economy, however great the economies of scale, might have a hard time matching the innate sustainability of on-farm biofuel systems which, like working animals, produce and expend the energy on site.

The contest between animal power and biofuels is a fascinating and important one, but it is bizarre how the cards are stacked. Animal traction is the dominant way of cultivating land throughout about half the world, while biofuels are barely used at all, except in Brazil, and experimentally in countries like the USA and Germany. Yet hardly any mainstream research is carried out into the comparative efficiency and sustainability of animal traction, whereas virtually every university in the western world has a faculty investigating the potential of biofuels. There are hundreds, if not thousands, of academic papers comparing different kinds of biofuels, but, aside from Jansen's study, I have yet to find a single one analysing the efficiency of animal power, which, after firewood, is the most frequently used bio-energy in the modern world. In such a prejudiced climate it seems inevitable that biofuels will emerge the winner.

The late Charlie Pinney, who for many years was a rare voice in the UK propagating the benefits of horse drawn cultivation, observed:

> The widespread nostalgic appeal of the draught horse often proves
> to be the biggest hindrance to the popular acceptance of its serious
> worth as a farm, forest and transport tool. Whenever you approach
> an individual or a government to suggest that the working horse can
> make a worthwhile contribution to a sustainable future for us all, you
> are almost invariably greeted with scorn or disbelief.[29]

The reluctance to examine animal power may also reflect a worry, on the part of scientists, that their efforts to turn biomass into energy through diverse highfaluting systems will prove no more efficient than the digestive systems of biological animals. The resistance of scientists is reinforced by the opposition of economists who view livestock as labour intensive: it takes a skilled man to handle a team of horses cultivating an acre or two a day, whereas the average farmworker these days drives a 140 horse power tractor capable of covering 20 times the area.

The social ecologist, however, will view the matter differently. Time may be money, but speed, though lucrative, is disproportionately expensive on energy; and as regards energy, it is land which is in short supply, not humans. More people on the land means less people on the streets; and people on land produce energy, whereas people on streets consume it.

Blessed Are the Meek

Whereas the second generation of biofuel is hopefully more efficient than its predecessor, the first generation of draught animals was more efficient than its successor. Horses are the most extravagant of draught animals. In the past they found favour as pullers of the plough only in a relatively small number of mainly wealthy countries, because they require high quality food (though mules were widespread in many dryer climates). Oxen were and are far more widely

---

29   Pinney (n.d.), *op cit*.16.

used throughout the world, largely because (like *Miscanthus*) they are cheaper to run. In the words of one writer, they were 'content with much more modest fare, more robust than horses, less likely to get injured, and when times got bad they could survive on simply dreadful food – dodgy straw and mouldy hay of a type that no horse would touch. At the end of their working life oxen also made good beef – after a few months in a fattening pen',[30] though there is no reason, other than sentiment, why the horse shouldn't end up in sausages as well.

Its not clear why a number of European countries moved from the ox to the horse – particularly since the horse might get requisitioned for war – hardly an advantage for the farmer, though it was for the breeder. It may have been because the horse was more prestigious than the ox, partly because it performed better on the road (it was faster on a good road, and on a road unfit for carts you could ride it) and partly because it is a more stimulating and exciting animal. Nobody, it seems, has ever raced oxen – but then there is no equine equivalent of bullfights either. According to one writer:

> It was the Enclosures Act of 1801, and with it the demise of a medieval open-fields approach to farming, that started a rapid decline in the use of oxen as beasts of burden. On the land a team of six or eight great beasts harnessed in pairs one behind the other was just too unwieldy to plough or harrow into corners of the new, smaller fields, but two or three horses harnessed side by side could do so with ease.[31]

However, it is by no means certain that a horse would do the work of two oxen. This is Lord Kames writing in 1776:

> As oxen have less air and spirit in moving than horses their motion is concluded to be slower. They are less expeditious than horses in galloping or trotting it is true but as farm work is performed in stepping, let the step of an ox and a horse be compared and the ox will be found not to be inferior especially where an ox is harnessed like a horse.
>
> Colonel Pool in Derbyshire ploughs as much ground in a day with three oxen as neighbouring farmers do with four or five horses. In summer they eat nothing but grass, in winter they have hay and turnips when much wrought, straw only when wrought moderately. About Bawtry in Yorkshire, four oxen in a plough do as much as the same number of horses. Near Beaconsfield, Mr Burke ploughs an acre in a day with four oxen and his neighbours do no more with four horses.[32]

Although less spirited than the horse, there nevertheless seems to have been some affection for the ox's meeker and more compliant temperament:

> Once a pair was selected they became each other's companions for life, working side-by-side and never far apart whether grazing in meadows or sleeping in the ox barn. Each ox had a name and within the pair one had a single syllable name and one had a longer name. So Quick and Nimble, Pert and Lively, Hawk and Pheasant all spent their working lives together.[33]

---

30   Williams, Michael (n.d.), *The Living Tractor*, http://www.foxearth.org.uk/oxen.html
31   Ibid.
32   Kames, Lord H H (1776), *The Gentleman Farmer, Being an Attempt to Improve Agriculture, by Subjecting it to the Text of Rational Principles*, W. Creech, http://www.foxearth.org.uk/oxen.html
33   Williams (n.d.), *op cit. 30.*

This convention, if it were ever true, is not universal. The oxen at Bhakti-vedanta Manor in Hertfordshire respond to polysyllabic Sanskrit names, and are not always worked together in the same team – though an ox which is used to working on the right hand side (ie walking in the furrow), will tend to be kept in that position, as will the ox habituated to working on the left (ie on the land). My very limited experience with oxen leads me to believe that they are probably easier than horses, because they are naturally slower and steadier, and with arable cultivation, slow and steady wins the race every time. At Bhaktivedanta I witnessed a three year old who had never been harnessed or broken before, yoked and guided round a yard by a novice within two hours.

In the open field farming system of medieval Europe, a *carucate*, the amount of land which could be cultivated by an eight ox team, was typically about 120 acres, or 48 hectares. A *virgate* was what two oxen would cultivate, 12 hectares, which is slightly more than what most writers allow for a pair of horses. However, the limitation upon what can be cultivated depends upon the length of the season more than the energy potential of the team. In the open fields of Anatolia, which until at least the 1950s operated a two course system similar to the English open field system, a man with a pair of oxen could only plough 10 acres, but this was because of time limits imposed by very short rainy seasons.[34]

An open field system still operates in some parts of the world, for example in parts of Ethiopia. In the Tigray region, according to a United Nations survey, a healthy oxen pair can plough one hectare in six to eight hours, which seems very fast. In some areas, extension workers have been introducing a one ox ploughing technique, though this, apparently, requires lighter soils.[35] Modern animal-drawn equipment is lighter than 19th and early 20th century equipment and organizations such as Intermediate Technology (now called Practical Action) have attempted to promote the use of more efficient modern horse drawn tools.

A further improvement in the efficiency of bovine traction can be achieved by using cows, as well as, or instead of oxen – the advantage being that they can also produce milk. Gandhi, asked to comment on the matter, remarked:

> This question was put to me as long ago as 1915. I felt then as now, that
> if the cows referred to were used for purposes of ploughing the land,
> it would not hurt them. On the contrary, it would make them strong
> and increase their yield of milk. But this benefit could only come about
> provided the cow was treated as a friend and not cruelly, as cattle in
> our land so often are … Every living being has to work within his or its
> limitations. Such work uplifts, never lowers either man or beast.[36]

A 1994 comparison of oxen and cow use in Ethiopia concluded that when farming families employed milking cows as draught animals, rather than the traditional oxen, production was improved by 67 per cent, because one animal was serving a dual purpose. Families had more surplus produce available for sale, but bought-in feed costs and labour costs were higher.[37]

34   My reference for this item has gone missing; for information on open fields in Anatolia see Stirling, P (1965), *Turkish Village*, 1965, University of Kent, Chapter 4; http://anthronexus.org/ Era_Resources/Era/Stirling/StirlingContents.html

35   Speiss, Hans (1994), *Report on Draught Animals under Drought Conditions, in Central, Eastern and Southern Zones of Region 1 (Tigray)*, United Nations Development Programme Emergencies Unit for Ethiopia, 1994, http://www.africa.upenn.edu/eue_web/Oxen94.htm

36   Gandhi, M K (1946), *How to Serve the Cow*, Navajivan Publishing House, Ahmedabad, 1954, p 51 (originally in *Harijan*, 15 Sept 1946).

37   Panin, Anthony and Brokken, Ray F (1989), *Economic Analysis of Cow Traction Farm Technology in the Ethiopian Highlands*, FAO, http://www.fao.org/Wairdocs/ILRI/x5483B/x5483b1g. htm#results%20and%20discussion

Ethiopia, since it so poor, is perhaps not the best advertisement for the use of oxen in the modern world. More relevant is the recent experience of Cuba, which is in many ways a fairly developed country. In respect of education, medical facilities and infant mortality, it is on a par with the United States. When category IV hurricane Dennis swept across the country in 2005, the Cuban authorities managed to evacuate 1.5 million people by bus from its path, while the Great Car Economy coped rather less well with hurricane Katrina when it hit New Orleans in the same year.

In the 1980s, Cuba reportedly had more tractors per hectare than California. But after the collapse of the Soviet bloc, which had been paying fair trade prices for Cuba's sugar, the country suddenly found itself undergoing an energy famine that may provide a foretaste of what is in store for all of us. Without oil, Cuban farmers had to resort to oxen. There were perhaps 50,000 teams of the animals left in Cuba in 1990, and maybe that many farmers who still knew how to use them, but that soon changed. Cuba's Agricultural Mechanization Research Institute developed improved machinery for ploughing, harrowing, ridging, and tilling, specially designed not "to invert the topsoil layer" and decrease fertility. Harness shops were set up to start producing reins and yokes, and the number of blacksmith shops quintupled. The Ministry of Agriculture stopped slaughtering oxen for food, and any in good physical condition were delivered to cooperative and state farms. By the end of the millennium there were 400,000 oxen teams plying the country's fields. One result was a dramatic reduction in soil compaction: 'Across the country we see dry soils turning healthier, loamier. Soon an ambitious young Cuban will be able to get a master's degree in oxen management.'[38]

This prediction from a Cuban professor, cited in a 2005 article by journalist Bill McGibben, has proved to be a bit optimistic. In 2008, Cuban environmentalist Roberto Sanchez Medina reported that with the end of the 'special period' (thanks to oil from Venezuela) the number of oxen teams had peaked – and that a major obstacle was the image problem. *Boyeros*, ox handlers, are held in low esteem and 'the biggest obstacles are cultural'.[39]

Could the reintroduction of teams of oxen ever be part of a panoply of responses to an energy crisis in the UK? Perhaps if the crisis got bad enough, but any move towards a revival of animal traction will probably start with horses. There are an estimated 1.35 million horses in the UK, perhaps occupying two million acres (about five per cent of our agricultural land), enough horsepower, were they all the right breed, to cultivate at least half of our arable land.[40] But only a tiny fraction of today's horse population performs any useful function, other than keeping a rural population that has been wrenched away from the food production process in contact with the animal kingdom, grass, muck, death and the forces of nature. More horses are towed around the country by gas-guzzling four-wheel-drives than are ever put between the shafts of a cart; and the NFU reports that obesity is becoming as serious a problem for horses as it is for their masters.[41]

If energy conservation becomes an imperative there will be increasing pressure to ensure that the UK horses pull their weight. And even though a proportion of the horse owning population are dripping with money, and

38   McKibben, Bill (2005), 'The Cuba Diet: What Will you be Eating when the Revolution Comes?' *Harper's Magazine*, April 2005.

39   Medina, R S (2008), 'The Cuban Ox Revival', *The Land* 5, Summer 2008.

40   BETA (2006), *National Equestrian Survey*, British Equestrian Trade Association; http://www.beta-uk.org/

41   NFU (2008), *NFU Countryside Magazine*, September 2006.

mistresses of the art of conspicuous consumption, they all have a bit of gumption and I dare say they will rise to the occasion. The recent trend for black and white gypsy cobs, which a few years ago were infra dig, suggests that popular taste in horses is veering back towards animals with more of a draught build. The British breeds of heavy horse are something of a Rolls Royce for the job (with a fuel consumption to match) and the reverence in which they are held will ensure that, if there is any economic or ergonomic profit to be had in employing them for the tasks for which they were originally bred, there will be no shortage of ploughmen and waggoners keen to take on that challenge.

But if pressure upon energy resources and land is so great that we have to dig deep for solutions, then there is no reason why we shouldn't turn to oxen. Many of our traditional breeds – Jerseys, Herefords, Welsh Black – are suitable stock for breeding working oxen. One of the factors that may help to give oxen a competitive edge over horses is the fact that we have less compunction about eating them. One exciting change will be that once again our cattle will be allowed to keep their horns. A horned ox is easier to handle and it has a reverse gear.

Veg Miles

One of the initial concerns which prompted me to write this book, but which now seems marginal, is the worry that a wealthy vegan agricultural economy might prove to be more energy intensive than a meat economy in respect of food transport. For a number of years, I lived in a rural community, with a fairly well developed subsistence economy, as well as a number of small commercial agricultural enterprises – which for convenience, and out of politeness I will call Happy Valley. The community, of a dozen or so adults plus a handful of children, included carnivores, vegans, and almost everything in between: lactovegetarians, piscovegetarians, lactophobes and triticophobes.

I was a stock-keeper there, looking after dairy cows, pigs and the horse. The dairy produce was consumed by over half the residents, and that subsistence consumption was invaluable in helping to make the cows viable, particularly after Environmental Health banned us from selling our cheese at market because it was made in our kitchen. But the pig products – pork, bacon, sausages and lard – were difficult to shift within the community. Since everybody at Happy Valley ate together, and took turns with the cooking, even the few who might have wanted to cook with lard or bacon were unable to. Communal food tends to be limited to the highest common factor of collective taste, and in this context, that meant veggie.

Although the pork could be sold elsewhere, this meant that the pigs and the cows were always regarded as separate enterprises, whereas the vegetables were communally managed. Yet the pigs were performing a community service, by disposing of waste in the most economical way possible, as indeed were the cows by grazing uncultivated land and providing manure. About 40 per cent of the pigs' feed came from waste produced on site, including whey from the dairy, the scrap food from the kitchen, apple pomace, chat potatoes and waste vegetables from the gardens. When there were no pigs, the whey was thrown away, and the food scraps, which had they been put on the compost heap would have attracted rats, were left in the woods to feed the burgeoning population of badgers who carried out daring raids upon our gardens, chicken-house and a highly fortified kitchen. The problem was eventually solved by siting the chickens right next to the house – in what the French call the *basse cour* – and feeding them a proportion of the waste.

But what I found most difficult of all was the food which was bought in to replace the pork, beef and the dairy that various members of Happy Valley wouldn't eat. My understanding of a subsistence-based rural community is that, by and large, you eat what you produce; and failing that, you eat what can be sourced in your locality or region. The food our community produced – vegetables, fruits, apple juice, potatoes, milk, yoghurt, cheese, butter, pork, beef, eggs, dripping and lard – together with other local or UK produce such as honey and mackerel, and staples such as wheat, oats, beet sugar and salt, was sufficient to provide a wholesome and varied diet, not that different from the one I was brought up on in the 1950s.

Others did not feel the same, because every two or three weeks the community ordered a consignment of food to the value of at least £200 from a wholefood wholesaler in Bristol which trades under the misnomer Essential Foods. The majority of foods delivered were foreign imports, usually of high nutritional value, and often from Third World countries. Chick peas, various kinds of beans and dals, soya milk, soya yogurt and other soya products, margarine, vegetable oil, olive oil, brazil nuts, peanut butter, tahini, coconut, molasses, dried fruits, canned tomatoes, rice, quinoa, hot chocolate and coffee were typical foreign products, and the most common source countries were the Indian subcontinent, China, Turkey and the USA.

I dare say these comments appear impertinent or censorious. People make personal decisions about what they eat: criticizing someone's diet is almost as offensive as criticizing their religion. Besides, I consume a number of foreign imports myself, mostly coffee and spices and citrus marmalade, for which there is no locally grown substitute. But it is instructive to compare Happy Valley's home produced fare with its imported food, because it provides a demonstration of the limitations of vegan and vegetarian diets in a northern country.

Particularly noticeable in the list of foods bought in by the community are fat products. All communal cooking at Happy Valley requires imported fat or oil because there aren't any edible vegetable oils grown in this country on any scale, aside from rape oil, which the community shunned in favour of sunflower and low grade olive oil.[42] Hemp oil and walnut oil are sometimes advanced as alternatives, but nobody is yet producing these at a price that is remotely competitive with rape oil, beef dripping or butter. For the time being, if a Northern country such as the UK stopped eating meat it would either have to consume prodigious amounts of rape oil, or else import virtually all its fat – and indeed the current prejudice against animal fats, particularly lard, suet and dripping, but also butter, full cream milk and oily fish, is such that we are not far off that situation already. After animal feeds, oils are the UK's largest import in terms of land-take from other countries.[43]

The situation is similar, though not quite so severe, for high protein foods. Take for example the haggises made in Scotland by James MacSween. The main ingredient is lamb offal, followed in descending order by beef fat, oatmeal, water, onion, salt, pepper and spices. Apart from the seasoning, all these ingredients are produced in Scotland, and could realistically all be produced on

42   Aside from rape oil which was strictly for industrial use until the middle of the last century, when the Canadians developed an arguably edible variety which they marketed under the brand name Canola.

43   The UK imports animal feeds grown on 1,750,000 hectares, and oils grown on 650,000 hectares.Mclaren, Duncan *et al* (1998), *Tomorrow's World*, FoE, Earthscan.

the same farm. But for the last 18 years James MacSween's company has also been making an equally tasty vegan haggis which contains, again in order of magnitude, oatmeal, water, vegetable margarine, kidney beans, lentils, peanuts, almonds, walnuts, carrots, turnips, onions, mushrooms, salt, pepper and spices. The oatmeal and the vegetables again could all be produced in Scotland. But the margarine is made primarily of palm oil, from Indonesia or Brazil, mixed with some UK rape oil; the beans and lentils are grown in countries such as Canada, Turkey and China; and the nuts are sourced from Argentina, India, China, the USA and elsewhere.[44] While the traditional meaty haggis is a local boy, the vegan substitute is a true cosmopolitan.

The range of high energy vegetarian foods is limited in the north, and becomes more limited the further north you go, whereas there is a huge variety which can be grown in warmer countries. This is why about ten years ago there was a fashion amongst vegan food growers to emigrate to Spain, where sunflowers, chick peas and tomatoes ripen fully, and olives, oranges and almonds grow on trees (though other reasons for emigrating were cheap land and a relaxed planning regime). The Spain drain seems to have slowed down, partly due to efforts by the Vegan Organic Network to promote stockless horticulture in the UK and to develop varieties of Mediterranean crops adapted to a north Atlantic climate. Vegans are becoming increasingly aware of the local food issue, and to their credit some are trying to address it. Some gardeners claim to have obtained decent yields of soya beans, olives and similar crops in the UK, but these are still a long way from being commercially viable. Even trees that grow semi-naturally in England, such as chestnuts and walnuts, have yet to produce commercial yields that can compete with imports from China or Italy. This could change if global warming raises the mean temperature in the UK by several degrees. But for the time being a shift towards a more vegan diet in a country such as the UK is likely to lead either to a more boring diet, or else to greater food miles.

More food miles means greater energy expenditure; but we should be wary of exaggerating this factor of the energy equation. In 2008, an influential study by Christopher Weber and H Scott Matthews concluded that transport accounted for 11 per cent of US food greenhouse gas emissions, while most of the rest was due to production processes, particularly of meat and dairy.[45] Surprisingly this is rather lower than the figure provided by the British Cabinet Office who state that 15.3 per cent of UK food emissions are caused by transport.[46] But Weber and Matthews' figures are better supported, and they use them to draw a challenging conclusion, namely that consumers concerned about the carbon impact of their food would do better to reduce meat and dairy consumption than to turn to a local diet:

> For the average American household, 'buying local' could achieve, at
> maximum, around a 4-5 per cent reduction in Greenhouse Gas (GHG)

44   James MacSween, personal communication, 2009.
45   Weber, Christopher and Matthews, H Scott (2008), 'Food Miles and the Relative Climate Impacts of Food Choices in the United States', *Environmental Science Technology*, Vol 42 No 10.
46   Cabinet Office, *Food Matters: Towards a Strategy for the 21st Century*, pp 13 and 16 citing *Hansard*, 2 Feb 2007, Column 613W, Ian Pearson MP: "Total direct greenhouse gas emissions arising from the UK livestock sector (sheep, pigs, bovines and poultry) were 4.5 per cent of the UK total emissions or 8.07 million tonnes of carbon equivalent (MtCe) in 2005. This was made up of: 5.03 MtCe of methane (of which 86 per cent was from enteric fermentation in the animals' digestive systems and about 14 per cent from manure management); 0.96 MtCe of nitrous oxide emission from manure management; and 2.08 MtCe of nitrous oxide emissions from grazing and crop production for livestock feed." http://www.parliament.the-stationery-office.co.uk/pa/cm200607/cmhansrd/cm070202/text/70202w0017.htm

emissions, due to large sources of both CO_2 and non-CO_2 emissions in the production of food. Shifting less than one day per week (ie one seventh of total calories) consumption of red meat and/or dairy to other protein sources, or a vegetable based diet could have the same climate impact as buying all household food from local providers.

In an accompanying table, the authors show that the average US household only needs to divert 17 per cent of its red meat or 38 per cent of its dairy consumption to vegan alternatives to achieve the same greenhouse gas savings that it would achieve by buying locally. Weber and Matthews' 'one day a week' recommendation provides the theoretical basis for the Paul McCartney's Meat Free Monday campaign (which, in the spirit of multiculturalism, has replaced the Christian tradition that Friday is the day for declining meat).

Weber and Matthews are right, the transport impacts of food production are slight compared to the total environmental impact of the food industry, and reducing consumption of industrially produced meat is an easier way for North Americans to lower the climate impact of their diet than going local. Nonetheless, they exaggerate their case by underestimating the potential for localization. When they talk about 'buying all household food from local providers', they don't mean buying food that is entirely locally produced; they mean food whose final delivery 'from farm or production facility to the retail store' is local. Upstream inputs such as deliveries of fertilizer or animal feed are still permissible within their definition of local; and in the case of composite products, such as a strawberry yoghurt, so is the transport of the strawberries, the milk and the sugar to the factory – only the delivery of the yoghurt from the factory to the retailer comes under the heading of 'final delivery'.[47]

In other words, the 'total localization' that they use in their comparison is not total at all but represents a reduction of just 39 per cent of all food transport GHG emissions. If it were total, the local consumer would come out looking a lot better in their final analysis: a meat and dairy consumer who went 100 per cent local and refused to buy foods with all these upstream transport impacts would make more or less the same greenhouse gas savings as the dairy consumer who gave up dairy and went vegan. Of course, going local and reducing meat consumption are not mutually exclusive – in many geographical situations they go hand in hand. Consumers who both reduced their animal protein intake to default levels and went 100 per cent local as well would be performing better still.

Going 100 per cent local is a bit extreme, but there is nothing inherently unrealistic about being up in the eighties or nineties, at least for consumers living in the countryside or market towns. Organic farms with default levels of livestock rely on low inputs from outside, they are less dependent upon fossil fuels than conventional dairy farms, and so are almost by definition highly localized. Unfortunately this scenario is not one that Weber and Matthews have covered in their analysis.

On top of that, there is a whole further section of the food transport system which is not even mentioned by Weber and Matthews – namely what happens to the food waste. Any nutrients which have not been absorbed by humans, whether or not they have passed through the body, have to find their way back to the land, and the further humans live from the land that produces their food, the further these nutrients have to travel. The recycling of fertility is the aspect of food transport that exercises the FAO most:

---

47   Christopher Weber, personal communication, 2 Nov 2009.

The cost of transporting nutrients to cropland is often prohibitive
… Longer food chains, driven by the concentration of consumers in
urban centres, mean that production systems have to bridge long
geographical distances between the site of feed production and the
consumer. Decreasing transport costs have allowed the relocation of
production and processing activities to minimize production costs.
Globally this process has helped to overcome local resource constraints
and allowed people in food-deficit areas to be fed. However it also
involves large-scale extraction and transfers of nutrients and virtual
water embedded in feed and animal products, with detrimental long
term consequences for ecosystems and soil fertility.[48]

Admittedly the FAO are here focussing on animal products, and their concern
is to ruralize factory farms by relocating them in the rural hinterland rather
than on the periphery of 'food deficit areas' (ie cities), closer to the land than
the consumer. But the same problem confronts an urbanized vegan economy:
how do you get all the waste from centralized food processing facilities, retail
outlets and peoples' dustbins to cascade back to the land from which it came?
The most efficient way discovered so far is by redistributing it in sacks of animal
feed that are sold to small farmers who feed it to animals and put their manure
back on the land. The least efficient way is to drive it out to a field and dump it,
but this is wasteful and the economies of scale involved usually means that there
is a problem with nitrogen leaching. For a vegan society, the most promising
possibility is anaerobic digestion, but again this takes food out of the food chain,
and there is still a transport toll involved in redistributing the heavy digestate
that remains after the energy has been removed. None of this crucial part of the
food distribution cycle is taken into account by Weber and Matthews.

The alternative is a local food economy whose citizens live where most of
the food can be grown, rather than in food deficit areas, and where waste, at
whatever stage it may be generated in the food cycle, has a only a short journey
back to the land from which it came – either via the route of an animal's gut and
with the assistance of its mobility, or not, as the case may be. This is what 'total
localization' involves and Weber and Matthews, perhaps because they live in
food deficit areas themselves, haven't grasped this.

That is not to say that total localization is either desirable or achievable. Grain
has to be traded to some extent to alleviate crop failures; and the environmental
cost of shipping non-perishable goods to regions where they cannot be grown is
slight. There are those who argue that Kenyans should be lifted out of poverty
through the otherwise fatuous business of flying fresh flowers and out of season
vegetables to Europe.[49] On the other hand, trucking goods across continents to
regions where they can be grown satisfactorily causes environmental damage
and results in unsustainable concentrations of biomass and nutrients where they
are not needed, for the sake of an economic advantage which is pocketed by
supermarkets, not by farmers. As for countries like New Zealand that have a
surplus of land and food, perhaps they should open their doors to immigrants
and parcel out some of their farms to landless peasants. If we are going to
globalize everything else, we should globalize land reform as well.

At the end of their paper, Weber and Matthews point out that GHG
emissions are 'only one dimension of the environmental impacts of food

48   FAO (2009), *Livestock in the Balance*, State of Food and Agriculture Report for 2009 (SOFA
2009), unpublished consultation draft, 11 August 2009, p 62.
49   Synergy (2005), *Trade Matters in the Fight against World Poverty*, Synergy, UK Department for
International Development.

production'. Food miles may not be over-extravagant in their energy use, but they are thickly implicated in a centralized distribution system which multiplies our energy expenditure at every opportunity and whose impacts include excessive packaging and refrigeration, waste, traffic congestion, road-building, noise, accidents, loss of local distinctiveness, exploitation and displacement of peasants, excessive immigration, urban slums, deforestation and habitat destruction, removal of biomass from third world countries, the undermining of local communities in the UK, the collapse of UK farming and the blood which is spilt over oil fields. It is idle to imagine that we can create local food systems in isolation from this globalized economy. To reinvent a truly local food economy we will have to make the whole system more decentralized, and relocalize the delivery not just of food, but of a panoply of goods such as energy, fibre, building materials, waste disposal and water.

13

GLOBAL WARMING: COWS OR CARS?

When I began this book, the global warming impact of livestock, mostly related to methane from cows, was considered a side issue. Since 2007, it has become the main argument against carnivory, and is now deployed as such by people who have long been advancing other reasons for not eating meat.[1] If you are a climate sceptic, then the global warming argument is irrelevant, and you can safely skip the next two chapters, unless you are interested in how the discourse has been altered by the rise of climate politics. I am not a climate sceptic, but that is not to say that I am convinced that 90 per cent of scientists must be right (any more than I believe that we live in an expanding universe born out of a big bang just because 90 per cent of physicists think so). I accept the global warming discourse, because of Pascal's Wager, otherwise known as the precautionary principle; and because I believe it is an appropriate ideology (or religion if you prefer) for humanity at a time when we are clearly placing too much pressure on the environment through excessive population and consumption.[2] In this chapter and the next I therefore take the climate change scenario, as modelled by the International Panel on Climate Change (IPCC), as a premise.

If you surfed around the Internet investigating meat and global warming in 2006, you were certain to alight upon this:

> Cut global warming by being vegetarian. Global warming could be controlled if we all became vegetarians and stopped eating meat. That's the view of British physicist Alan Calverd, who thinks that giving up pork chops, lamb cutlets and chicken burgers would do more for the environment than burning less oil and gas. Writing in this month's *Physics World*, Calvert [*sic*] calculates that the animals we eat emit 21 per cent of all the carbon dioxide that can be attributed to human activity. We could therefore slash man-made emissions of carbon dioxide simply by abolishing all livestock.

This press release was posted on the website of *Physics World*, a magazine which published Calverd's article on 7 July 2005. Unfortunately, to read the whole article, you have to subscribe to the magazine which costs £240, or £599 if you want access to its archive of back issues. I tried e-mailing *Physics World*, but the e-mail bounced back. So I keyed in 'Alan Calverd global warming' in the hope of locating a pirated copy. I had no luck, but instead got well over 100

---

1   E.g. Goodland, R (1988), 'Environmental Sustainability in Agriculture: Bioethical and Religious Arguments Against Carnivory' in Lemons, J, Westra, L and Goodland, R (eds), *Ecological Sustainability*, Kluwer Academic, pp 235-5.

2   Fairlie, S (2002), *The Prospect of Cornutopia, or If Global Warming Didn't Exist We Would Have To Invent It*, Chapter 7 Publications, also published as 'The Prospect of Cornucopia', in *Irish Pages*, Autumn Winter 2002/3, p 224; Orr, David (1992), 'Pascal's Wager and Economics in a Hotter Time', *The Ecologist*, 22:2, pp 42-3.

citations to the article, including coverage in the *Financial Times*, the *Daily Mail*, *The Sun* and the *Sydney Morning Herald*. I opened up about 50 of these and on only one did I find any evidence that anyone had actually read the article: a contributor to the Oregon Public Network site quoted a sentence from the article which could not be found on *Physics World*'s press release.[3] Every other citation I found did nothing more than repeat the press release word for word.

I also found out that Alan Calverd was an expert in 'photon beam therapy level graphic calorimetry', and discovered an e-mail address for him, but my message to that address bounced back as well. I ordered his article through the interlibrary loans service, but the response came back that it could not be located. Since this hardly ever happens, I was beginning to wonder whether the magazine, the article, and Mr Calverd weren't just a figment of some ingenious blogger's imagination.

They weren't, and eventually I located the article through contacts in the US academic library service, and even they had some difficulty. Instead of being, as I had anticipated, a lengthy scientific paper, it was a one page opinion piece, adorned with a photo of the author, a genial soul, blowing on a euphonium, who has 'a passion for good food, radical thinking and playing instruments in the bass clef'. Calverd's argument was simple: all animals generate carbon dioxide by breathing, and therefore we could make sizable reductions in our CO_2 emissions – around 21 per cent in his view – 'by abolishing all livestock and eating plants instead'.[4]

Of course it is not as though nobody has ever thought of this before. The same idea is expressed in facile bumper stickers on the lines of 'if you think CO_2 is a problem, hold your breath'. But most people discard the thought almost immediately because any carbon released into the atmosphere by a living animal is carbon that has previously and helpfully been withdrawn from the atmosphere by what is known as the short term carbon cycle. The carbon dioxide which any animal breathes out has to come from somewhere, and unless it has been fed fossil fuels, it comes from the vegetable matter it eats (or if it is a predator, from the vegetable matter consumed by the prey). Vegetable matter obtains its CO_2 from the atmosphere through photosynthesis, so when animals breathe out, they are merely returning to the atmosphere carbon dioxide that their food has previously withdrawn from it. The sum total of animals in existence at any one time in fact serves as a carbon sink, and if we got rid of them all, in which case carbon would return to the atmosphere through burning or through methane caused by rotting vegetation, there would arguably be slightly more carbon in the atmosphere. Calverd is no doubt fully aware of this; and any joker who makes headlines in the national press, and convinces virtually the entire vegan web-community to reprint a load of codswallop, is quite entitled to go 'oompah!'

Eighteen Per Cent

No sooner had I solved the Calverd mystery than another and far bigger 'meat causes global warming' wave burst over cyberspace. This time the figure quoted was not 21 per cent of CO_2 attributable to meat, but 18 per cent of all greenhouse gases. Typical headlines were 'Cattle Causes Most Warming'; and in particular 'Livestock More Damaging than Vehicles'.

3   www.lists.oppn.org/pipermail/org.opn.lists.garbanzo/2005-August/000405.html, now unobtainable.
4   Calverd, A (2005), 'A Radical Approach to Kyoto', *Physics World*, July 2005.

This time the perpetrator was not some wag; it was a team of top livestock economists and agronomists at the United Nations Food and Agriculture Organization (FAO), headed by Henning Steinfeld, in their wide-ranging overview of global livestock, called *Livestock's Long Shadow,* published in 2006.[5] Steinfeld and his friends have been writing papers for the FAO on livestock issues for a good many years, without ever attracting much public attention. But *Livestock's Long Shadow* was issued with a media fanfare and an FAO press release which began:

> Which causes more greenhouse gas emissions, rearing cattle or
> driving cars? Surprise! According to a new report published by the
> United Nations Food and Agriculture Organization, the livestock
> sector generates more greenhouse gas emissions as measured in CO_2
> equivalent – 18 percent – than transport.

The timing was perfect. The 18 per cent figure was picked up and bounced around cyber-space by ideological tub-thumpers of all persuasion. Vegans crowed with delight, right wingers ribbed Al Gore for eating meat and the motor industry breathed a sigh of relief. Normally responsible journalists reported the figure without questioning its provenance or its accuracy. Geoffrey Lean in the *Independent* began his article 'Meet the world's top destroyer of the environment. It is not the car, or the plane, or even George Bush: it is the cow ... Livestock are responsible for 18 per cent of the greenhouse gases that cause global warming, more than cars, planes and all other forms of transport put together.'[6]

Over the next few years the figure has been tossed around by various environmental celebrities as though it was an undisputed fact. Jonathon Porritt used it in advertisements for Compassion in World Farming; the Green Party MEP, Caroline Lucas, cited it in speeches and radio interviews (though after I had telephoned her about it, she did acknowledge publicly that the figure had been challenged). And in September 2008, Rajendra Pachauri, the chair of the Intergovernmental Panel on Climate Change, endorsed the 18 per cent figure at a talk in London hosted (once again) by Compassion in World Farming.

The IPCC is the Nobel prize-winning body of scientists whose word is normally taken as gospel on matters relating to global warming. Virtually all of its statistics are hedged by provisos, subject to intensive peer review and backed up by volumes of impenetrable technical data, so it was strangely cavalier of Mr Pachauri to be volunteering a figure which far exceeded most other estimates made by reputable scientific organizations, including the IPCC itself, which claims that the whole of agriculture only contributes 10–12 per cent of global GHG emissions.[7] The World Resources Institute's global warming flow chart, which is based on 1996 IPCC statistics, allocates just 5.1 per cent of global greenhouse gases to 'livestock and manure'.[8]

This discrepancy does not necessarily mean that one figure is wrong and the other right. It is mainly a matter of which sector various kinds of emissions are allocated to. For instance, the widely quoted figure of around 14 per cent of global emissions attributed to transport, only includes direct fossil fuel consumption; emissions caused by vehicle manufacture, road building, airports, oil extraction,

5   Steinfeld, H *et al* (2006), *Livestock's Long Shadow*, FAO.
6   Lean, Geoffrey (2006), 'Cow "Emissions" More Damaging to Planet than CO_2 from Cars', *The Independent*, 10 December 2006.
7   Metz, B *et al* (eds) (2007), *Climate Change 2007: Mitigation. Contribution of Working Group to the Fourth Assessment Report of the IPCC*, IIPCC, Cambridge University Press.
8   Baumert, K *et al* (2005), *Navigating the Numbers: Greenhouse Gas Data and International Climate Policy*, World Resources Institute, 2005, http://www.wri.org/publication/navigating-the-numbers

the electricity to power trains and a host of other costs are slotted under 'industry' or 'iron and steel' or 'cement' or 'commercial buildings'. If you include all these emissions in the transport sector then the figure would rise considerably; I have been unable to find a convincing assessment, but it is easy to compute from existing figures that the oil extraction and refining alone would add 1.5 per cent to the global transport figure. A Canadian study estimated that whereas fuel burnt in vehicles represents 31 per cent of all Canadian GHG emissions, once oil refining, vehicle manufacture and road building are included, the figure rises to 52 per cent.[9]

'There are many assumptions that one needs to make when quantifying emissions,' states Gidon Eshel, a US professor of environmental studies, who has estimated that global livestock emissions account for about 10 or 11 per cent of US emissions. 'It's not that any one assumption is correct. Almost all of them are defensible.'[10] The assumptions that different analysts make will (whether they mean to or not) reflect their ideological position. Statistics are never black and white, they always come in a certain colour.

Amidst the publicity about *Livestock's Long Shadow*, few people bothered to read the 400 page report to examine what assumptions the main author, Henning Steinfeld, and his team had made to arrive at their figure of 18 per cent. It doesn't take a very long perusal of these figures to see that the FAO find fossil fuel use associated with livestock to be of negligible importance, whereas the main 'organic' sources account for 97 per cent of all livestock emissions. Less than three per cent of livestock emissions are caused by burning fossil fuels for nitrogen fertilizer, farm machinery and processing. Three gases are responsible in roughly equal degrees for all the rest: a third of the total consists of carbon dioxide from deforestation, another third methane produced by animals' guts, and the remainder nitrous oxide from manure and fertilizer.

Moreover Steinfeld and his colleagues make another distinction. In subsidiary columns in their main table they show that 70 per cent of these emissions are attributable to extensive livestock (ie grazing) and only 30 per cent are attributable to intensive livestock (industrial farming).

Perhaps, given what has already been revealed about *Livestock's Long Shadow* in Chapter Six, you can detect an agenda emerging here. The conclusion the FAO economists are heading for is not that we should all eat less meat, still less that North Americans should stop stuffing cows and pigs with grain that hungry people could eat. On the contrary, they anticipate that meat consumption will more than double, from 229 million tonnes in 2001, to 465 million tonnes in 2050. The conclusion of the report is that:

> It is hard to see an alternative to intensification of livestock production
> … The principle means of limiting livestock's impact on the
> environment must be to reduce land requirements for livestock. This
> involves the intensification of the most productive arable and grassland
> used to produce feed or pasture.

In other words reduce grazing and move towards factory farming. This will come as no surprise to long term observers of the FAO or of Henning Steinfeld.

---

9   Gagnon, Luc (2006), *Comparing Energy Options: Greenhouse Gas Emissions from Transportation Options*, Hydro Quebec; http://www.hydroquebec.com/sustainable-development/documentation/pdf/options_energetiques/transport_en_2006.pdf For similar figures on USA see: EPA (2006) *Greenhouse Gas Emissions from the U.S. Transportation Sector 1990-2003*, March 2006, http://www.epa.gov/otaq/climate/420r06003.pdf

10   Adler, Ben (2008), 'Are Cows Worse than Cars?', *The American Prospect*, 3 Dec 2008, http://www.prospect.org/cs/articles?article=are_cows_worse_than_cars

Back in the early 1990s, *The Ecologist* waged a fierce campaign against the FAO claiming that its programme was targeting vulnerable peasants, and blaming them for environmental damage which was rooted in the pressures upon their livelihoods coming from industrial farmers. In another 1998 FAO document, Steinfeld wrote:

> We cannot afford the common nostalgic desire to maintain or revive mixed farming systems with closed nutrient and energy cycles … To avoid overuse of immediate natural resources, mixed farmers and pastoral people alike need to substitute them with external inputs … Grain-fed livestock should be promoted 'even though, prima facie, there may be competition between food and feed uses of some commodities … The trend of further intensification and specialization of demand-driven production is inescapable. Attempts to change the direction are doomed to failure.[11]

It is the same argument as that put forward by the Global Opponents of Organic Farming, whom we met in Chapter 8: 'spare, don't share'. Maximize production and human impact on some land so as to relieve pressure elsewhere. Its is an approach that inevitably tends to favour urbanized, grain-fed pigs and poultry over wide-ranging fibre-eating cows and goats.

Once you appreciate that this is where Steinfeld is coming from it is easier to understand where he is going. Throughout much of *Livestock's Long Shadow*, the assumptions made, the criteria chosen and the matters left out are those that one would make and choose and leave out if one wanted to show that intensive livestock farming was less harmful to the climate than extensive livestock farming, or as the authors themselves put it, 'by far the largest share of emissions come from more extensive systems where poor livestock holders often extract marginal livelihoods from dwindling resources'.

There are many people within the FAO who spend their working lives trying to assist and defend peasant livestock farmers, so it is not surprising to hear that the stigmatization of their 'marginal livelihoods' in *Livestock's Long Shadow*, has aroused some anger within the organization. This reaction is almost certainly responsible for the tenor of FAO's 2009 State of Food and Agriculture Report (SOFA), the bulk of which is focussed on the need to protect peasant livelihoods from rampant industrialization of the meat industry. The SOFA report (at least the draft version) does much to redress the balance in favour of smallholders, though it still regards the march of industrialization and urbanization as 'unstoppable'. And it repeats, without any re-examination, the livestock GHG figure of 18 per cent, which in the rest of this chapter I shall argue is misleading and ideologically biased in favour of intensive farming.

CO$_2$ from Deforestation

The FAO's figures are laid out in Table 6. Take first the 34 per cent of livestock emissions – more than six per cent of all the world's greenhouse gases – attributed to deforestation. When areas of forest are cleared for any reason, very large quantities of carbon stored in the timber are released into the atmosphere, mostly from burning. According to the FAO, only in Latin America is land cleared on any scale for livestock (in Africa the tsetse fly makes this difficult),

---

11   Fresco, L and Steinfeld, H (1998), *A Food Security Perspective to Livestock and the Environment*, FAO 1998, http://www.fao.org/WAIRDOCS/LEAD/X6131E/X6131E00.HTM

Table 6: Role of Livestock in Carbon Dioxide, Methane and Nitrous Oxide Emissions

Gas	Source	Mainly related to extensive systems (billion tonnes CO_2 eq).	Mainly related to intensive systems (billion tonnes CO_2 eq).	Percentage contribution to total animal food GHG emissions	Percentage contribution to total anthropogenic GHG emissions
CO_2	**Total anthropogenic CO_2 Emissions**	**24(~31)**			
	Total from livestock activities	**~0.16 (~2.7)**			**~0.4 (~6.75)**
	N fertilizer production		0.04	0.6	0.11
	On farm fossil fuel, feed		~0.06	0.8	0.14
	On farm fossil fuel, livestock-related		~0.03	0.4	0.07
	Deforestation	(~1.7)	(~0.7)	34	6.04
	Cultivated soils, tillage		(~0.02)	0.3	0.05
	Cultivated soils, liming		(~0.01)	0.1	0.02
	Desertification of pasture	(~0.1)		1.4	0.25
	Processing		0.01 - 0.05	0.4	0.07
	Transport		~0.001		
CH_4	**Total anthropogenic methane emissions**	**5.9**			
	Total from livestock activities	**2.2**			**5.5**
	Enteric fermentation	1.6	0.20	25	4.44
	On farm fossil fuel, feed	0.17	0.20	5.2	0.92
N_2O	**Total anthropogenic N_2O emissions**	**3.4**			
	Total from livestock activities	**2.2**			**5.5**
	N fertilizer application		~0.1	1.4	0.25
	Indirect fertilizer emission		~0.1	1.4	0.25
	Leguminous feed cropping		~0.2	2.8	0.5
	Manure management	0.24	0.09	4.6	0.87
	Manure application/deposition	0.67	0.17	12	2.13
	Indirect manure emission	~0.48	0.14	8.7	1.54
Total emissions from livestock activities		**~4.6 (~7.1)**			**(18%)**
Grand total of anthropogenic emissions		**33(~40)**			**(100%)**
Total extensive v. intensive emissions		**3.2 (~5.0)**	**1.4(~2.1)**		
Percentage of total anthropogenic emissions		**10 (~13)%**	**4 (~5)%**		

Taken from *Livestock's Long Shadow*, except the column on the right has been calculated by SF from the other data. Values in brackets are or include emissions from land use, landuse change and forestry (LULUCF). Relatively imprecise estimates are presented by a tilde. The FAO's publicized figure of 18% for total livestock contribution to total anthropogenic GHGs is actually 17.75% if it is reached by dividing 7.1 by 40.

Credit: FAO

and two thirds of this clearance is for beef and most of the rest for soya bean production. Nobody denies that this is a grave problem. Nobody denies that livestock are heavily implicated. But most other climate analysts do not allocate these emissions to 'livestock' because to do so distorts the picture. For a number of reasons they are normally allocated under a sector cumbersomely entitled Land Use, Land Use Change and Forestry (LULUCF).

(i) Deforestation has declined since *Livestock's Long Shadow* was written

The first of these reasons is that the amount of deforestation changes from year to year, often quite significantly, and so any calculation has to be based on projections, which can be wrong, and in the case of *Livestock's Long Shadow*, are proving to be wrong. The FAO anticipate that total deforestation in Latin America over the period 2001 to 2010 will average 3.7 million hectares a year – exactly what it was over the previous decade. About 60 per cent of the 1990s deforestation, or 2.3 million hectares, was in the Brazilian Amazon.[12] When Steinfeld and his colleagues were preparing their report, deforestation in Brazil was around this rate, peaking at 2.7 million per year in 2004. But since then deforestation rates have dropped dramatically, and in 2008/2009 there were just 0.7 million hectares cleared in Brazil – equivalent to less than a third of the amount projected by the FAO.[13] If deforestation rates remain this low, it will mean that the six per cent of global warming attributed by the FAO to rainforest beef and soya is closer to two per cent, and their 18 per cent figure drops to 14 per cent at one fell swoop – even though global meat consumption is continuing to rise over the same period.

Of course it is by no means certain that deforestation will remain at the current level or continue to decline; annual rates have been astonishingly volatile in the past. But there are reasons for hoping that it may tail off. The surge between 2002 and 2004 was due to peculiar economic conditions, including the existence of foot and mouth in Europe while it was absent in Brazil. By 2006 the Woods Hole Research Centre was reporting that the conduct of the increasingly respectable beef industry in the Amazon was veering towards regulation and even conservation.[14] And in October 2006, over 90 per cent of the soy processing industry signed an agreement not to source beans from recently cleared land.[15]

The soy ban has been effective, helping to halve the rate of deforestation, but it has done nothing to stop the ranchers, who are now visibly responsible for most of the new clearings. Then in 2009, in response to pressure from Greenpeace, several supermarket chains including Walmart and Carrefour announced that they would suspend contracts with beef suppliers found to be involved in Amazon deforestation. To back that up the Brazilian government filed law suits against the cattle industry for environmental damage; and in October 2009, four of Brazil's largest beef trading companies agreed to a moratorium on buying cattle from newly deforested land.[16] The situation now looks hopeful, not least because enforcing against deforestation is the most obvious way for Brazil to meet any commitments it might have to make in ongoing climate negotiations. But pressure for land in the south could easily reverse the progress that has been made.

12   The figure of 3.7 is equivalent to the annual rate of change in South America in 1990-2000. Brazil contains 60 per cent of the Amazon and in that decade Brazil was responsible for 62 per cent of deforestation. FAO (2003), *The State of the World's Forests*, found at http://www.mongabay.com/deforestation_tropical.htm

13   Mongabay (2009a), *Brazil Releases Official Amazon Deforestation Figures for 2009*, Mongabay.com, 13 November 2009, http://news.mongabay.com/2009/1113-brazil_amazon_deforestation.html

14   Nepstad, D *et al* (2006), 'Globalization of the Amazon Soy and Beef Industries: Opportunities for Conservation', *Conservation Biology*, 2006. http://www.whrc.org/policy/COP/Brazil/Nepstad_et_al_2006_Cons%20Biol.pdf

15   Mongabay (2008a), *Amazon Soy Ban Seems to be Effective in Reducing Explicit Deforestation*, Mongabay.com, 3 April 2008. http://news.mongabay.com/2008/0403-amazon_soy_ban.html

16   Mongabay (2009b), *Brazilian Beef Giants Agree to Moratorium on Amazon Deforestation*, Mongabay.com, 7 Oct 2009, http://news.mongabay.com/2009/1007-greenpeace_cattle.html

(ii) Deforestation is not directly related to levels of meat consumption

A second reason why emissions from deforestation are not normally added to the livestock account is this: they do not reflect annual livestock production, but annual expansion of the livestock industry, which is not the same thing. To include them is to confuse capital with income, thereby distorting the arithmetic and making the percentage figure meaningless. That is why deforestation has declined, while meat consumption has increased; the two are not directly related. If forest is cleared for beef or for soya production, this is a one-off event, and the emissions resulting from the burning of biomass do not recur annually. If we succeed in halting deforestation in the near future, the beef and soya currently sourced from cleared forest areas could be produced for years to come, indefinitely, without any further carbon emissions – in other words the world would be enjoying the same amount of meat, for less than 12 per cent of global GHG emissions.

True, if we restored some of the forest by reducing the number of livestock, then there would be an increase in vegetation which would act as a carbon sink, and so we can potentially assign 'carbon sinks forgone' emissions to all livestock. But this is true of nearly everything that humans do on this earth. If you are going to account for potential carbon sinks in deforested areas the size of Yorkshire in Brazil, then you have to account for Yorkshire as well, and add a 'carbon sinks forgone' toll to the acreages of wheat, suburbia and motorway that make up a considerable part of that county – a matter that people in Brazil are keen to point out in climate negotiations.

(iii) The causes of deforestation are complex

The third reason why attributing emissions from deforestation to livestock is problematic is that the causes of deforestation are too complex to assign with any degree of statistical confidence. The popularly held view, which equates piles of hamburgers with football fields of felled forest, is supported by figures showing that at any one time about three quarters of the cleared forest is being ranched for beef.[17] But just because livestock are in the vanguard of deforestation, that does not mean they are causing it. A common view among analysts for many years has been that deforestation in the Amazon is not primarily driven by beef farmers, but that cattle are used because they are the most efficient way of keeping land clear and the most visible way of asserting title. As Steinfeld himself wrote in 1996:

> There is increasing evidence that livestock ranching in deforested areas is the most obvious symptom of a much more complex degradation process, with a variety of driving forces … The key phenomenon was the land grab in the rainforests, which occurred as a result of a general increase in prices. A large part of the expansion of pasture land may therefore have had more to do with land speculation than with cattle raising'[18]

In 1996, Steinfeld took about 1,000 words and cited ten different papers to prove that cattle ranching was not a root cause of deforestation. In *Livestock's Long Shadow,* he makes a complete volte-face citing a 2004 World Bank report to show that 'deforestation for cattle ranching has proven to be profitable in

17   Greenpeace Brazil (2009), *Amazon Cattle Footprint, Mato Grosso: State of Destruction*, 29 January 2009, http://www.greenpeace.org/international/press/reports/amazon-cattle-footprint-mato

18   De Haan, C V, Steinfeld, H and Blackburn, H (1996), *Livestock and the Environment: Finding a Balance*, FAO 1996, Chapter 2 'Driving Forces'; http://www.fao.org/docrep/x5303e/x5303e00.htm

itself'.[19] The projections which conclude that two thirds of deforestation over the first ten years of the century will be due to ranching are derived from a paper co-authored by himself.[20] It is certainly true that the steps made to eradicate foot and mouth have increased the amount of Brazilian beef going for export, and hence the profitability of the industry. However more recent papers note that the surge in Amazon deforestation levels, peaking in 2004, was exacerbated by an increase in soya production, one study concluding that 'the growing importance of larger and faster conversion of forest to cropland [mainly soya] defines a new paradigm of forest loss in Amazonia and refutes the claim that agricultural intensification does not lead to new deforestation'.[21]

Early in 2008, observers began to express concern at a sudden spurt in forest deforestation, which threatened to undo the progress made over the previous three years, and was ostensibly due to beef ranching, since the moratorium on soya planting remained in place. This surge, which fortunately only lasted a few months, was in response to the rise in commodity prices, and most observers concluded that pressure for soya in previously cleared areas was pushing ranchers further into the forest. The conservation group Mongabay wrote:

> Typically rainforest lands are cleared for low-intensity cattle ranching then sold to soy producers some two to three years later. With land prices appreciating and soy cultivation expanding in previously cleared areas and the neighbouring *cerrado* grassland ecosystem, ranching is increasingly displaced to frontier areas, spurring deforestation ... 'The powerful Brazilian soy lobby has been a driving force behind initiatives to expand Amazonian highway networks, which greatly increase access to forests for ranchers, farmers, loggers and land speculators', said Dr William F Laurance, a researcher at the Smithsonian Tropical Research Institute.[22]

Over the past ten years the fluctuations in the levels of deforestation have shown a much closer correlation to movements of soya prices than to movements of beef prices.[23] The increase in soya production prior to 2003 (as we have already seen on p 46) was largely livestock-driven – to supply the EU with high protein animal feeds that it could no longer source from slaughterhouse wastes because of the BSE fiasco. But the more recent spike in deforestation in 2007–2008 was driven by both oil and meal prices more than doubling, with oil prices arguably rising more than meal in Brazil, whereas meat prices only rose by 10 to 20 per cent in 2007 and 2008. Moreover most commentators agreed that the rise in meal and oil prices was at least in part caused by the demand for biofuels, especially from US corn.[24] Dr Laurance observes:

> American taxpayers are spending $11 billion a year to subsidize corn producers – and this is having some surprising global consequences.

19   Margulis, S (2004), *Causes of Deforestation in the Brazilian Amazon*, World Bank.

20   Wassenaar, T *et al*, 'Projecting Land Use Changes in the Neo Tropics: the Geography of Pasture Expansion into Forest', *Global Environmental Change*, 17, pp 86-104, 2006.

21   Morton, D *et al*, 'Cropland Expansion Changes Deforestation Dynamics in the Southern Amazon', PNAS, Sept 2006, http://www.pnas.org/cgi/content/abstract/103/39/14637

22   Mongabay (2008b), *Amazon Soy Ban Seems to be Effective in Reducing Explicit Deforestation*, Mongabay.com, 3 April 2008, http://news.mongabay.com/2008/0403-amazon_soy_ban.html

23   Butler, Rhett A (2007a), *Amazon Fires Amongst Worst Ever*, Mongabay.com, Oct 2007, http://news.mongabay.com/2007/1021-amazon.html ; Butler, Rhett A (2007b) *Influence of Soy Prices on Deforestation in the Amazon*, http://photos.mongabay.com/07/soy_prices_defor-568.jpg

24   ABIOVE, *Soybean Complex Average Prices*, Brazilian Association of Soybean Industries, 2009 http://www.abiove.com.br/english/cotacoes_us.html; Butler, Rhett A (2008), *U.S. Biofuels Policy Drives Deforestation in Indonesia, the Amazon*, mongabay.com January 17, 2008.

Amazon fires and forest destruction have spiked over the last several months, especially in the main soy-producing states in Brazil. Just about everyone there attributes this to rising soy and beef prices .... We're seeing that these predictions – first made last summer by the Woods Hole Research Centre's Daniel Nepstad and colleagues – are being borne out. The evidence of a corn connection to the Amazon is circumstantial, but it's about as close as you ever get to a smoking gun.[25]

There is also evidence that the spread of sugarcane plantations in Brazil, again mostly to supply biofuels, has been pushing small farmers and beef ranchers towards the Amazon.[26] Much of the 2008 increase in deforestation, it seems, was not driven by the need to feed beef, but the need to fuel motor cars.

The FAO in *Livestock's Long Shadow* claim that 65 per cent of deforestation is for beef pasture and 27 per cent for soya. But only half of the area cleared for soya is attributed to livestock, because nearly half the value of the soya is in the oil, which doesn't go to animals. If you take the view (as Henning Steinfeld did in the 1990s) that cattle are just the foot soldiers of deforestation and that now demand for soya and other crops is the main driving force, then the amount of deforestation attributable to livestock would go down accordingly. At the current reduced rate of deforestation, livestock would be responsible for little more than one per cent of the world's CO_2 emissions, rather than the six per cent claimed by the FAO.

If you conclude that soya cultivation is the main culprit, it also follows that many more of the emissions would be attributable to intensive agriculture, and many less to extensive cattle farmers, a matter I will return to later. Blame cattle and you are blaming a handful of cowboys and the mainly Brazilian consumers who buy the beef; blame soya and you are blaming global agribusiness – not least British agribusiness, which imports 90 per cent of its soya from Brazil and Argentina.[27] This decision, how to partition the responsibility for deforestation between the driving force and the advance guard, has a direct bearing on one third of all the carbon emissions that the FAO attribute to livestock. Yet it is as much an ideological decision as it is a scientific one, and as such it provides an unreliable basis for attaching a percentage of the blame.

(iv) Rainforest cattle are not typical

If, however, you agree with the FAO's view that two thirds of rainforest clearance is due to cattle, then there is a fourth reason to question their calculation. The rainforest cattle which they hold responsible for a quarter of all livestock GHG emissions represent a very small proportion of the world's livestock: perhaps five per cent of the world's cattle, and about 1.5 per cent of the world's animal products.[28] The authors of *Livestock's Long Shadow*, when it was published, were

25   Butler, Rhett A (2007c), *U.S. Corn Subsidies Drive Amazon Destruction*, Mongabay.com, 13 December 2007, http://news.mongabay.com/2007/1213-amazon_corn_sub.html

26   Sparovek *et al* (2009), 'Environmental, Land-Use and Economic Implications of Brazilian Sugar-Cane Expansion', 1996-2006. *Mitig Adapt Strateg Glob Change*, 14: 285-98.

27   According to the Brazilian government only two per cent of rainforest beef is exported: Eduardo Roxo (Brazilian Ambassador to UK), 'Brazil's Actions are Combatting Amazon Deforestation', letter, *The Guardian*, 5 June 2009; UK figures: DEFRA and Food Standards Agency, *GM Crops and Foods: Follow-up to the Food Matters Report* by Defra and the FSA, August 2009.

28   According to Greenpeace there are 56 million cows in the Brazilian Amazon, which is 36.6 per cent of the entire Brazilian herd (2006 data). Brazil produced 7.9 million tonnes of beef in 2007 so on a pro rata basis that is 2.9 million tonnes, or 4.7 per cent of world beef production, or about one per cent of world meat production. I cannot find any figures for rainforest beef from other Latin American countries, but judging by FAO figures for total production in countries with rainforest, it is unlikely to be more than a third as much as in Brazil. Greenpeace (2009), *Slaughtering the Amazon*,

alone in choosing to take the massive emissions from this tiny rogue sector, which they acknowledge is 'atypical', and using them to bump up, by around 30 per cent, a figure which they then proceed to publicize as typical emissions from livestock throughout the world.

Some might argue that it is valid to tar all meat with the rainforest brush in this way because demand for Brazilian beef reflects increasing global demand for meat in general. According to Greenpeace, the recent rise in rainforest beef exports is partly a result of increasing demand for beef from the EU, particularly the UK and Russia. The Soil Association, however, points out that beef production in the UK has declined by 20 per cent in recent years (because fewer calves are coming out of an increasingly intensive dairy industry), implying that it may not be demand for meat, but price advantage that is driving the increase in exports. They therefore claim that increasing beef production in the UK would reduce pressure upon the Amazon.

Perhaps it is legitimate to cast the blame for deforestation upon meat in general, but if so, what purpose does that serve? The most direct solution is not to lower the consumption of meat in general, but to put an end to the consumption of the offending items. Whatever the FAO's motives, some vegan and vegetarian campaigners are only too happy to propagate the 'hamburger connection' as a reason for giving up meat. But to invoke the emissions from rainforest beef as an argument for halting meat consumption, makes as much sense as arguing that we should boycott all vegetable oil, because palm oil is responsible for deforestation in Indonesia, and soya oil responsible in the Amazon.

* * *

None of the four points above should be taken to suggest that deforestation is not a grave problem; nor that livestock are not in large measure responsible. I have raised them to indicate how misleading it can be to apply a figure like six per cent to the emissions from deforestation attributable to livestock. There is a direct relationship between the amount of petrol a consumer puts in his (or her) car and the amount of carbon he emits into the atmosphere. There is no such relationship between the amount of meat he eats and the amount of deforestation he is responsible for. A consumer who switched from a diet of New Zealand butter, English lamb and corn fed pork to one based on margarine and meat analogues made from Brazilian soya would probably be increasing his contribution to deforestation, not diminishing it.

The FAO are well aware of these difficulties, and acknowledge them discreetly in the text. They even put brackets around the LULUCF figures in their table, suggesting that they should be considered separately, which means that their 18 per cent is also in brackets and that the total printed at the bottom of the table without brackets is the equivalent of 11.66 per cent of manmade GHGs. Why, one wonders, have they given such prominence to the higher figure of 18 per cent in their publicity?

Nitrous Oxide

Another third of the FAO's 18 per cent – representing 5.5 per cent of all the world's GHGs – derives from nitrous oxide emissions. Nitrous oxide, or N_2O, is a potent greenhouse gas which leaks out of the agricultural nitrogen cycle at various stages and finds its way into the atmosphere, where it remains for

2009; FAO SOFA (2009), *op cit.*

a disturbingly long time. When nitrogen fertilizers are applied to crops about one per cent of the nitrogen ends up as N_2O, and this rate is assumed by the IPCC to be the same for synthetic fertilizers, organic manure and green manure. However the storage of manure and the grazing of animals also leads to N_2O emissions; and there are further 'indirect' N_2O emissions as nitrogen leaches or spreads into waterways and the wider environment.

Table 6 shows how the FAO have derived their N_2O emissions from these various sources, using IPCC 1997 guidelines. There are some impressive looking equations in these guidelines, up to seven lines long, though on scrutiny these are not nearly as daunting to the non-scientist as they first appear. They have since been updated in 2006, and some default 'emission factors', key to the FAO's calculation, have been lowered significantly, suggesting that their N_2O figures are now a considerable overestimate.[29] Bear in mind that all the FAO's figures are based on IPCC emission factors that are continually being reassessed as scientific opinion shifts about matters that are still poorly understood.

But there are other reasons to suspect that the total of 4.6 million tonnes of N_2O emitted by, through or on behalf of animals, equivalent to 5.5 per cent of the world's GHGs, may not be an accurate reflection of the toll exacted on the world's climate by livestock. We also have to examine the 'rebound' emissions, or what economists call 'opportunity costs'. If we got rid of all our livestock, would the 4.6 million tonnes of N_2O disappear?

In Table 6 the figures given for both direct and indirect N_2O emissions from fertilizer application refer only to synthetic fertilizers applied to animal feed crops, representing about 20 per cent of all synthetic nitrogen. If the human race were to get rid of its domestic animals, or even just those which are fed on chemically grown feed crops, then those emissions would be eliminated.

But the much larger figures given for 'manure application and deposition' and 'indirect manure emission' refer to manure applied not only to feed crops but also, and probably mainly, to human food crops. If we dispensed with livestock we would have to find another source of nitrogen for these food crops, either synthetic nitrogen or green manures, which would (according to the IPCC methodology) be emitting equivalent amounts of N_2O both directly and indirectly.

The same will be true of many of the leguminous feed crops which, in the case of lucerne and clover, produce a surplus of nitrogen in their roots and residues available for subsequent food crops. And it will be true of a proportion of the grazing animals, about two thirds of which are reared in mixed farming systems, where pasturing builds up fertility that can subsequently be used for cropping.[30] You can take the animals out of these systems but a proportion of the nitrogen is necessary for human food production, whether or not it passes through the gut of an animal.

---

29   The FAO do not make a clear distinction between emissions from manure and emissions from grazing; in regard to the indirect emissions they remark that 'this methodology is beset with high uncertainties, and may lead to an overestimation because manure during grazing is considered'.

The FAO's figures are based on IPCC 1997 methodology. The updated 2006 methodology lowers a number of default emission factors quite significantly including EF1 (relating to the amount of N_2O per kilo of N fertilizer or manure applied to the land) from 1.25 to 1.00; and EF5, for nitrogen leaching in run-off water from 0.025 to 0.0075, a reduction of 70 per cent. On the other hand the FAO use a Global Warming Potential for methane of 21, which has been superseded, for research purposes, by a GWP of 23, and in 2007 of 25. An updated version of the FAO's table would therefore display marginally increased emissions from methane and considerably lower emissions from N_2O. It would also need to account for an increase in total numbers of livestock, and in fossil fuel emissions.

30   FAO (2006), *op cit. 5*, p 53.

On top of this, since animal products supply a third of all the world's protein, if we eliminated livestock, we would have to produce half as much again of vegetable protein crops to replace the meat foregone (though less if we only aimed to replace the energy in the meat forgone).[31] This means that we would still need to apply the nitrogen fertilizer that we had previously been spreading on feed crops onto replacement food crops; and we would need to continue to produce leguminous crops such as soya beans for human consumption.

And in addition to all of this, animal faeces are not going to disappear from the face of the earth if we stop keeping domestic animals. Figures in *Livestock's Long Shadow* suggest that currently three tonnes of volatile nitrogen are produced by wild animals for every 23 tonnes produced by domestic animals.[32] It seems more than likely that if humans stopped grazing livestock, then the population of wild animals would increase. Many might regard this as a good thing, but it would eat into the N_2O emissions savings, though probably not enormously.

In short, many of these nitrous oxide emissions are a consequence not of livestock, but of an agricultural ecology in which livestock plays an incidental role, transmitting rather than creating nitrogen. That is why the World Resources Institute chart (see Fig 3 p 177), for example, only classes a small fraction of its nitrous emissions under 'Livestock and Manure', and places the majority of them under the heading 'Agricultural Soils'.

Perhaps it would be possible to work out what would be the net savings in N_2O emissions if we dispensed with livestock and made global agriculture stock free – but I am not going to do anything more than hazard a broad guess. To summarize, if we went stock free,

(a) we would still need at least the area currently devoted to feed crops for our own nourishment;

(b) we would need to substitute synthetic nitrogen or green manures for perhaps two thirds of the manure we currently use;

(c) there would be a small, perhaps negligible, amount of increased emissions from wild animals.

Taking this into account, together with the fact that the IPCC have recently lowered some of the default emission factors, I suspect that the net N_2O emissions from meat might only be around half as high as those given by the FAO.

Methane

Similar issues arise when we consider the final third of the FAO's 18 per cent of GHGs (5.4 per cent to be precise) which consist of methane, mostly emanating from the digestive systems of ruminants. In one sense it is the most crucial of the three greenhouse gases because it is intrinsic to farm animal metabolism, and less a consequence of the way they are farmed. Sheep are normally kept indoors overnight in France, and anyone who has ever slept the night in a *bergerie* containing a 500 strong flock will be familiar with the music of ovine flatulence. The soothing cadence of mastication, farts, belches and showers of piss never ceases from dusk till dawn; only when there is cause for alarm – usually a lamb lost from its mother – will an animal resort to using its vocal chords. Digestion, in the world of the ruminant, seems to serve as a form of peaceful communication.

31 Ibid, p 34.
32 Ibid, p 104. The figures actually relate to ammonia emissions from manure.

Methane is integral to the metabolism of the ruminant, and indeed to the entire biological world. It is released under anaerobic conditions, and most animals, insects, plants and trees emit some methane at some point in their life or death cycle. Ruminant animals are reputed to be the second largest source, after natural wetlands – 'marsh gas' being a traditional term for methane, and the 'will o' the wisp' a manifestation.

Methane also happens to be a potent greenhouse gas whose levels in the atmosphere have risen by about 150 per cent since pre-industrial times. After carbon dioxide it is second highest element in the GHG budget, and you will often read that 'methane is a potent greenhouse gas, 25 times as powerful as carbon dioxide'. While, gram for gram, this is an agreed exchange rate, in most contexts this is about as useful as saying that a tonne of iron weighs more than a tonne of hay. The value of the two gases is measured on a common scale, known as Global Warming Potential (GWP) using units known as CO_2 equivalents (CO_2eq). Currently every gram of methane is viewed as having a GWP of 25 – ie it has the same effect as 25 grams of CO_2.

Human induced methane emissions are currently thought to be responsible for between 14 and 16 per cent of all man-made greenhouse gas emissions – a decline from a figure of 20 per cent three decades ago, but still quite a chunk.[33] However, there is considerable disagreement amongst academics as to which human activities are responsible, and the IPCC, the authority for climate statistics, unhelpfully doesn't commit itself on this matter. In the 1980s and 1990s, the majority of studies estimated that ruminants were responsible for between 16 and 24 per cent of man-made methane emissions[34] but this estimate has gone up, partly because methane emissions from other sources have gone down. The UK Environmental Change Institute state that livestock are responsible for 23 per cent of global human-made methane emissions, whereas the US Environmental Protection Agency reckons the figure is 34 per cent.[35] The EPA's figure is the equivalent of 5.4 per cent of all man-made emissions, which is similar to that given in *Livestock's Long Shadow*.

To eliminate the bulk of these animal-sourced methane emissions, the human race would not need to eliminate all its livestock; just its cows, sheep, goats and other ruminants – the fibre eaters – representing about half the nourishment derived from livestock. This is because methane is a byproduct of digestive systems that have evolved to process protein from high fibre diets. So whereas beef cattle produce anything from 18 to 66 kg of methane per year, and dairy cattle up to 115 kg, pigs produce just 1 to 1.5 kg per year and humans a pathetic 50 grams. Horses, which are not ruminants, but can digest grass thanks to their long colon, are reported as averaging 18 kg of methane, so if we decided to get rid of cows because of their methane emissions, we perhaps ought to dispense with horses as well.[36]

33   IPCC (2007), *Climate Change 2007, Synthesis Report*, http://www.ipcc.ch/pdf/assessment-report/ar4/syr/ar4_syr_spm.pdf; EPA, Methane to Markets, 2008 http://www.epa.gov/methanetomarkets/docs/methanemarkets-factsheet.pdf; Stern, N (2007) *The Economics of Climate Change*, Cambridge, p 223.

34   See, for example, a range of estimates given in Environmental Change Institute (2006), *Methane UK*, Chapter 2; http://www.eci.ox.ac.uk/research/energy/methaneuk.php

35   *Ibid.*, citing Khalil, M (2000), *Atmospheric Methane: Its Role In The Global Environment*. The EPA in 2009 raised their figure from 32 per cent to 34 per cent. EPA (2009), *The Methane to Markets Partnership*, US Environmental Protection Agency. http://www.epa.gov/methanetomarkets/pdf/2009-accomplish-report/the_methane_to_markets_partnership.pdf

36   Crutzen, P J (1986), 'Methane Production by Domestic Animals', *Tellus B Chemical and Physical Meteorology*, 1986, Vol 38, p 271.

But once again, as with nitrous oxide, before coming to any conclusions about the environmental impact of these grass eating animals, we need to examine the opportunity costs of eliminating beef and dairy from our diets.

Without beef, sheep and dairy protein, we would need to consume an extra 25 per cent of our current production of grains, pulses and vegetables to compensate. The IPCC estimates that plant crops produce 17 per cent of current man-made methane emissions, exactly half of the amount emitted by livestock, so in theory this would increase correspondingly, eating into the methane savings made by getting rid of our ruminants.

However most of this (ten out of the 17 per cent) comes from rice grown in paddy fields, which produce methane, like natural wetlands. Twenty years ago most analysts were of the view that global rice cultivation produced more methane than the beef and dairy industries, for approximately the same global total of protein (though this generated less media interest than cows, probably because rice neither farts nor belches).[37] But in recent years, estimates for the amount of methane from rice have reduced by about half. According to M Khalil and M Shearer, this is because China over the last two decades has reduced its annual methane emissions from rice from 30 million tonnes per year to ten million tonnes, mainly by replacing organic fertilizer with mineral fertilizers.[38] This raises the question of what has happened to its organic fertilizers, particularly since the amount of available manure must have risen enormously in China over the last 20 years. Have some of these methane emissions been transferred from the rice sector to the animal sector through inefficient recycling of nutrients? Or do they appear as nitrous oxide emissions through the application of manure to dryland crops? Khalil and Shearer don't say, although they do acknowledge that the chemically fertilized rice is probably emitting more nitrous oxide.

Even after these improvements in rice performance, each kilo of rice now is said to generate about 100 grams of methane, while a kilo of milk from a productive dairy cow generates about 13 to 26 grams of methane (so rice pudding is to be avoided).[39] Since the protein content of milk is about a third of that of rice, weight for weight, and the energy about a tenth, the milk provides somewhat more protein per unit of methane than the rice, but less energy. There would not be a lot to be gained by eliminating dairy production and replacing it by rice

37   Bolle (1986) allocates 18 per cent of emissions to ruminants and 31 per cent to paddy fields. Reeburgh (1993) assigns 16 per cent to animals (80 million tonnes) and 20 per cent to rice (100 million tonnes). Bolle, H J *et al* (1986), 'Other Greenhouse Gases and Aerosols: Assessing their Role for Atmospheric Radiative Transfer', in Bolib, B *et al* (eds), *The Greenhouse Effect, Climatic Change and Ecosystems*, Wiley, NY, 1986. Reeburgh, W S *et al* (1993), 'The Role of Methylotrophy in the Global CH4 budget', in Murrell, J C and Kelly, D P (eds), *Microbial Growth on C-1 Compounds*; Intercept, UK 1993. www.igac.noaa.gov/newsletter/highlights/1996/ch4.php

38   Khalil, M A K and Shearer, M J (2005), *Decreasing Emissions of Methane from Rice Agriculture*, International Congress on Greenhouse Gases and Animal Agriculture, GGAA Working papers, p 307-15.

39   Rice figures from Le Mer, J and Roger, P (2001), 'Production, Oxidation, Emission and Consumption of Methane by Soils: A Review', *European Journal of Soil Biology*, 37, 25, cited in Environmental Change Institute, *op cit. 34 p 16*, which tally with my own calculations. Milk figures based on my calculations from FAO (2006), *op cit. 5*, and Nix, J, *Farm Management Pocketbook, Imperial College, 2007* and corroborated by others, e.g. L Chase (n.d.), *Methane Emissions from Dairy Cattle*, Cornell University, http://www.ag.iastate.edu/wastemgmt/Mitigation_Conference_proceedings/CD_proceedings/Animal_Housing_Diet/Chase-Methane_Emissions.pdf ; and M J Auldist *et al* (n.d.), *Methane Emissions from Grazing Dairy Cows in Victoria*, University of Melbourne, http://www.asap.asn.au/ASAP1-page_example.pdf. These figures may not include the enteric emissions from calves and heifers, nor the meat and leather value of the culled cow. Methane emissions per litre of milk in India, where yields are lower, have been estimated at 53 grams per kilo of milk; Singhal, K K *et al* (2005), 'Methane Emission Estimates from Enteric Fermentation in Indian Livestock: Dry Matter Intake Approach', *Current Science*, Vol 88, No1 Jan 2005.

cultivation. However this is unlikely to happen to any great extent, because most of the countries which produce a lot of milk do not grow much rice, and the main rice eating countries are not great milk producers, preferring pigs and poultry – the exception being India. If beans, nuts and dryland grains were the main substitutes for beef and dairy, the methane emissions they generated would be relatively small.

It is also worth comparing the methane toll of milk against that of cars. At 25 grams of methane per kilo of milk, someone consuming a pint of milk a day would be emitting about 130 kg of CO_2 eq a year. This is about three per cent of the 3.9 tonnes emitted by driving 10,000 miles per year in a small car averaging 37 miles per gallon.[40] Since about five per cent of total emissions from fossil fuels are methane, the car driver is arguably responsible for as much methane as the milk drinker, quite apart from emitting 30 times as much CO_2.[41]

A more significant concern for anyone proposing to reduce or eliminate the global cattle herd is the extent to which it would be replaced by methane-emitting wild animals. Whenever domestic animals are removed from grassland, other species try to move in. In Scotland, for example, when sheep are removed, deer take over: if you want trees to grow you need a deer fence not a sheep fence.[42] In wet areas like the UK we could, if we wanted, turn most of what is presently cow and sheep pasture into woodland, by fencing out wild animals. But large areas of the world consist of natural grasslands which are too dry to maintain tree cover and it is hard to imagine anything else replacing cattle or sheep, except their wild predecessors.

When Tanzania's Serengeti National Park was established in 1951, resident Maasai tribes were removed from the area, together with their cattle herds which had already become greatly depleted by rinderpest and tsetse fly. Since then the population of wild animals, particularly ruminants, has multiplied: in the early 1960s there were an estimated 250,000 wildebeest in the park; by 1979 there were 720,000, and in all about two million large ruminants, including buffalo, gazelles and giraffe, plus 240,000 zebra. Using the 1979 census, the climate scientist PJ Crutzen calculated that their annual methane emissions totalled 19,500 tonnes.[43] By global standards this was a tiny amount – the total amount of methane emitted by domestic animals at the time, 74 million tonnes, was 3750 times as large.

However, the Serengeti is small by global standards, 14,700 sq km, almost the same size as Northern Ireland: you could fit 2,600 Serengetis into the 38 million square kilometres of the world's land which the FAO classes as grassland.[44] If all this land were reclaimed by wild ruminants at the same concentrations as in Serengeti in 1979, they would emit 52 million tonnes of methane, or about five eighths the amount which domestic livestock are now presumed to be emitting.[45] Since Crutzen carried out his study, the herds of wildebeest are reported to

---

40   These figures are derived from a breakdown of vehicle emissions at the carbon calculator site Carbon Independent,http://www.carbonindependent.org/sources_car.htm. They broadly accord with figures given in McKay, D (2009), *Sustainable Energy – Without the Hot Air*, UIT, but the calculation via McKay is more complicated, because he works in kilowatt hours, and he allocates the embodied energy of cars and petrol to other sectors.

41   I say 'arguably' because the methane emissions from oil are less than from gas or coal.

42   Fenton, James (2004), 'A New Paradigm for the Uplands'. ECOS 25:1, p 4.

43   Crutzen, P J (1986), *op cit.*

44   FAO, *op cit.*, p 363. The WRI give 40 per cent of the earth's surface excluding Greenland and Antarctica. Grassland Ecosystems, WRI, http://www.wri.org/biodiv/pubs_content_text.cfm?ContentID=281.

45   EPA and FAO global estimates, FAO (2006), *op cit.*, p 97.

have doubled again in size to 1.5 million.[46] The current wildebeest population in Serengeti suggests that some of the world's grasslands might end up being so heavily stocked with wildlife that they produce as much methane as they currently do with domestic animals.

Of course, it is unscientific to extrapolate from one example in this fashion, but Serengeti does illustrate the rapidity with which abandoned rangeland can be repopulated with wild ruminants. Before the arrival of 110 million cattle in the USA, there were 60 million bison on the great plains and about 100 million small antelope, belching out proportionate amounts of methane.[47] If the cows disappeared, it is possible that the bison would return, not least because there is a significant movement in the US to establish a 'Buffalo Commons'.[48] Bison emit the same amount of methane in relation to their weight and dry matter intake as cows, suggesting that their total emissions might have been close to half those of the present day US cows.[49] In an intriguing study called 'Methane from the House of Tudor and the Ming Dynasty', Susan Subak concludes that in 1500 'emissions from bison alone were as much as three million tonnes' and that with deer and other animals included this might have risen to four million tonnes – compared with current livestock emissions from North America of 5.05 million tonnes.[50]

Who knows whether this is an accurate reflection of what might happen if the dry rangelands of North, Central and South America and Sub Saharan Africa were relieved of their cows? No one seems to have produced any models, though there is a lot of noise about it on the internet from cowboy types. However it is safe to say that if cattle, sheep and other domestic ruminants were eliminated, only a minority of their emissions would resurface in a resurgent wild ruminant population. Less than a third of the world's cattle and sheep are fed on rangeland; and if Australia removed its cows it would most likely see the resurgence of the kangaroo, whose methane emissions are negligible.

However, there are many other natural sources of methane including wetlands, small mammals, termites and other insects, trees and plants, most of which are very poorly understood, and some of which have only recently been discovered. The total dry weight of humans, together with domestic animals, amounted to only 20 per cent of that of all terrestrial species of animals in 1980.[51] Wetlands are currently agreed to be the largest source of methane – either natural or anthropogenic – and it seems likely that if cattle or sheep farming in wet areas were abandoned, drainage systems might be neglected, or even deliberately abandoned for biodiversity reasons, leading to increased methane emissions. Termites are held responsible for about three per cent of global methane emissions; who knows if they won't multiply if we remove all livestock from

46   This is the figure most commonly cited, e.g. Basset, Peter (2004), *Wildebeest on the Serengeti, World on the Move*, BBC Radio 4 http://www.bbc.co.uk/radio4/worldonthemove/reports/wildebeest-on-the-serengeti/; Travellers Point (2004), *The Herds of God: The Great Wildebeest Migrations*, www.travellerspoint.com/forum.cfm?thread=3663; the range given in a variety of mostly promotional literature is from 1.3 million to 2 million.

47   Avery, D (2002), *Would Organic Farming Unleash a Billion Cattle on US Wildlands?* Center for Global Issues, http://www.highyieldconservation.org/background.html

48   Popper, D E and F J (1999), 'The Buffalo Commons: Metaphor as Method', *Geographical Review*, 89(4), 1999, pp 491-510, draft available at http://www.gprc.org/buffalocommons_method.html

49   Clauss, Marcus and Hummel, Jürgen (2005), 'The Digestive Performance of Mammalian Herbivores: Why Big may not be Better', *Mammal Review*, Vol 35, 2005, pp 174-87.

50   Subak, Susan (2004), *Methane from the House of Tudor and the Ming Dynasty*, CSERGE/ University of East Anglia, http://www.uea.ac.uk/env/cserge/pub/wp/gec/gec_1994_06.htm

51   Westing, Arthur H (1976), 'A World in Balance', *Environmental Conservation*, 8 (3), pp 177-83; cited in Coppinger, R and Smith, C (1985) *The Domestication of Evolution*, Environmental Conservation, 10 (4), pp 283-92.

our farming systems? The controversial wildlife biologist Allan Savory observes that in the conditions which arise when pastures are burnt, or irregularly grazed, 'termite numbers can increase to such an extent that they consume all the soil-covering litter'.[52]

Even more disconcerting for anyone seeking a reliable livestock methane emissions figure was the recent discovery in 2006, by the Max Planck Institute, that plant tissues can generate methane under aerobic conditions, while they are alive, whereas previously it was thought that they only produced methane in the absence of oxygen, for example when immersed in a swamp or bog.[53] The Max Planck *in vitro* experiments were corroborated by PF Crutzen, with 18 year old data relating to forest and savannah in Venezuela which had been puzzling him.[54] This discovery could explain two questions that have perplexed scientists: why strong fields of methane have been observed over tropical forests; and why methane levels in the atmosphere have not increased as much as expected – the possible explanation being deforestation. 'Initial extrapolations of this new source' states S Houweling, another scientist currently examining the matter, 'point to a global source strength of 63–236 million tonnes of methane a year, which is roughly comparable to wetlands (commonly considered the largest source of methane).'[55]

If this source were confirmed it would suggest that reductions in methane emissions made by reducing livestock numbers might be partially offset by an increase in plant biomass. However even if methane emissions from cows were completely offset in this way, there would probably still be a global warming advantage in replacing grazing cows by forest in many areas with enough rainfall to grow trees, because of the superior ability of the trees to act as a carbon sink.

In 2008 another department of the FAO reported:

> Since 1999 atmospheric methane concentrations have levelled off while the world population of ruminants has increased at an accelerated rate … Prior to 1999 there was a strong relationship between change in atmospheric methane concentrations and the world ruminant populations. However, since 1999 this strong relation has disappeared. This change in relationship between the atmosphere and ruminant numbers suggests that the role of ruminants in greenhouse gases may be less significant than originally thought, with other sources and sinks playing a larger role in global methane accounting.[56]

The fact is that scientists still do not have a complete idea of how the biological methane cycle works, and even less idea of what might happen if we removed an entire order of creation out of the human food chain which now dominates the biosphere. The FAO's figure of 5.4 per cent represents the maximum possible GHG savings that might be expected from dispensing with livestock; since it

52   Savory, Allan with Butterfield, Jody (1999), *Holistic Management*, Island Press, p 198.

53   Keppler, F *et al* (2006), 'Methane Emissions from Terrestrial Plants under Aerobic Conditions', *Nature*, 439, 187-91.

54   Crutzen, P J *et al* (2006), 'Methane Production from Mixed Tropical Savanna and Forest Vegetation in Venezuela, *Atmospheric Chemistry and Physics Discussions*, 6, 3093-7, 2006.

55   Houweling, S *et al* (2006), 'Atmospheric Constraints on Methane Emissions from Vegetation', *Geophysical Research Letters*, 33 Art No L15821 . However subsequent studies have not corroborated this view and suggest that methane production by plants is "relatively moderate": Kirschbaum et al (2007) 'How Important is Aerobic Methane Release by Plants', *Functional Plant Science and Biology*, 1:1, pp 138-145.

56   FAO/IAEA (2008), *Belching Ruminants, a Minor Player in Atmospheric Methane*, Joint FAO/ IAEA Programme: Nuclear Techniques in Food and Agriculture, http://www-naweb.iaea.org/nafa/ aph/stories/2008-atmospheric-methane.html

is higher than the EPA's or any other estimate, and since it takes no account whatsoever of substitute methane emissions from increased cropping or from wildlife, nor of the uncertainty that surrounds the entire methane budget, I suspect it is an overestimate, albeit perhaps not a large one.

Carbon Colonialism

To summarize, if we place to one side the 'atypical' rainforest beef, for all the rest of the world's meat and dairy we are looking at net emissions of nowhere near 18 per cent of global GHGs, more like 10 per cent, and possibly less. These emissions are comprised of:

(i) 0.32 per cent of global GHGs for fossil fuels, none of which are strictly necessary for livestock production;

(ii) 1.5 per cent from soya in the Amazon (though less than half this if we adopt 2006 to 2009 rates of deforestation);

(iii) an uncertain estimate of 2.8 per cent from nitrous oxide not essential to the production of food;

(iv) an unknown amount from methane, which might turn out to be anything up to five per cent, but according to current scientific thinking is nearer the top end, mainly from cows and sheep.

The total is no higher than 9.6 per cent – plus those rainforest beef, whose encroachments have been halved in recent years. This upper figure is still in excess of other less well publicized estimates for meat emissions, of around five per cent for global livestock, and 10–12 per cent for global agriculture; and in line with those for industrial countries which range from about 4.5 per cent to 11 per cent of national GHG emissions.[57]

The FAO's 18 per cent is the odd one out (or at least it was when it came out) and it is so largely because they choose to include the LULUCF figures for Amazon deforestation and nitrous oxide emissions for agricultural soils, when others find it wiser to treat them separately. Why does the FAO choose to do this? I couldn't find any reasoning in the 400 word report, so I can only guess at the motive. One possibility is that the FAO's promotional department decided that a high figure would bring publicity for them and their discipline. 'Hey people, listen to us, we're bigger than transport.' If so the PR people were right: one leading UK green campaigner told me that if the FAO had distorted the figures upward, so much the better because it had raised the profile of the meat issue. Truth, it seems, is a casualty, not only of war, but of environmental campaigns as well.

However, more than one person has told me that Henning Steinfeld came in for a lot of criticism from other sections of the FAO on account of this publication.[58] Another equally conjectural reason is that including emissions for deforestation and nitrous oxide reinforces the main ideological subtext of *Livestock's Long Shadow*. At the foot of the table which lays out all their GHG emission calculations, is a row which compares 'total extensive vs intensive livestock

57   There is a useful discussion of some of these figures in Garnett, Tara (2008), *Cooking Up a Storm, Food Greenhouse Gases and our Changing Climate*, Food Climate Research Network, 2008.

58   That certainly explains the subsequent FAO document on livestock, belatedly published in 2010, called *Livestock in the Balance*, which focuses primarily on protecting smallholders from the hasty introduction of intensive farming methods. FAO, (2010), *Livestock in the Balance*, State of Food and Agriculture Report 2009.

emissions', showing that the total emissions from extensive systems are five billion tonnes of CO_2 equivalent, while those from intensive systems are 2.1 billion tonnes. The text goes on to state 'by far the largest share of emissions come from more extensive systems where poor livestock holders often extract marginal livelihoods from dwindling resources'.

Nowhere in the report can I find any explanation of how the categories 'extensive' and 'intensive' are defined, nor are we told what proportion of the world's food is produced by extensive and intensive systems respectively. However, a moment's perusal of the two columns headed 'extensive' and 'intensive ' reveals one main difference between the two. Nearly all the intensive emissions are related to activities that are heavily fossil fuel dependent – synthetic fertilizer, tillage, soya cropping, processing, transport, and factory farms. The extensive emissions relate to just two activities, grazing and organic agriculture, which are not intrinsically dependent upon fossil fuels, and have been carried out since long before fossil fuels were discovered. In short, the intensive emissions are representative of an industrial fossil fuel economy, while the extensive emissions are the survivors of a pre-industrial non-fossil economy (and perhaps forerunners of a post-industrial one).

What is strange is that the emissions of the extensive systems are twice as high as the emissions of the intensive, which is not what one would expect, since industrial systems are known to generate higher emissions than pre-industrial systems. It could be that the extensive

Total: 44,153 MtCO$_2$ eq.
Sector

ENERGY

Transportation   14.3%

Electricity & Heat   24.9%

Other Fuel   8.6%
Combustion

Industry   14.7%

Fugitive Emissions 4.0%

Industrial Processes 4.3%

Land Use Change* 12.2%

Agriculture   13.8%

Waste   3.2%

emissions are higher because there are more extensive farmers, or because they produce more food, in which case we ought to be told. It is certainly the case that the extensive emissions are inflated by the inclusion of Amazon deforestation which is mostly attributed to 'extensive' beef farmers (even though many of these have been pushed into the forest by the spread of intensive agriculture). But another possibility is that the reductionist approach taken by the FAO fails

Figure 3. World Greenhouse Gas Emissions in 2005

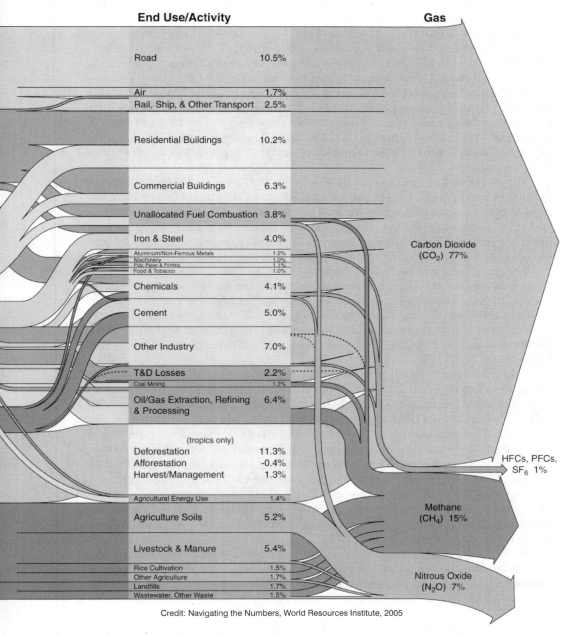

Credit: Navigating the Numbers, World Resources Institute, 2005

Sources and Notes: All data are for 2005. All calculations are based on CO_2 equivalents, using 100-year global warming potentials from the IPCC (1996), based on a total global estimate of 44, 153 $MtCO_2$ equivalent. See Appendix 2 of *Navigating the Numbers: Greenhouse Gas Data & International Climate Policy* (WRI, 2005) for a detailed description of sector and end use/activity definitions, as well as data sources. Dotted lines represent flows of less tha 0.1% of total GHG emissions.
* Land Use Change includes both emissions and absorptions, and is based on analysis that uses revised methodologies compared to previous versions of this chart. These data are subject to significant uncertainties.

to take account of the full basket of emissions generated by each system. The average intensive farmer in the US or Europe emits far more emissions than the average 'poor livestock holder extracting a marginal livelihood', and this is no doubt true of most industrial farmers in Brazil or India as well, so how come this isn't reflected in the FAO's figures?

The reason is that there is a bias in the FAO's figure which leads them to focus on biospheric sources of emissions and to downplay fossil fuels. The WRI chart (Fig 3 on page 177)[59] shows clearly the provenance of different gases: under the 'End Use Activity' column, everything from 'Oil Gas Extraction, Refining and Processing' upwards is fossil-fuel based, ie dug out of the ground and, after use, dumped in the atmosphere – and as yet there is no easy way to put them back where they came from. All these activities, which can be said to represent 'industrial ' emissions, generate CO_2, though oil and gas extraction also gives off significant amounts of methane; and in total they represent 62 per cent of all GHG emissions.

The 38 per cent of emissions which lie below 'Oil and Gas Extraction' are not fossil-fuel based; half of these are LULUCF carbon dioxide emissions (mostly caused by deforestation) while the rest consist of the majority of the methane and nitrous oxide emissions. These represent fluxes that occur above ground; they form part of the biosphere's life cycle, circulating between the atmosphere, the oceans, the soil and terrestrial biomass at different speeds.[60]

The biospheric emissions from nitrous oxide, methane and CO_2 are common to both extensive and intensive systems, but they are more associated with the former, because ruminants generate methane and as high fibre eaters they are better adapted for grazing. The fossil fuel emissions are almost entirely associated with intensive agriculture. Extensive livestock farmers in local economies don't need fossil fuels in any quantity, a difference that is clearly reflected in the FAO figures.

Take for example the figure for transport given in the FAO's table: less than one million tonnes of CO_2eq – an amount so low that it doesn't even figure in the column headed 'percentage of animal food GHGs emissions'. Whilst it is entirely believable that Maasai herdsmen or backyard poultry keepers in South East Asia use negligible amounts of fossil fuel for transporting their goods, the figure seems absurdly low for a global industry which mainly serves the industrialized countries; and it is completely at variance with other estimates. For example, the UK's Cabinet Office reckons that transport for the UK food industry generates the equivalent of 19 million tonnes of CO_2 eq a year.[61] If we assume that a quarter of UK food is meat or dairy, that means that in the UK alone we emit nearly five times as much carbon transporting livestock products as the FAO considers the entire world does.

How does the FAO arrive at this glaring underestimate? They explain that they calculated the emissions generated by a few major international shipping routes involved in meat and feed commerce:

---

59   Baumert, K *et al* (2005), *Navigating the Numbers*, World Resources Institute pp 4-5. http://pdf. wri.org/navigating_numbers.pdf

60   That is why, in the LULUCF section of Fig 3, items such as Afforestation and Reforestation register as a negative value – they represent newly established carbon sinks which absorb some of the carbon that is released into the atmosphere when other sinks are destroyed. LULUCF emissions and the matter of carbon sequestration follow a different dynamic from the other elements of the carbon budget and are dealt with in subsequent chapters, which focus on land use.

61   The Strategy Unit (2008), *Food Matters: Towards a Strategy for the 21st Century*, Cabinet Office, p 13.

These flows represent some 60 per cent of international meat trade. Annually they produce some 500,000 tonnes of CO_2. This represents more than 60 per cent of total CO_2 emissions induced by meat-related sea transport, because the trade flow selection is biased towards the long distance exchange. On the other hand, surface transport to and from the harbour has not been considered. Assuming for simplicity, that the latter two effects compensate each other, the total annual meat transport-induced CO_2 emission would be in the order of 800–850,000 tonnes of CO_2.[62]

In other words, 'for simplicity', the FAO has omitted to take into account all surface transport and airfreight anywhere in the world – all the cattle trucks and milk tankers, the lorries loaded with hay and straw, the vets' visits to farms and the relief milkers' trips to work, the deliveries of feed and fertilizer, the Spanish artics and the Tesco vans, the just-in-time deliveries of McDonalds burgers and M & S chilled beef vindaloo – all of this is discounted in order to cancel out a modest overestimate made in its calculation about international shipping. Sea transport represents a minor part of food transport emissions. The UK imports around 40 per cent of its food, but international shipping represents just 13.5 per cent of transport emissions, whereas road transport constitutes 73 per cent and aviation 13.5 per cent.[63]

It is hard to pin down what other assumptions the FAO may have made to 'simplify' their calculations. For example their breakdown makes no mention of refrigeration, other than in connection with shipping by sea, so we have no way of knowing to what extent on-farm, wholesale, retail and domestic refrigeration is a component of the figure given for processing. In total, the FAO estimate that just 2.2 per cent of all livestock emissions come from fossil fuels, and 11 per cent of intensive livestock emissions come from fossil fuels. It is hard to reconcile this with sets of statistics which show that, in the UK, fertilizer production, food processing and distribution account for nearly half of all food emissions.[64] To a large extent these are emissions that the backyard poultry keeper or the peasant making a 'marginal livelihood' from a dairy cow, does not have.

The FAO analysis accounts for CO_2 from tractors, fertilizers, on-farm processing and a miniscule amount for transport – but it doesn't register all the high impact baggage that tends to come with intensive farming, in order to accommodate its so called economies of scale: 4x4s, concrete yards, paved roads, electric lights, air conditioning, refrigeration, burglar alarms, slaughterhouse costs, animal waste disposal, health and safety measures, carcase incineration, livestock registration and identification, product tracking, computerized accounts, conferences, packaging, advertising, middlemen, retail chains, just-in-time delivery, supermarket journeys, processing waste disposal, domestic waste disposal, journeys to work etc. You don't buy into the intensive farmers' club

---

62   FAO op cit 5.

63   *Ibid.*

64   Tara Garnett's figures show that emissions from fertilizer manufacture, distribution and processing (FDP) represent 49 per cent of UK food consumption emissions; a similar preliminary estimate from DEFRA gives 35 per cent. The bulk of these emissions will be from fossil fuels. By contrast, the FAO's calculations for livestock production give 0.07 billion tonnes for FDP (plus or minus 0.02); ie just 5 per cent of the figure of 1.4 billion tonnes due to intensive livestock production *alone* (not counting LULUCF emissions); and only 1.5 per cent of the 4.6 billion tonnes from extensive and intensive livestock combined (again not counting LULUCF). There is clearly a huge discrepancy between the methodology adopted by DEFRA and Garnett on the one hand, and by the FAO on the other. DEFRA figures and Garnett's are compared at Garnett, T (2008), *Cooking Up a Storm*, Food Climate Research Network, p 23.

with just a tractor and sack of fertilizer – intensive farming brings with it all the paraphernalia of an industrialized urbanized lifestyle, which the peasant in a local economy for the most part manages without. That is why nearly two thirds of global warming is caused by fossil fuels. The FAO, by failing to account for all this have contrived to stigmatize lifestyles which, taken in their entirety, and providing there is enough land to accommodate them, ought to be applauded for their low impact.

The FAO is not alone in its bias. In the UK, a Cranfield University report concludes that 'the global warming potential from arable cropping is dominated by N_2O, not by CO_2 from fuel use … In agriculture N_2O dominates, with substantial contributions too from methane.'[65] The N_2O factor is indeed significant, but it is fairly constant to any given level of productivity, whereas there is a wide difference in fossil fuel use between a muscle-powered, organic, locally-centred arable economy, and a highly mechanized, export-orientated agriculture serving distant urban consumers.[66]

An analysis that focuses unduly on biospheric emissions and is blind to wider fossil-fuel penalties could, for example, conclude that the extra fossil fuel required to power an industrialized system of rice cultivation that replaced a peasant system, was negligible compared to the methane emissions that emanated from both of them. Similarly current advice to improve the yield of dairy cows in order to reduce methane emissions fails to take into account increased fossil fuel emissions that may result from intensification. This biospheric bias is most pronounced with livestock. Since animals concentrate nutrients at the top end of the food chain, it stands to reason that they concentrate methane and nitrous oxide emissions at the top end of the carbon chain, while the fact that they can move means that their fossil fuel demands are intrinsically low, or in the case of a grass fed, subsistence dairy cow or goat, zero.

The more that agricultural climate change policy is focussed on biospheric emissions, the more the Asian rice farmer, the UK arable farmer and livestock farmers everywhere will be expected to reduce methane and nitrous oxide emissions, yet allowed to gloss over the relatively low emissions emanating directly from intensive fossil-fuel based food production (even though the inevitable corollary is an energy-intensive system of food distribution which is central to a society whose carbon emissions are 62 per cent derived from fossil fuels). Poultry factories producing packs of frozen turkey breasts will be viewed as more sustainable than neighbourhood milk co-operatives, because all the energy-intensive industrial infrastructure and consumer paraphernalia that is integral to the production and distribution of turkey breasts is not factored within the farm budget, but clocked up under other headings such as 'transport', 'industry' or 'built environment'. Such a policy is suspect in itself; but it is especially prejudicial for those farming families who have few fossil fuel expenses, and whose entire carbon budget on this planet is counted in the natural currencies of nitrous oxide and methane – the 'poor livestock holders extracting marginal livelihoods from dwindling resources'. These are the people targeted by *Livestock's Long Shadow* – yet they expend far fewer carbon emissions than you or I.

65   Cranfield (2006), *Determining the Environmental Burdens and Resource Use in the Production of Agricultural and Horticutural Commodities*, DEFRA project report ISO295, Natural Resource Management Institute, Cranfield University Silsoe Research Institute, August 2006, p 96.

66   The IPPC uses the same default factors for a given amount of nitrogen, whether it comes from artificial fertilizer, manure or green manure. See also Crutzen P J *et al* (2007), 'N_2O Release from Agro-Biofuel Production Negates Global Warming Reduction by Replacing Fossil Fuels', *Atmos Chem Phys Discuss* 7, pp 11191-205

A further matter of potential concern for livestock farmers is the growing interest in targeting methane as a 'quick fix.' A salient feature of methane is that it doesn't stay in the atmosphere very long: within 12 years it is nearly all converted into less harmful chemicals, whereas carbon dioxide has an atmospheric lifetime of about 200 years. This means that a given value of methane exerts most of its global warming impact at a high intensity within 12 years, while the same value of CO_2 has its effect spread out over two centuries. The CO_2 equivalent unit has been established by estimating the total Global Warming Potential (GWP) of each gas over a period of 100 years. The currently favoured exchange rate of 25:1 between methane and carbon dioxide is therefore not a scientific fact; it is an assessment, based in part on the rate at which we discount the future.[67] It has changed twice in recent years, from 21:1 to 23:1 between 2001 and 2006, and in 2007 the IPCC recommended a figure of 25:1.[68] If we were to estimate the GWP over a shorter timescale of 20 years, rather than 100 years, then the exchange rate would change drastically to around 72:1 rather than 25:1, and as a result methane would then be held to be as great a cause of global warming as carbon dioxide.

In a context where addressing global warming appears to be a matter of urgency, and might soon become a matter of emergency, methane's short lifetime makes it an attractive target for policy makers and governments committed to reductions in greenhouse gas emissions, because the savings will be realized sooner than with CO_2. Oxford University's Environmental Change Institute observes:

> Reducing emissions of short-lifetime potent gases such as methane
> is therefore a valuable means of rapidly slowing global temperature
> rise. This gives reduction of methane emissions a high economic value
> (perhaps even greater than that reflected in the GWP of 23) as they are
> effective at slowing the rate of global warming. They are also likely to
> be even more important at some point in the future should the effects
> of climate change become critical and fast-acting measures need to be
> adopted.[69]

In other words, if pressure from climate change becomes more intense we are likely to see the GWP of methane jacked up even further with the result that methane will occupy a more imposing proportion of the global carbon budget. In the language of economists, the higher the discount rate, the more commanding a position methane takes in the global carbon budget. There are increasing calls from vegan and anti-livestock commentators such as Peter Singer and Robert Goodland for the GWP of methane to be raised from 25 to 72. Singer, together with two colleagues Geoff Russell and Barry Brooks, has employed a useful analogy to explain the effect of GWP ratings:

> A tonne of methane has 100 times the warming during the first five
> years of its lifetime as a tonne of CO_2, but under current Kyoto rules,
> its comparative potency is set at 21. This is because the relative impacts
> of ALL greenhouse gases are averaged over the same period of 100
> years, regardless of their atmospheric lifetime. This is like applying
> a blow torch to your leg for ten seconds but calculating its average

---

67   Clarkson, R and Deyes, Kathryn (2002), *Estimating the Social Cost of Carbon Emissions*, Government Economic Service Working paper 140, HM Treasury/DEFRA, 2002.

68   IPPC (2001), *Climate Change 2001: The Scientific Basis*; IPCC Working Group 1 AR4 Report, Chapter 2, p 214.

69   Environmental Change Institute (2006), *op cit*.34.

temperature as just 48 degrees because that's what it is when averaged over 20 minutes, with 20 minutes being used because that happens to be some agreed international standard. The implication of course being that a blow torch for ten seconds and a 48 degree hot water bottle have the same effect.[70]

The analogy is potentially illuminating, but it is misleading because the methane blowtorches and the CO_2 hot water bottles aren't being applied to anything as sensitive as a leg, but are heating up the atmosphere in a room; and it is also incomplete, because there isn't just one blowtorch and one hot water bottle, there are hundreds of them being brought into the room continuously. Although the blow torches are individually intensely hot, they go out within a matter of seconds, whereas the hot water bottles keep piling up until their collective heat far outstrips that of the relatively few blowtorches that remain ignited at any one time.

At the point where the heat becomes unbearable, the obvious first course of action is to reduce the flow of blowtorches into the room. That will be the quickest way of reducing the temperature back to the level it was just before it became unbearable. Reducing the flow of hot water bottles will have comparatively little immediate effect. But removing the blowtorches won't prevent the hot water bottles continuing to pile up until the heat becomes unbearable again; that will eventually happen even if the flow of blowtorches is completely stopped, and when it does happen it will be much harder to lower the temperature again.

In fact, the short lifetime of methane also speaks in its defence. In order to maintain the blowtorch heat, the blowtorches have to come into the room thick and fast. In other words, in order to maintain a given level of methane in the atmosphere we have to keep pumping it out regularly otherwise the number of parts per million will fall away rapidly. According to the IPCC's 2001 figures, a reduction of global emissions by just 22 MT, or 6.1 per cent of all the methane currently emitted by human activities, is sufficient to prevent any further rise in levels of methane in the atmosphere.[71] This could be achieved by reducing livestock emissions by just 20 per cent, by halving methane emissions from landfill sites, or even by reducing fossil fuel use by about a quarter. Any reductions in annual methane output beyond this would actually be reducing the amount of methane in the atmosphere, until it settled at a new lower threshold. On the other hand, if humans stopped burning fossil fuels tomorrow, the burden of the CO_2 emitted in the 20th century would linger on for several generations.

This means that if the world decided to stabilize all greenhouse gases at their current level, we would only need to reduce global methane output by 6.1 per cent, but CO_2 emissions by more than 80 per cent.[72] Any 'targeting' of methane to compensate for the manifest failure to reduce CO_2 emissions, especially if it resulted in reductions of substantially more than 6.1 per cent, would in effect be

70   Russell, Geoff (2009), *Kyoto's $20 Trillion Greenhouse Gas Blunder*, Crikey.com,7 July 2009, http://vegclimatealliance.org/kyotos-20-trillion-greenhouse-gas-factor-blunder/. The blowtorch analogy first apeared in Russell, G, Singer, P and Brook, B (2008), *The Missing Link in the Garnaut Report*, 10 July 2008, The Age.com.au; http://www.theage.com.au/opinion/the-missing-link-in-the-garnaut-report-20080709-3cjh.html?page=-1
71   This figure is dependent on some variables such as the amount of OH in the atmosphere. Environmental Change Institute (2006), *op cit.* 34; IPCC (2001), *Climate Change 2001: The Scientific Basis. Contribution of Working Group I to the Third Assessment Report of the Intergovernmental Panel on Climate Change*, 4.2, Cambridge University Press, 2001. IPCC (1995), *Climate Change 1994, Radiative Forcing of Climate Change. Working Group 1. Summary for Policymakers*. International Panel on Climate Change, Cambridge University Press, UNEP, 1995.
72   Stern, N (2007), *The Economics of Climate Change*, Cambridge, p 223.

scapegoating methane to bale out CO_2; or put another way, it would be extracting a subsidy from methane emitters for the benefit of fossil fuel users.

In fact the level of methane in the atmosphere, contrary to expectations and despite a rise in the numbers of livestock, has remained almost stable since 1999, according to IPCC assessments, so it is possible that such a reduction in methane output has already taken place.[73] It is very odd that so much attention should have been focussed upon methane in the last decade while its presence in the atmosphere has remained stable.

That stability could be a temporary blip, and it doesn't alter the fact that reducing methane emissions has an immediate effect. But while targeting methane might be desirable or even necessary to avert catastrophe, it would be politically invidious if it required disproportionate reductions in emissions from livestock rearers, rice producers, fuelwood users or anyone else benefiting from 'natural' sources of methane. It would be making rural occupations pay for the sins of the city, and poor people in third world countries pay for the sins of rich fossil-fuel burners. Whereas UK and US methane emissions comprise 7.5 per cent and 10.5 per cent of their GHG emissions respectively, India's methane emissions, two thirds of which come from cows and rice, are reported to comprise 35 per cent of her total.[74] India, like most poor countries, burns far less fossil fuel per head than the USA or the UK, and for many of her poor, a goat or a cow may represent almost the entirety of their greenhouse gas emissions. Targeting methane emissions such as these to compensate for a failure to reduce CO_2 emissions is another facet of the neo-colonialism that has pervaded international climate negotiations.

Something of this trade-off between methane emissions and CO_2 emissions can already be seen in specific proposals for mitigating methane emissions from animals. At the farm level the FAO states that 'the most promising approach for reducing methane emissions from livestock is by improving the productivity and efficiency of livestock production through better nutrition and genetics. 'Higher yielding cows produce more milk for less methane. This advice is widely repeated throughout the agricultural press, for example *Farmers Weekly* writes:

> Want to reduce your herd's carbon footprint? Then the easiest way is to increase milk yields ... A 9000-litre cow produces 125 kg of methane a year, a 5000-litre cow produces 103 kg of methane.[75]

While there may well be advantages in increasing yields from dairy animals, there are also drawbacks: a high yielding cow will be less likely to survive on default produce from the farm, and will require both more and higher grade nourishment that can usually only easily be provided by carbon intensive farming practices such as tillage, irrigation or use of synthetic fertilizers. Dairy cows would evolve into feedlot cows, turning human food into milk and methane inefficiently, rather than earning their keep by digesting fibrous material. A tax on emissions from cows, already mooted in New Zealand, the USA and Denmark, would have severe repercussions in poor countries. As the FAO acknowledges 'it

---

73   IPCC (2007), *Contribution of Working Group I to the Fourth Assessment Report of the Intergovernmental Panel on Climate Change*, Cambridge University Press, Chapter 2, pp 140-142 http://www.ipcc.ch/pdf/assessment-report/ar4/wg1/ar4-wg1-chapter2.pdf
74   DEFRA (2008), *UK Climate Change Sustainability Indicator: 2006 Greenhouse Gas Emissions Final Figure*, http://www.defra.gov.uk/news/2008/080131a.htm; EIA (2008), *Emissions of Greenhouse Gases in the US 2008*, ftp://ftp.eia.doe.gov/pub/oiaf/1605/cdrom/pdf/ggrpt/057308.pdf; Padma, T V, *India and Climate Change: Facts and Figures*, Sci Dev Net, 31 August 2006, http://www.scidev.net/en/climate-change-and-energy/mitigation/features/india-climate-change-facts-and-figures.html
75   *Farmers Weekly* (2007), 'Milk Yield Holds the Key to Lower Carbon Footprint', *Farmers Weekly*, 20 August 2007.

will be difficult to apply the [polluter pays] principle to methane emissions from single cows owned on an Indian mixed farm of half a hectare.'[76]

On the macro scale, a major shift from ruminants to monogastrics, again favoured by the FAO, would also require intensified production methods, involving increased reliance on fossil fuel dependent, N_2O-emitting arable crops, and the loss of soil carbon through the ploughing up of pasture. As the FAO remarks 'ruminant production, both meat and milk, tends to be much more rural-based', because of the need for bulky fibrous feeds. If large numbers of 'inefficient' cows were culled, millions of poor rural dwellers would lose the ability to harness their local biomass and add to the swell of refugees flocking into megacities. There is a very good case for reducing the numbers of feedlot cattle and of rainforest cattle, which are extravagant and destructive in other respects; but beyond that, large scale reductions in ruminants would have uncertain returns in terms of methane emission because we cannot be sure how the biosphere will respond, would increase pressure for fossil fuel consumption, and would have severe social repercussions.

Even a cursory glance at the global GHG budget is sufficient to see that the main problem is CO_2 – cows have been around for thousands of years, while global warming takes off with the discovery of oil. Moreover IPCC figures show that between 1970 and 2004, emissions of CO_2 emissions from fossil fuels have doubled, while emissions of nitrous oxide and methane have increased by 50 per cent and 40 per cent respectively.[77] Fossil fuel CO_2 emissions over the last 40 years have increased at a much higher rate than non-fossil fuel CO_2 emissions. Since 1970:

> Emissions from the energy supply sector have grown by over 145 per cent, while those from the transport sector have grown by over 120 per cent; as such, these two sectors show the largest growth in GHG emissions. The industry sector's emissions have grown by close to 65 per cent, LULUCF by 40 per cent, while the agriculture sector (27 per cent) and residential sector (27 per cent) have experienced the slowest growth.[78]

A child could deduce that if we were serious about preventing global warming, the most obvious and reliable course of action would be to leave all fossil fuels in the ground. If we did that, then not only would the CO_2 burden begin to stabilize, but the amount of methane in the atmosphere would decline dramatically, since more than a quarter of all anthropogenic methane comes from fossil fuels. The exaggerated emphasis on the alleged four or five per cent of GHGs emitted by cattle, and the mendacious rhetoric about cows causing more global warming than cars, look suspiciously like an attempt to shift some of the blame for global warming from below ground to above ground, from fossil fuels to the natural biosphere, from the town to the country and from the rich nations to the poor.

Calverd's Return

Remember Alan Calverd, the euphonium player mentioned at the beginning of this chapter who warned of the dangers of animals breathing? Once I had finally obtained a copy of his article, and satisfied myself as to its content, I felt I could

---

76   FAO (2006), *op cit.*5, p 277.

77   IPCC (2007), *Climate Change 2007: Mitigation, Contribution of Working Group III to the Fourth Assessment Report (AR4) of the IPCC* (Introduction) eds B Metz *et al*, Cambridge University Press, 2007, pp 103-5.

78   *Ibid.*, p 104.

safely lay it aside to gather dust. Even the authors of *Livestock's Long Shadow*, however keen they might be to fix blame for global warming on livestock, state categorically that 'respiration by livestock is not a net source of CO_2'.

However, in 2009, the Calverd thesis was resurrected, this time in all seriousness by Robert Goodland and Jeff Anhang. Goodland is a colleague of David Pimentel and a former adviser to The World Bank who has written a paper putting the environmental, ethical and religious case against meat-eating (though there was no mention of global warming in this).[79] Anhang is a research officer at the International Finance Corporation, an arm of the World Bank which facilitates private sector investment in developing countries' economies. In a document published by the World Watch Institute entiled *Livestock and Climate Change*,[80] Goodland and Anhang (whom henceforth I shall call G&A) begin by rubbishing the FAO's affirmation that animal breathing has no net effect upon global warming. Animal respiration, G&A claim, is responsible for more CO_2 emissions than all the rest of the FAO's 18 per cent, effectively doubling the amount of CO_2 attributable to livestock:

> Livestock (like automobiles) are a human invention and convenience, not part of pre-human times, and a molecule of CO_2 exhaled by livestock is no more natural than one from an auto tailpipe. Moreover, while over time an equilibrium of CO_2 may exist between the amount photosynthesized by plants that equilibrium has never been static. Today tens of billions more livestock are exhaling more CO_2 than in pre-industrial days, while earth's photosynthetic capacity (its capacity to keep carbon out of the atmosphere by absorbing it in plant mass) has declined sharply as forest has been cleared.

While there is little point in pursuing semantic arguments about the relative 'naturalness' of photosynthesis and fossil fuels, the last sentence requires scrutiny. First of all, the tens of billions of extra livestock, where do they come from? If there are more of them in the world today than there were 200 years ago, that has to be because there is more of the kind of vegetation they like to eat, both through the replacement of relatively inedible forests by edible grassland, and through the increased luxuriance of vegetation thanks to additional inputs of nitrogen, phosphate and water. There is no other way these animals can absorb the carbon that they exhale. The process is neutral and short term: animals eat and exhale every day, and plants go on absorbing carbon throughout most of the year.

That leaves us with G&A's assertion that the equilibrium has been disturbed as a result of the decline in photosynthetic capacity caused by forest clearance. Forests photosynthesize carbon, and since much of this carbon is not eaten by animals, instead it is stored in the trees, and released over a longer time scale, when the forest is burnt or the trees rot. The clearance of forests releases this carbon prematurely into the atmosphere, causing an increase in CO_2 levels in the atmosphere. All this we know already. But carbon release through deforestation caused by livestock is already accounted for by the FAO: by adding their figure for livestock respiration to the FAO's figure for livestock-induced deforestation, G&A are double counting – clocking up carbon at one part of its cycle as it comes

79   Goodland, Robert (1998), 'Environmental Sutainability in Agriculture: Bioethical and Religious Arguments Against Carnivory', in J Lemons *et al* (eds), *Ecological Sustainability and Integrity*, Kluwer, 1998, pp 235-65.
80   Goodland, R and Anhang, J (2009), *Livestock and Climate Change: What if the Key Actors in Climate Change are Cows, Pigs and Chickens?* Worldwatch, http://www.worldwatch.org/node/6294

out of an animals' lungs to enter the atmosphere, and a second time when it remains in the atmosphere because it fails to be sequestrated for a period of years in a forest that no longer exists. As for deforestation that is carried out for other purposes than livestock, it is hard to see what connection that could possibly have with animals whose CO_2 output is entirely reliant on the photosynthesized CO_2 that they ingest.

G&A cite only one source to back up their calculations about animal respiration – Alan Calverd's article in *Physics World*.[81] He is their only source because, in their words, 'Calverd's estimate is the only original estimate of its type'. They do not inform their readers, who will have considerable difficulty locating Calverd's study, that this is a one page opinion piece written by an author whose mugshot shows him playing the euphonium.

G&A do not stop there but continue with an unprecedented tirade of blame heaped against the animal kingdom. Livestock, they claim, are responsible for 32,564 million tonnes of CO_2eq, more than four times as much as the 7516 tonnes calculated by the FAO. This would bump the FAO's figure of 18 per cent up to 80 per cent of the total of 40 billion tonnes of greenhouse gases which the FAO and the World Resources Institute estimate are attributable to human activity. Eighty per cent is a bit hard to swallow, but G&A also argue that the figure for anthropogenic GHGs should be increased from 40 billion tonnes to 63.9 billion tonnes, with the result that livestock represent a more modest 51 per cent of anthropogenic GHGs. This proportion conveniently allows them to argue that domestic animals account for 'at least half of all human-caused GHGs' and therefore that replacing livestock products with alternatives would be 'the best strategy', since it would have 'far more rapid effects ... than actions to replace fossil fuels with renewable energy'. Not only are cows worse than cars, but fauna causes more damage than fossil fuels.

Aside from respiration, G&A provide two other main reasons why they consider the FAO's 18 per cent to be a severe underestimate. Like Peter Singer and his colleagues, they argue that the Global Warming Potential of methane should be increased from 25 to 72, with the result that by a stroke of the pen livestock methane emissions are tripled, and become a more prominent target than fossil fuels. As I have already pointed out, targeting methane in this way may be the most effective strategy in an emergency, but it involves scapegoating methane emitters to compensate for our failure to address the long-term effects of fossil fuels.

The third factor which bumps up the livestock GHG tally by a further 2.6 billion tonnes is one that I have also touched on in the discussion on *Livestock's Long Shadow* under the heading 'carbon sinks forgone'. The 26 per cent of land worldwide used for grazing, together with the 33 per cent of arable land used for feed crops, could, G&A argue, be more effectively used as a carbon sink by being converted back to forest, or alternatively used to reduce fossil fuel emissions by growing biofuels. No doubt in many cases, particularly in respect of animal feed crops, G&A are right; but to assign a greenhouse gas penalty to these putative carbon sinks and add that to the toll of actual emissions attributed to livestock leads to statistical confusion. If theoretical GHG emissions are to be assigned to grazing and feedcrop land on the grounds that it could be covered in trees or biofuel crops, then even more non-existent emissions should be assigned to the 67 per cent of arable land upon which we grow food for humans, not to mention all the land taken up by houses, roads and other infrastructure. On top of that

81   Calverd (2005), *op cit. 4*.

there is uncertainty about what levels of carbon sequestration can be achieved, and disagreement as to whether or not grazing land can be an effective sink for soil carbon – a matter I cover in the next chapter.

The authors of *Livestock and Climate Change*, emboldened perhaps by the success of the FAO's 18 per cent, appear to have whipped themselves into a state of statistical hysteria, where the GHG emissions they attribute to livestock have swelled to such proportions that those attributed by the IPCC to fossil fuels and other sources have magically shrunk into insignificance. The undisguised message is that cutting emissions from fossil fuels is not worth bothering about:

> A substantial body of theory, beliefs, and even vested interest has been built up around the idea of slowing climate change through renewable energy and energy efficiency. However, after many years of international climate talks and practical efforts, only relatively modest amounts of renewable energy have been developed ... GHG emissions have increased since the Kyoto Protocol was signed ... Action to replace livestock products not only can achieve quick reductions in atmospheric GHGs but can also reverse the ongoing world and water crises.

The action G&A propose is the production and high-pressure marketing of analogue meat products made from 'superfoods' like spun soya protein. They advise promoters of these products, like McDonalds, to orient their advertising heavily towards children who are the 'most susceptible' consumers, and the illustrated examples of imitation meat products such as Morning Star 'Bacon Strips' and Quorn Naked Chikn Cutlets [*sic*] fit this bill. G&A show minimal interest in reviving traditional, slow food diets founded on whole grains and pulses, fresh vegetables and imaginative use of low levels of default meat, offal and dairy. The heavily processed and packaged products they advocate are symptomatic of urbanized societies reliant on high levels of transport, refrigeration, waste and energy use – but what else should one expect from an author who works for the International Finance Corporation, a body that has long been criticized by NGOs for investing in 100,000 hectare latifundias, factory farms, luxury hotels, water privatization, soya bean and palm oil plantations and the like? The world that these World Bank representatives describe may turn out to be the one our children will inherit. But if the human race can only be saved from global warming by living on a diet of turkey-less twizzlers, one wonders if it is really worth saving.

14

HOLISTIC COWBOYS AND
CARBON FARMERS

In his book, *The Carbon Fields,* under the heading 'No More Climate Change', Graham Harvey writes:

> Our food supply hides a big, fat life-denying secret. It's something no one in the food and farming business ever wants to talk about. Yet it has the potential to transform the lives of everyone on this planet as well as the lives of future generations. It's the power of soils to take carbon dioxide out of the atmosphere and to end for all time the threat of global warming.

Harvey is the agricultural story editor of the British radio soap opera *The Archers,* and if you are wondering why David and Ruth Archer spent the winter of 2008/9 fencing their pastures off into small paddocks, then this chapter on grass farming and the sequestration of carbon in the soil offers an explanation. Be warned, however, that this may be all that is successfully explained. The carbon fields, I have discovered, are a minefield, and of all the chapters in this book, this is the one that has given me most trouble and that delivers the most verbiage for the least amount of solid conclusion. You can safely skip it without impairing your understanding of subsequent chapters. On the other hand, if you are intrigued by the agricultural heresies that can be loosely grouped under the heading 'grass farming', read on. Harvey continues:

> Though you'll seldom hear it mentioned, the world's soils are the largest terrestrial reservoir of carbon. A sizable part of the damaging extra load of greenhouse gases in the atmosphere today comes from soil carbon released when we switched from traditional farming to intensive grain growing. The good news is that it is a process we could easily reverse. By moving to sustainable ways of growing our food – particularly through the use of grazing animals – we could quickly put the excess carbon back in the soil.[1]

Harvey's promise of instant eco-salvation, through what is known as carbon sequestration, sounds a bit too good to be true; and the odour of conspiracy theory with which he surrounds it makes it seem all the more suspect. It is manifestly untrue to say that this matter is one nobody wants to talk about. There are reams and reams of learned articles on soil carbon sequestration – so many that they almost outnumber the customary plethora of blogs and website rants on the subject. Throughout the USA and Australia in particular, universities are conducting symposiums on carbon farming and interest groups and companies

---

1   Harvey, G (2008), *The Carbon Fields,* Grassroots, p 41. This quotation has been shortened.

hold conferences with titles like 'Soil Carbon: The Next Cash Crop'.[2] In the UK the Nuffield Trust held a Carbon Farming Conference at the showground at Stoneleigh in 2008, while the Permaculture Association has been hosting conferences on Low Carbon Farming.[3] The Soil Association has been plugging carbon sequestration as one of the main advantages of organic farming; and at the Oxford Farming Conference in January 2009, environment secretary Hilary Benn revealed that the government was willing to sponsor a new low carbon farming award to help promote green best practices.[4] Carbon farming is sliding into the seat of honour that biodiesel slunk out of in disgrace.

And like biofuels, carbon farming is potentially big business. Article 3.4 of the Kyoto Protocol allows carbon emissions to be offset by the improvement of agricultural soils, thereby setting the stage for soil carbon to become a tradable commodity. The USA, at the time of writing, has yet to sign the Protocol, but it already has a burgeoning carbon farming offset market: the Chicago Climate Exchange Market (in co-operation with the National Farmers Union and Iowa Farm Bureau) is paying farmers up to $3 per acre for sequestering carbon with funds derived from selling carbon credits to industrial companies.[5] A company trading under the stirring title *Carbon Farmers of America* pays farmers $6 for every tonne of carbon they sequester in their soil. The money is raised by selling 'Carbon Sinks' costing $25 apiece to individuals and businesses. For $2,500 you can buy a 'Family Carbon Sink Package'; for a mere $175 you can 'turn your vehicle into a carbon sink'; for an unspecified sum you can certify your company by buying a Business Carbon Sink Package.[6]

Amongst scientists, there is widespread, probably universal, agreement that agricultural soils can sequester carbon in much the same way that tree cover can, and that the potential for them to do so in some circumstances is not negligible. Beyond that there is uncertainty and dispute. For every scientific paper showing that a land use change such as converting pasture to woodland, using minimum tillage on cropland, or excluding livestock from pasture increases the amount of carbon sequestered in the soil, there is another showing that, in other circumstances, the effect is the opposite.[7] Moreover an increase in soil carbon on one site, for example by converting it from arable to pasture, may result in a decrease on another where pasture is converted to arable. An increase in soil carbon may also result in the release of other greenhouse gases: pasture sequestrates carbon, but grazing animals release methane. Similarly, adding nitrogen, by planting legumes for example, increases vegetation and hence the amount of carbon which the soil can potentially assimilate; but it also releases nitrous oxide into the atmosphere.

Within the scientific mainstream there is little consensus as to whether soil carbon sequestration can only have a minor impact upon our overall greenhouse

2   Carbon Farming Expo and Conference, Orange NSW, 18-19 November 2008; http://carbonfarming.blogspot.com/
3   *The Carbon Footprint of British Agriculture*, The Nuffield Carbon Farming Conference, Stoneleigh, 30 April 2008; *Low Carbon Farming*, The Greenhouse, Bristol, 25 November 2008.
4   Sams, Craig (2007), *Farming Carbon*, Soil Association Conference 2007; Benn, Hilary (2009), *Challenges and Opportunities: Farming for the Future*, Oxford Farming Conference, 6 January 2009, DEFRA, http://www.defra.gov.uk/corporate/about/who/ministers/speeches/hilary-benn/hb090106.htm
5   Chicago Climate Exchange (2008), *Soil Carbon Management Offsets*, http://www.chicagoclimatex.com/docs/offsets/CCX_Soil_Carbon_Offsets.pdf
6   Carbon Farmers of America (n.d.), *Carbon Sinks*, http://www.carbonfarmersofamerica.com/products.htm
7   For a circumspect view of minimum tillage and some other approaches to SOC sequestration, see Goulding, K *et al* (2009), *The Potential for Soil Carbon Sequestration, but with a Full Greenhouse Gas Budget*, Rothamsted 2009.

gas emissions, or whether it has the potential to solve all our problems. One scientific paper, authored by nine scientists, states that 'the IPCC estimates for the global mitigation potential of carbon sequestration in agricultural soils are 0.4 to 0.6 billion tonnes per year (over 100 years) – which is less than ten per cent of our current annual carbon emissions from fossil fuels ... From this perspective, soil carbon sequestration can make only modest contributions to the overall need for mitigation of atmospheric CO_2 build-up.' [8] Yet a year later, one of the nine authors, Dr Rattan Lal – who is the world's number one guru on soil carbon, and frequently cited by the IPCC – stated that 'the maximum potential rate of Soil Organic Carbon (SOC) sequestration of three billion tonnes of carbon per year is high enough to almost nullify the annual increase in atmospheric concentration of CO_2 at 3.4 billion tonnes per year.'[9]

A perusal of the mainstream scientific literature on soil carbon sequestration suggests that livestock may have something to offer, but perhaps not a lot. Land which is grazed usually sequesters more carbon than land which is cultivated – hence Harvey's enthusiasm for replacing intensive grain growing with grazing. But grazed land is usually less productive than cultivated land, so the conversion of arable land to pasture would either require the reverse taking place somewhere else, or else a decline in livestock numbers and meat-eating. By and large (though by no means always) woodland sequestrates more carbon than pasture. According to studies carried out by Pete Smith and others at Aberdeen University, a system of farming similar to that described in the Organic Livestock scenario described in Chapter 9 in which a third of the EU's arable land was put into ley-arable rotation, would increase total soil carbon stocks by 39 million tonnes per year. Putting a similar area of land over to forest would increase soil and wood carbon stocks by 49 million tonnes. But these figures represent only 4.27 and 5.41 per cent of all the CO_2 produced annually in Western Europe.[10]

The FAO in *Livestock's Long Shadow* conclude 'the potential for incremental accumulation of organic carbon in soils is huge and adapting extensive livestock systems is the key to unlocking this potential.' [11] They even recommend including 'soil carbon as a class of offset credits for greenhouse gas emissions trading.'[12] But on further reading it looks as though 'adapting extensive livestock' means 'reorienting extensive grazing towards the provision of environmental services', which in turn may be a euphemism for restricting it:

> Carbon sequestration, through adjustments in grazing management or abandonment of pastures will also be difficult, but given the potential of the world's vast grazing lands to sequester large amounts of carbon and reduce emissions, mechanisms must be developed and deployed to use this potentially cost-effective avenue to address climate change ... Grazing access will have to be restricted and managed, often in a way that makes livestock production a secondary output and environmental services a primary one. [13]

8   Paustian, K *et al* (1997), 'Agricultural Soils as a Sink to Mitigate CO_2 Emissions', *Soil Use and Management*, 13, 230-44.

9   Lal, R (1998) 'Myths and Facts About Soils and the Greenhouse Effect', in Rattan Lal (ed.), *Soil Carbon Sequestration and the Greenhouse Effect*, Proceedings of a Symposium held by the the the SSSA in Baltimore MD 18-22 Oct 1998, Soil Science Society of America Inc, Madison WI, 2001.

10   Smith, Pete *et al*, 'Potential for Carbon Sequestration in European Soils: Preliminary Estimates for Five Scenarios Using Results from Long term Experiments', *Global Change Biology* 3, 67-79, 1997.

11   FAO, 2006, *Livestock's Long Shadow*, FAO, p 114.

12   *Ibid.*; see also Conservation Technology Information Center (2008), *Mitigating Climate Change, Conservation Agriculture Carbon Offset Consultation*, West Fayette Indiana, 28-30 October 2008, http://www.amazingcarbon.com/

13   FAO (2006), *op cit*. 11, p 280.

This echoes the 'spare not share' agenda which is advocated throughout *Livestock's Long Shadow*, and by the GOOFs – intensification of arable agriculture and withdrawal from biomass-based natural processes – though the FAO pick their words carefully. It derives from a widespread view that extensive grazing lands have been overgrazed. This perception has already led to a halving of cattle numbers on public rangeland in the US, and in the UK it is reflected in recent policy to reduce livestock populations, particularly sheep, through withdrawal of subsidies, and through stewardship agreements that impose a maximum stocking rate. Yet there is no conclusive evidence that reducing or eliminating grazing on pasture land sequesters carbon; indeed there are some studies showing the reverse, that both heavy and moderate grazing can *increase* carbon sequestration.[14]

Faced with this level of uncertainty amongst the scientific community, hundreds of scientific papers showing often conflicting results, and the widespread view that only a fraction of our total CO_2 emissions can be sequestered with any degree of confidence, my instinct is to dump the whole vexed issue of soil sequestration into a parenthetical box or footnote, and forget about it until the scientists come up with as much consensus about soil carbon as they have about fossil fuel carbon. However I am dissuaded from that course because the heretical viewpoint – as disseminated in the UK by Graham Harvey – represents a permaculture approach to livestock management which is attracting some interest, and which, though it yet has to prove itself, deserves an airing in this book.

Amazing Grazing

Graham Harvey speaks for a different breed of carbon farmers who see livestock as a pro-active way of improving carbon sequestration. For these farmers, a herd of grazing animals is not a voracious destroyer of soil and landscape, but an indispensable element of an integrated semi-natural environment. Enhancement of the carbon carrying-capacity of the soil is inseparable from the enhancement of both the plant and the animal carrying-capacity of the land. Insofar as these elements are all seen as symbiotic, livestock carbon farming represents one possible permaculture approach to carbon sequestration.

This movement can be traced back to a single book, written in the 1950s with the unprepossessing title *Grass Productivity*. Though ostensibly dry and scientific, this is no textbook.[15] The author, André Voisin, was a French biochemist and small farmer, who, the blurb tells us, 'was known to spend hours watching the cows graze on his farm in Normandy. It was then that he realized that simple observation of the cow at grass could teach more about ecological relationships than the most sophisticated research of the time.' One attraction of Voisin's book is that he goes some way towards explaining how to 'think like a cow' with less of the anthropomorphism that pervades some other attempts to unveil the secret thoughts of bovines. But the reason why the book has a cult following is that it is the first attempt to analyse in detail the principles behind what Voisin called 'rational grazing', and has since come to be known as 'pulse grazing', 'mob grazing', 'short duration grazing' or even 'controlled overgrazing'.

Pulse grazing involves carefully rotating livestock around paddocks so that they eat grass at the optimum moment in its growth, and for the optimum period

14   Schumann, G E *et al* (2001), 'The Dynamics of Soil Carbon in Rangelands', in Follett, R F, Kimble, M and Lal, R (2001), *The Potential of US Grazing Lands to Sequester Carbon and Mitigate the Greenhouse Effect*, Lewis Publishers.

15   Voisin, André, *Grass Productivity*, Island Press, 1988.

of time. If they are left for more than two or three days days in the same paddock, they will return to the favoured plants that were eaten on day one and have now begun to grow back to take a second bite, while ignoring the less palatable grasses, resulting in excessive grazing pressure on the higher quality grass, and inefficient use of the lesser. This is hardly a radical observation, but Voisin has a keen eye for detail and he goes on to propose some sophisticated methods of managing and fencing land to ensure 'rational grazing'.

Voisin's book was largely ignored until it was picked up in the 1970s by a biologist, Allan Savory, who was working with buffalo, elephant and other big game in what was then Rhodesia, and was dismayed that cattle were 'overgrazing my beloved Africa to death'. Savory was approached for help by some cattle ranchers who themselves were becoming worried by the deterioration in their rangelands:

> Although my antagonism to cattle was well known, what I was saying publicly at the time made sense to them … In tackling this new challenge to manage cattle and wildlife together while improving the land I dusted off my copy of *Grass Productivity* and was astounded to find that Voisin had already solved the riddle of *time*. He had proven that overgrazing had little relationship to the number of animals but rather to the time plants were exposed to the animals, the time of exposure being determined by the growth rate of plants. If animals remained in any place for too long, or returned to it too soon, they overgrazed certain plants. Suddenly I could see how trampling also could be either good or bad. Time determined that too. The disturbance needed for the health of the soil became an evil if prolonged too much or repeated too soon.[16]

Savory surmised that the ecology of natural rangelands had evolved to favour pulsed grazing. In wide open grassland, large herbivores herd together for safety from predators who escalate the numbers game by hunting in packs. Wildebeest and zebra in Africa do not amble across the savannah in family groups sampling the vegetation, picking out the choicest morsels. They move like an army upon one location kicking up dust and spewing out methane, and graze close to the ground before moving on. It sounds like a scorched earth policy, but the volume of dung deposited on the areas they have grazed deters them from returning until the grass has recovered. Their migration patterns are as much a carefully timed rotation as Voisin's tidy fences, while their hooves break up soil crusts and trample dead vegetation and seed into the aerated surface, like a rotary harrow. By contrast, a herd of sheep or cows left to their own devices in a fenced field with sufficient grass for a season and no fear of predators, will spread out, graze in a haphazard fashion, and scatter their dung here, there and everywhere.

Migratory mob-stocking works for wildebeest (as the rise in their numbers in Serengeti demonstrates); could it work for cattle? Savory began to put his theories into practice in Zimbabwe but his outspoken opposition to Ian Smith's regime forced him to leave. He moved to the United States in 1978, set up a Centre for Holistic Management in Albuquerque, and began to spread the gospel that 'western rangelands are understocked and overgrazed'.

Savory collected his observations into a lengthy volume now entitled *Holistic Management*, which, though it never mentions the word 'permaculture' could justly be subtitled 'a permaculture approach to rangeland management'.[17] As his

---

16   Savory, Allan, Introduction to the Island Press Edition of *Grass Productivity*, *ibid*.
17   Savory, Allan with Butterfield, Jody (1999), *Holistic Management*, Island Press, p 198.

message has spread, it has raised a storm of controversy between various factions in the western ranges. Conservationists from the Bureau of Land Management and the Sierra Club have long argued that there are far too many cattle on the public and private lands of the West; and a good many agree with the sentiments that the hero of *Earth First!*, Edward Abbey, voiced at a redneck-packed public meeting at the University of Montana:

> Overgrazing is much too weak a term. Most of the public lands in the West, and especially in the Southwest, are what you might call 'cowburnt'. Almost anywhere and everywhere you go in the American West you find hordes of these ugly, clumsy, stupid, bawling, stinking, fly-covered, manure-smeared, disease-spreading brutes. They are a pest and a plague. They pollute our springs and streams and rivers. They infest our canyons, valleys, meadows and forests. They graze off the native bluestem and grama and bunch grasses, leaving behind jungles of prickly pear. They trample down the native forbs and shrubs and cacti. They spread the exotic cheatgrass, the Russian thistle and the crested wheat grass. Weeds. Even when the cattle are not physically present, you'll see the dung and the flies and the mud and the dust and the general destruction. If you don't see it, you'll smell it. The whole American West stinks of cattle.[18]

Not any more, replies the new breed of holistic cowboy. Cattle have been mismanaged, not overstocked. Ever since the dust bowl years there has been a widespread view that extensive grazing lands have been overgrazed, and this perception has led the Bureau of Land Management in the USA to reduce grazing on 124 million acres of public lands from about 22 million authorized livestock units in 1941 to 12.5 million in 2008 (though not all these permits are used and in 2008 the actual stocking level was 8.6 million).[19] There were once 60 million bison on the Great Plains, now there are an estimated 60 million cattle on the USA's 336 million hectares of grazing land.[20]

By contrast, in some parts of Africa, for example KwaZululand and communal areas of Zimbabwe and Botswana, experts have been claiming for more than 50 years that land is overstocked, while local pastoralists (who believe otherwise) have maintained stock levels, without any apparent decline in carrying capacity. 'Why are there so many animals in the communal areas of Zimbabwe?' asks agricultural ecologist Ian Scoones. Because communal herders, unlike the farmers with fenced private ranches, can move their cattle to suit to seasonal and climate conditions and 'exploit resource patches within a heterogeneous landscape in a flexible opportunistic way'.[21] The movements of nomadic herders are not that far removed from the migrations of wildebeest and bison.

Range cattle, say the disciples of Voisin and Savory, need to be pulse grazed, much as vast herds bison were pulse grazed by the Blackfoot Indians who shadowed them, playing the same role as packs of lions in Africa. Dan Dagget's book *Gardeners of Eden* takes us on a tour of ranches where cattle, or in some cases bison, have been employed to bring rangeland back from a state of parched exhaustion to one of floral abundance, pictured in glossy Arcadian

---

18   Abbey, Edward (1988), 'Free Speech: The Cowboy and His Cow', in Abbey, Edward (1988), *One Life at a Time Please*, Henry Holt and Co.
19   Bureau of Land Management (2009), *Factsheet on the BLM's Mangement of Livestock Grazing*, US Dept of the Interior, Jan 2009, http://www.blm.gov/wo/st/en/prog/grazing.html
20   Herman Mayeux, Foreword to Follett *et al* (2001), *op cit. 14*.
21   Behnke, Roy H Jr, Scoones, Ian and Kerven, Carol (1993), *Range Ecology at Disequilibrium*, Overseas Development Institute, 1993.

photographs of flower-strewn meadows and mist-shrouded pastures. Several of his photographs borrow a device also used by Savory and beloved by advertisers of soap powders: this side of the fence (brown and sparse) is undergrazed public land – the ranch on the other side (green and lush) used holistic management.

On one of Dagget's farm visits, to the Goven Ranch in North Dakota, a State University Scientist takes a spade and digs a hole in the pasture: "'See how easy that shovel cut into the dirt" he said, "that shows us we've got dirt here instead of concrete. The organic content of this soil, the carbon, makes it easier to dig."' Over the other side of the fence lies Federal Conservation Reserve Program Land, where grazing has been restricted for fifteen years. In the middle of a field the scientist pulls out his spade and starts digging: 'Bouncing up and down on his shovel as if it were a pogo stick, he finally managed to jackhammer a hole a few inches deep. He then showed us that the soil here was a light gray rather than the rich black we had seen in the pasture. He also showed us that there were fewer roots in what soil he could dig up.'[22]

Targeted intensive grazing also provides a way of dealing with the cheatgrass (*Bromus tectorum*) mentioned by Edward Abbey. This European invader (an annual known as drooping brome in the UK) has colonized between 17 and 25 million acres of formerly biodiverse native grassland. Richard Manning in his book *Grassland* called it a 'natural revenge on the cattlemen'.[23] Tolerated initially because, when young, it is palatable and nutritious for cattle, it is now widely (though not universally) regarded as a scourge: 'It's called cheatgrass because it cheats the farmer out of the water and nutrients on his land and slowly converts the range area from a grazed land to barren land', says a BASF consultant as part of her patter for selling weedkiller. Cheatgrass spreads because it doesn't 'cure'.[24] Most native American grasses remain nutritious and palatable to livestock once they have dried on the stem, whereas *Bromus tectorum*, and most other European grasses don't – that is why cattle farmers in Europe make hay, whereas bison graze throughout winter on native grasses (unlike European cattle, they know how to scrape the snow away with their muzzles).[25] Moreover, cheatgrass seeds have sharp awns that injure the mouths of cattle, making it still more unpalatable. A stand of cheatgrass inadequately grazed in spring becomes a blanket of tinder-dry vegetation which is six times more likely to cause a wildfire than native sagebrush. After a fire, the following spring, the fast growing cheatgrass germinates with enthusiasm and smothers slower-growing native plants, progressively establishing its own private fire-assisted monoculture. 'There are no biological controls for cheatgrass,' the BASF consultant advises; but Dagget is supported by most of the academic literature in his contention that timed, intensive grazing is the best way of controlling cheat grass, whereas non-intervention allows it to spread.[26]

On the one hand, Savory's theories seem to make sense: why shouldn't humans be able to imitate with cows what nature achieves with buffalo? On the other hand, many of Dagget's findings sound too good to be true (an impression reinforced by the folksy journalese by which they are described). And neither's

22   Dagget, Dan (2005), *Gardeners of Eden*, Thatcher Charitable Trust, Santa Barbara CA, 2005, pp 72-4.

23   Manning, Richard, *Grassland*, Penguin, p 179.

24   Spainhower, Richard, *Cheatin' Heart*, BASF Advertorial, The Dickinson Press, Dickinson, N Dakota 2003, http://www.vmanswers.com/pdfs/CheatgrassCanRob.pdf

25   Manning, *op cit*. 23, p 127.

26   Young, James A and Clements, Charlie D 'Cheatgrass' (2007), USDA Agricultural Research Centre Reno; University of Idaho (2006), *Cheatgrass or Downy Brome, Targeted Grazing*, http://www.cnr.uidaho.edu/rx-grazing/Grasses/Cheatgrass.htm

case is helped by an analysis of pulse grazing carried out by New Mexico University agronomist Jerry Holechek, and published in a 1999 edition of the journal *Rangelands*.[27] None of the 13 research studies reviewed by Holechek showed that 'short duration grazing' offered any advantages over continuous grazing: there was no evidence that it could support a higher stocking rate, and the assertion that it improved the species mix was 'strongly rejected by the authors'. Holechek attributes the enthusiasm for short rotation grazing in the late 1980s and early 1990s to a fortuitous abundance of rain, remarking 'history shows that it's human nature to pursue a good story rather than pursue the truth'. The article elicited a debate in the letters pages of *Rangelands*, but only a weak reply from Savory; the Holistic Management website has no rebuttal of Holechek's criticisms and Dagget's book makes no mention of them.[28]

Another paper by Briske *et al*, published in 2008, analysed 28 different different studies and came to the same conclusion as Holechek, that 'these experimental results conclusively demonstrate that rotational grazing is not superior to continuous grazing across numerous rangeland ecosystems'.[29] Briske did, however, suggest some reasons for the wide gulf between the perceptions of ranchers who used rotational grazing methods, and the findings of scientists. He acknowledges that in experiments, 'grazing treatments are often applied on a more rigid schedule to ensure experimental integrity and repeatability compared to commercial systems that are adaptively managed'. Briske provides a diagram with a grid showing equal periods of grazing time allocated to paddocks throughout both wet and dry periods in the grazing season, from which he concludes that rotational grazing cannot adapt to seasonal fluctuations in grass growth – but the diagram more likely demonstrates that scientific experiments cannot adapt to them. Voisin is adamant that the time spent in paddocks must be varied according to the rate of grass growth, and emphasises in bold letters that: 'Flexibility in management is essential … It is not a case of rigidly obeying figures: one must follow the grass … Figures are only guides: in the end it is the eye of the grazier that decides.' One wonders how many of these scientific experiments have been managed by someone with the eye of a grazier.

Nonetheless the results of all these studies are depressing for anyone who (like me) finds Voisin's and Savory's theories inspiring. The dispute is similar to that which persists between enthusiasts of herbal and homeopathic treatments and members of the medical profession who only place confidence in 'evidence-based medicine' and dismiss everything else as 'anecdotal'. In this respect at least, the holistic cowboy Dan Dagget finds himself on the same side of the fence as cow-hater Edward Abbey, who assured his Montana audience: 'I'm not going to bombard you with graphs and statistics, which don't make much of an impression on intelligent people anyway. Anyone who goes beyond the city limits … can see for himself.'

---

27   Holechek, Jerry, 'Short Duration Grazing: The Facts in 1999', *Rangelands* 22(1), 2000, http://uvalde.tamu.edu/rangel/feb00/holechek.pdf

28   Described as 'weak' by another correspondent to the magazine. I have so far been unable to obtain a copy of Savory's reply. In November 2009 I attended a talk by Savory in London, and asked him what his response was to Holechek's criticism. He referred me to an article on his website entitled 'Doing the Right Thing'. However this article answers none of Holechek's criticisms, and the name Holechek doesn't give any results on the site's search facility. Rio de la Vista (2001), *Doing the Right Thing – The McNeil Ranch*, Holistic Management in Practice, July/August 2001.http://www.holisticmanagement.org/n9/Education/InPractice_Archives/Doing_the_Right_Thing.pdf

29   D. D. Briske et al, (2008), 'Rotational Grazing on Rangelands: Reconciliation of Perception and Experimental Evidence', Rangeland Ecol Manage 61:3–17 | January 2008.http://essm.tamu.edu/people/briske/documents/REMSynthesis08.pdf

A Sink for All Carbon?

While this battle is being waged out west, André Voisin's theories of rational grazing have also been spreading amongst farmers in the more temperate zones of North America, Australia and New Zealand. Increasing numbers of ecologically concerned livestock rearers are feeding their cattle or sheep entirely on grass, with little or no fattening on corn or soyabeans. Many of them consciously call themselves 'grass farmers' rather than livestock farmers. The movement has been spearheaded in the USA by the magazine *Stockman Grass Farmer*, edited by Allan Nation, who described Voisin's *Grass Productivity* as 'the real McCoy, the book that started the worldwide revolution in grassland management'. The revolution's extent can be seen on the website *Eatwild.com*, which includes a directory of grass farms across the USA, listing 61 in the state of Pennsylvania alone.

Stockrearers are drawn to grass-farming for a host of reasons: because, in their view, it is ecologically benign, because it has low overheads, because the food tastes better, because it uses less fossil fuel. Many, if not most, will be aware of Voisin's theories and a considerable number of them follow his prescriptions about rotation. Joel Salatin, whose Polyface Farm is a star feature of Michael Pollan's book *The Omnivore's Dilemma*, describes his 'mob and move' system in language borrowed from both Voisin and Savory:

> What we're trying to do here is mimic on a domestic scale what herbivore populations do all over the world. Whether it is wildebeests on the Serengeti, caribou in Alaska or bison on the American plains, multi stomached herds are always moving onto fresh ground, following the cycles of grass.[30]

Salatin moves his cattle every day – 'my neighbours think I'm insane'. He points out that as well as favouring grass growth 'short duration stays allow the animals to follow their instincts to seek fresh ground that hasn't been fouled by their own droppings, which are incubators for parasites'. Three days after the cows have left the patch, Salatin moves on his chickens to pick out the maggots and the worms from the cowpats. His entire farm (like any good farm) is a web of symbiotic relationships. The chickens also introduce fertility, but that is another matter that I will come to in a minute.

Pollan goes on to explain how the regular grazing system brings more sunlight to groundhugging clovers, and hence more nitrogen into the soil, increasing productivity:

> The productivity means Joel's pastures will, like his woodlots, remove thousands of pounds of carbon from the atmosphere each year, instead of sequestering all that carbon in trees. However, grasslands store most of it underground, in the form of soil humus. In fact grassing over that portion of the world's cropland now being used to grow grain to feed ruminants would offset fossil fuel emissions appreciably. For example, if the sixteen million acres (6.4 m ha) now being used to grow corn to feed cows in the US became well-managed pasture, that would remove fourteen billion pounds (6.4 Mt) of carbon from the atmosphere each year, the equivalent of taking four million cars off the road.

The penultimate chapter of Dagget's *Gardeners of Eden* homes in on the same matter. The black, spade-friendly, root-packed humus on the holistic side of the

---

30   Pollan, Michael (2006), *Omnivore's Dilemma*, Bloomsbury, p 192.

fence sequestrates far more carbon than the compacted pan on the ungrazed side. Targeted grazing of cheatgrass means that it is converted into manure and humus; continuous grazing or minimal intervention means that it goes up in smoke. Dagget interviews a Utah rancher and consultant called Greg Simonds who is developing a package of satellite mapping and computer monitoring techniques for estimating the amount of carbon sequestrated on ranches. 'I've seen where carbon sequestration has sold for about $25 a ton' says Simonds. 'A ranch, even in this dry country can sequester about a ton of carbon per acre. Twenty-five dollars an acre, that's a very good return for a rancher in this kind of country.'[31]

The only book I have found which attempts to explain the holistic cowboy's approach to carbon farming comes from an Australian called Allan Yeomans (a lot of these mavericks are called Allan for some reason). His father, PA Yeomans, developed a mathematically ingenious system of contour ploughing, known as the Keyline system, and also manufactured Australia's first chisel plough for breaking up the soil without turning it over – an important element of arable soil carbon sequestration. Allan Yeomans' 2005 book *Priority One* is a rollicking good read if you like challenging ideas; but his contention that *all* of the excess carbon in the atmosphere can be sequestered in the soil through organic carbon farming would be more convincing if some of the facts and figures were backed up by references. It would be even more convincing if the longest chapter in the book did not consist of a 150 page-long conspiracy theory explaining how virtually every environmental campaign, from land rights struggles against hydroelectric dams to the use of bicycles in cities, is funded and promoted by the fossil fuel industry, in order to detract attention from the real short term solution – carbon sequestration. His attack on 'the slogan dominated cult of rainforest protection' and his proposal to replace a massive chunk of the Amazon with sufficient sugar cane plantations to power the entire American car-fleet sits uneasily with his uncompromising insistence on organic farming. Politically he is to the right, brandishing a contempt for land rights and a conviction that the imperial nations did not exploit the resources of other countries but 'pulled themselves up by their own bootstraps'. Perhaps it is because he is an embarrassment that Graham Harvey, who seems to share Allan Yeomans' belief that soil sequestration on its own can solve global warming, doesn't mention him in his own book ,*The Carbon Fields*, referring instead to Yeomans' brother, Ken, and their father.

Allan Yeomans argues that to reduce the concentration of carbon dioxide in the atmosphere from the current 380 parts per million to below 300 ppm we would have to remove about 167 billion tonnes of carbon. For this to be sequestered in the world's five billion hectares of arable and grazing land, each hectare would have to absorb, on average, an extra 32.8 tonnes of carbon. This is the equivalent of converting 1.6 per cent of the top 12 inches of soil into organic matter.[32] These

---

31   Dagget (2005), *op cit. 22*, pp 128-9.

32   Yeomans, Allan J A (2007), *Priority One*, Biosphere Media, http://www.yeomansplow. au/priority-one.htm. Yeomans arrives at his 1.6 per cent through the clumsy device of imagining columns of atmosphere and layers of soot, and does so on the grounds that 'quoting huge numbers with lots of noughts is confusing and usually becomes meaningless, that's why it's fed to us.' Actually it's his layers of soot that are confusing – like those football fields in the rainforest. Fortunately, Peter Donovan of the Canadian group, the Soil Carbon Coalition, has taken the trouble to transcribe Yeoman's figures clearly into conventional mathematical units that can be compared easily with everybody else's. His calculations are as follows: The atmosphere currently contains about 800 Gt (gigatons or billion metric tons) of carbon. The vast majority of this is in the form of carbon dioxide, which is currently about 383 parts per million (Yeomans used 380 ppm for his calculation). Each of these parts per million = 800/383 = 2.089 Gt C. So, to take out 80 ppm we are talking about removing 80 x 2.089 or 167 Gt (167,000,000,000 metric tons) of carbon from the

figures are consistent with those given by mainstream experts such as Dr Rattan Lal, of Ohio University.[33] It doesn't sound like a massive amount, and according to Yeomans it is easily achievable through a three pronged approach:

- organic farming, because chemical nitrogen not only fails to introduce organic matter in the form of manure, but also restricts the fungal and bacterial activity that provides humus;[34]

- 'controlled overgrazing', in the manner described by Voisin and Savory;

- the use of subsoil ploughs because 'the one thing the animals can't do in the development of soil is to loosen up the ground to let in air and rain as easily as can a wise farmer with a chisel plough or subsoiler.'[35]

The potential benefits of organic farming and minimum tillage in respect of sequestering carbon are widely (though by no means universally) acknowledged, and the only element in Yeoman's recipe that might be viewed as wacky is the 'controlled overgrazing'. In this respect, the subsoiling makes sense, since one of the main criticisms of Savory's grazing methods is that it causes compaction of the soil. But how does this assist carbon capture? And what part do the animals play? Yeomans explains:

> Healthy grazing animals, nutritional grasses and fertile soils have evolved and combined over time to form highly productive symbiotic relationships. The cycle begins when the grazing animal bites the top of the plant. The plant no longer has leaves, and with no chlorophyll, can't utilize the sun's energy for growth. The plant immediately switches energy sources and utilizes the carbohydrates stored in its root structures to rapidly regrow its leaves. The plant roots, after being utilized in this regeneration process, are shed like autumn leaves. Microbial soil life thus gets a crop of dead root material to feast on. Each partly eaten plant that goes through this cycle generates a net increase in soil organic matter.[36]

This organic matter is 58 per cent carbon (one of the few undisputed figures you will find in the entire field of soil carbon) and the more carbon there is stored

---

atmosphere. Soil density is usually between 1.2 and 1.4 on a dry basis. That is in relation to the density of water which is 1.0. A hectare of soil (100 m x 100 m), 12 inches or 30.48 cm deep, has a volume of 3,048 cubic metres. At a soil density of 1.2, this foot-deep hectare of soil weighs 3,658 metric tons. One per cent of this weighs 36.58 metric tons, and if this 1 per cent is organic matter (58 per cent carbon by weight), it contains 21.21 tons of carbon. According to the World Resources Institute (*World Resources 2005: The Wealth of the Poor-Managing Ecosystems to Fight Poverty*, table 11, p 216), there are 5,096,000,000 hectares of crop and grazing land worldwide. Dividing the 167 billion tons we aim to remove by this figure, each hectare has to take, on average, 32.8 tons of carbon. Since each per cent of organic matter in the top foot of soil contains 21.21 tons of carbon, this comes to roughly 1.56% as the amount of soil that we must convert to organic matter on the top foot of these lands, to lower atmospheric concentrations by 80 ppm. In other words, if the organic matter content is currently 1 per cent, we need to raise it to 2.56 per cent, if 3 per cent we raise it to 4.56 per cent, and so on. Donovan, Peter (n.d.), *The Calculation*, Soil Carbon Coalition, http://soilcarboncoalition.org/calculation

33   Lal, R (2003), *Agricultural Sequestration to Address Climate Change through Reducing Atmosphering [sic] Levels of Climate Change*, Address to the Subcommittee on Clean Air, Climate Change and Nuclear Safety, US Senate Committtee on Environment and Public Works, July 8, 2003.

34   Yeomans, *op cit. 32*, chapter 6.

35   Yeomans, *ibid,* p 148.

36   Yeomans, *ibid,* p 148.

in the soil, the less there will be in the atmosphere. Pollan describes exactly the same process taking place as one of Salatin's cows browses: 'The shorn grass plant, endeavouring to restore a rough balance between its roots and its leaves, will proceed to shed as much root mass as it's just lost in leaf mass. When the discarded roots die, the soil's resident population of bacteria, fungi and earthworms will get to work breaking them down into rich brown humus.' Graham Harvey also explains the 'rebalancing process that keeps the roots and aerial parts in equal proportion', concluding that 'the key to efficient carbon capture is good grazing'. Dagget explains that the rich black earth on the Goven ranch is due to the very same process. Savory agrees that 'any time a plant is severely defoliated root growth ceases as energy is shunted from root growing to regrowing leaves'. However he is more concerned that excessive root death might be a symptom of excessive grazing: 'Overgrazing is "grazing of the roots".'[37]

As carbon sequestration climbs up the agenda, this process of plants shedding roots in sympathy with their lost leaves seems to have become an article of faith amongst carbon farmers. None of the five authors I have cited gives any scientific reference for this biological phenomenon, though it is unlikely to be verifiable by any means except scientific experiment. There is not a word about it in my standard textbook, *Improved Grassland Management* by John Frame, but then even the word 'roots' doesn't occur in its 14 page index.[38] Robert Kourik's enticingly titled but disappointingly suburban *Roots Demystified*, has a chapter on buffalo grass in lawns, but not a word about the buffalo.[39] It took a lot of thumbing through the abstracts of scientific papers before I came across a reference, in a book edited by Rattan Lal and others, to a paper by Hodgkinson and Baas Becking, written in 1977, which reported that 'when shoots are grazed, root mortality frequently increases and root extension and branches decreases. This root mortality could result in significant increases in Soil Organic Carbon (SOC) over a relatively short time.'[40] Here, at last, was peer-reviewed scientific confirmation of the fact that grazing triggers root death. However, later I came across a 2009 research paper by Klumpp *et al*, entitled 'Grazing Triggers Soil Carbon Losses by Altering Plant Roots.'[41] Klumpp and her colleagues had conducted laboratory tests on 'monoliths' or samples extracted from temperate permanent pastures, which showed that when grazing was increased from a very low frequency to five times a year, there was indeed increased root mortality, but that this led to decreased root biomass and lower levels of soil carbon. This happens all the time in the murky world of soil carbon science – no sooner do you read a paper claiming one thing, than you find another showing that in slightly differing circumstances the very opposite occurs. However there is one matter on which almost everybody is in agreement – that roots are more important than shoots. The above ground biomass (leaves, stalks and so on) even when ploughed into the ground and decomposed by microbes, make considerably less contribution to the carbon content of the soil than the roots. According to one study, this is because it is more stable and lasts over twice as long in the soil.[42]

37   Savory (1999), *op cit.* 17, p 219.
38   Frame, John (1992), *Improved Grassland Management*, Farming Press.
39   Kourik, R (2008), *Roots Demystified*, Metamorphic Press.
40   Hodgkinson, K C and Baas Becking, H G, 'Effect of Defoliation on Root Growth of Some Arid Zone Perennial Plants', *Aust J Agric Res*, 29:31-42; cited in Schumann G E *et al* (2001), *op cit.* 14.
41   Klumpp K *et al* (2009), 'Grazing Triggers Soil Carbon Losses by Altering Plant Roots and their Control on Soil Microbial Community', *Journal of Ecology*, 97, pp 876-85, 2009.
42   E.g. Rasse *et al* conclude from their experiments that 'the relative root contribution to SOC has an average value 2.4 times that of shoots'. D Rasse *et al* (2005), 'Is Soil Carbon Mostly Root Carbon? Mechanisms for a Specific Stabilization', *Plant and Soil*, 269, pp 341-56.

The much bigger question mark raised by *Priority One* is the scale of the rise in organic matter that Yeomans anticipates, which is of a level of magnitude higher than those reported by mainstream scientists. In 2000, the IPCC listed carbon sequestration figures from a large number of studies covering a wide number of methods applied to grazing and croplands. The majority of reports listed improvements in the range of 0.1 to 0.5 tonnes per hectare per year, while the mean for minimum tillage systems was 0.3 tonnes.[43] At this rate (and providing the carbon level doesn't reach saturation point) it would take between 65 and 328 years to reach Yeoman's target of 32.8 tonnes. The only improvements noted by the IPCC which anywhere matched the rate anticipated by Yeomans were when legumes or other new species were introduced into native grasslands – resulting in annual increases of between 0.36 and 3.34 tonnes. One survey of the planting of deep rooting grasses and legumes in Colombian savannah reported an increase of up to 14.4 tonnes, far higher than anywhere else.[44] However, the report of this study in *Nature* prompted a response from scientists from Woods Hole Research Centre, who voiced concerns about the length of time that this sequestration could be maintained, and the decline in biodiversity that would result from improving pastures in this manner.[45]

Admittedly none of these studies examine land managed under a combination of two or all of the three conditions that Yeomans stipulates. None of the IPCC studies were listed as being organic (though perhaps some of them were). The highly productive Rodale experiment (see Box on page 81) over a period of 21 years raised the proportion of carbon in the soil by about 0.6 per cent in the organic livestock system, and 0.5 per cent in the stockless organic system, whereas there was barely any increase of carbon in the chemical system. This is by no means negligible, but at that rate (if it could be maintained) it would take 56 and 67 years respectively to reach Yeoman's projected 1.6 per cent increase. Global warming, by most accounts, will be well under way by then.[46]

Nonetheless, the fact that organic, minimum tillage and grazing systems all score quite well in mainstream literature does suggest that Yeomans might be on the right track. Unfortunately, the only evidence he provides to show that the levels of carbon sequestration he proposes are actually achievable is a single eye witness report of his father's soil building prowess:

> Sir Stanton Hicks writes in his book *The Nutritional Requirement of Living Things*: 'After three years he (PA Yeomans) was able to demonstrate that black soil had formed to a depth of 12 inches where scarcely any soil had previously existed on this rockstrewn countryside. Thousands of farmers and many distinguished visitors from overseas have witnessed the results of this transformation of a barren tract of land into a parklike region, carrying fine livestock all year round.'[47]

At no point does Yeomans produce any before-and-after measurements of carbon on either his land or anyone else's. I have witnessed soil of the kind he describes built up on land that had previously been stripped of its organic matter,

43   IPCC (2000), *Land Use, Land Use Change and Forestry*, http://www.grida.no/publications/other/ipcc%5Fsr/?src=/Climate/ipcc/land_use/index.htm

44   Fisher, M J *et al* (1994), Carbon Storage by Introduced Deep-Rooted Grasses in the South American Savannas', *Nature*, vol 371, 15 September 1994. Actually the highest figure reported was 11.73 ± 2.58, but the IPCC cite the top end of the standard error range.

45   Davidson *et al* (1995), 'Pasture Soils as Carbon Sink', *Nature*, Vol 376, 10 August 1995.

46   Pimentel, D *et al* (2005), *Organic and Conventional Farming Systems: Environmental and Economic Issues*, Cornell University, 2005.

47   Yeomans, *op cit.32*, 146.

for example by growing and mulching comfrey – but only on a small scale, in a temperate landscape and probably with considerable inputs of compost. Much as those of us with organic or permaculture leanings would like to believe his theory, more credible evidence would be necessary to prove the point.

Graham Harvey's book on the carbon fields of Britain doesn't shed much light on the matter either. He cites the 1.6 figure, which he credits, not to Yeomans, but to the Carbon Farmers of America, though they feature it in an excerpt from Yeomans' book on their website. Harvey prefers to rely on another Australian, Dr Christine Jones, who runs an accreditation scheme, which currently has a pilot project in Western Australia generously paying farmers 90 Australian dollars (about 57 US dollars) for every tonne of carbon their soils absorb. The fund is subsidized by Rio Tinto Coal, a subsidiary of the Anglo-Australian mining company Rio Tinto – which rather scuppers Yeomans' notion that the fossil fuel industry wants to divert attention away from carbon sequestration.[48]

I got off to a bad start with Dr Jones when I picked up on this comment in Graham Harvey's book: 'Christine Jones … calculates that raising the level of soil organic matter by one per cent across just 15 million hectares of land would capture the carbon equivalent of the earth's total greenhouse emissions.'

'Some mistake, surely?' I thought when I read this and turned to Dr Jones' website, where I found this variant: 'it would only require a one per cent increase in soil carbon on 15 million hectares of land to sequester 8Gt (billion tonnes) of carbon dioxide, which is equivalent to the greenhouse emissions of the whole planet.' I struggled long and hard with this statistic. First I had to work out that when she says 'one per cent increase in soil carbon' (which suggests that to achieve the desired increase, the 15 million hectares would have to start out a metre deep in solid soot) what she means is 'the conversion of one per cent of the soil to carbon.' Next was the matter of the world's greenhouse emissions. The annual global anthropogenic emissions of carbon are about 8Gt – but that is the equivalent of 30 Gt of carbon dioxide, while Jones' 8Gt of carbon dioxide are around 2.16 Gt of carbon. Eleven tonnes of CO_2 are equivalent to about three tonnes of carbon. Perhaps she got muddled up, its a very easy mistake, but not one you'd expect from a PhD in Soil Science. But even if we take the lower figure, her one per cent increase consists of an annual increment of 144 tonnes of carbon per hectare – exactly ten times as much as the highest potential increase recorded by the IPCC in its overview of carbon sequestration studies, and well over 100 times as much as the vast majority of the increases it records.[49]

Perplexed, not only by this, but by some other statistical inconsistencies on her website, I emailed Christine Jones asking for clarification. She wrote back: 'We are in the middle of an extremely busy fieldwork period here. On top of that, hundreds of e-mails come in from all over the world every day. I suggest you READ the articles on the Amazing Carbon website – you will find all your answers there.' I had read nearly half the website before e-mailing, and so set down to read the rest.

After this inauspicious beginning, and given the aura of flakiness and quackery surrounding a wing of the carbon farming movement, I was not overoptimistic. However after rooting through a lot of very repetitive papers from her website and other sources, I managed to obtain a fairly good picture of how the Australian Soil Carbon Accreditation Scheme (ASCAS) proposes to save the planet.

48   Jones, C (2007), Australian Soil Carbon Accreditation Scheme, Renewable Soil.com, 29 March 2007, http://renewablesoil.com/australian-soil-carbon-accreditation-scheme.html
49   IPCC (2000), *op cit. 43.*

Since the European settlement of Australia, 50 to 80 per cent of carbon has been lost from most farmed soils through unsustainable farming practices reliant on heavy duty tillage and blanket use of chemical fertilizers. This can be put back, Jones argues, by using regenerative farming methods that use minimal tillage, keep a ground cover at all times and have a high proportion of perennials. The method she particularly champions – in fact in most documents it is the only one she cites – is the 'pasture cropping' or 'perennial cover cropping' system developed by a farmer called Colin Seis. This version of zero tillage involves drilling grain crops into perennial pasture at a time of year when it is dormant. It is reliant on fertilizer; and yields of grain crops are respectable for Australia, although well below UK levels. Seis acknowledges that it may only be applicable in certain areas.[50] It is certainly hard to imagine it working in the south west of the UK, where pasture is never dormant, except when it is frozen. There have been a number of experiments in the UK involving wheat or other cereals sown into clover, which almost certainly increase soil carbon content. Unfortunately most of the trials are more concerned with yields, which are respectable by organic standards, but lower than compared to conventional chemical monoculture, and it is hard to find results for soil carbon content in reports of these experiments.[51]

The level of carbon sequestration that Jones thinks is generally 'achievable by landholders practising regenerative cropping and grazing practices' is 0.15 per cent of the soil turned to carbon per year – the equivalent of adding 23.1 tonnes of carbon to every hectare, each year. This is broadly consistent with another planet-saving statistic that she likes to repeat, that 'it would require only a 0.5 per cent increase in soil carbon on two per cent of [Australia's] agricultural land to sequester all of Australia's emissions of carbon dioxide.' (If this seems implausible, it is worth bearing in mind that Australia is huge and underpopulated: Jones is talking about an area half the size of Britain's agricultural land for a population about a third the size of Britain's.)

Twenty three tonnes per hectare is about 20 times the amount of carbon that the IPCC considers can normally be gained from a variety of land use practices, such as minimal tillage or pasture improvement.[52] Dr Jones' projections seem to be hypothetical; I could find no evidence of trials showing that this level of carbon sequestration had been achieved on any regular basis. The mainstream guru of carbon sequestration, Rattan Lal, estimates that the USA has the potential to sequestrate between 122 and 357 million tonnes of carbon per year on 491 million hectares – about ten per cent of the world's arable and grazing land.[53] This is way short of Dr Jones' estimate of 185 million tonnes from 8.9 million hectares; the rate of sequestration anticipated by Jones is 40 times the average anticipated by Lal, who, by mainstream standards is a soil carbon 'optimist'. Moreover, this rate of sequestration cannot be kept up indefinitely, since soil, like forests, reaches a state of carbon saturation – it only offers humanity a breathing space while we find an alternative to fossil fuels.

50   The information about Christine Jones' work is culled from the following documents, available on www.amazingcarbon.com: Jones, C (2008), 'Liquid Carbon Pathway Unrecognized', *Australian Farm Journal*, 1 July 2008; Porteous, James and Smith, Frank (2008), *Farming, A Climate Change Solution*, ECOS (Australia) Feb/Mar 2008; Jones, C (2007), *op cit.*; Jones, C (2006a), *Carbon and Catchments*. Also Jones, C (2006b), *Soil and Carbon Credits*, 5 March 2006, http://www.soilcarboncredits.blogspot.com/

51   Whitefield, P (2004), *The Earthcare Manual*, Permanent Publications, pp 269-70. The importance of legumes to soil carbon sequestration is convincingly demonstrated in experiments carried out by G De Deyn *et al* (2009), 'Vegetation Composition Promotes Carbon and Nitrogen Storage in Model Grassland Communites of Contrasting Soil Fertility', *Journal of Ecology*, 97, 864-75.

52   IPCC (2000), *op cit.* 43.

53   Lal, R (2003), *op cit.* 33.

So who is right? Jones claims that 'sequestration rates under regenerative agricultural regimes may be quite a bit higher than estimated by current models', in other words that the modelling system used in Australia and the UK, called Roth C, developed at the UK research institute at Rothampsted, is wrong. The Roth C model, Jones claims, 'is based on the assumption that most carbon enters soil as 'biomass inputs', that is from decomposition of plant leaves, plant roots and crop stubbles.'[54] But it does 'not take into account the process of humification of root exudates or contributions from mycorrhizal fungi.'[55] The role of these fungi has risen in scientific estimation since 1996 when the USDA researcher Sara Wright discovered the substance glomalin, a protein produced by mycorrhizal fungi that 'may account for one third of the organic carbon stored in agricultural soils'.[56]

The process is explained more clearly in an article about Jones' work that appeared in the Australian magazine *ECOS*: 'Through photosynthesis, the activities of symbiotic bacteria and fungi, associated with roots and fed by the sugars, enable the exuded carbon to be combined with soil minerals and made into stable humus which locks the carbon away ... 'This can't happen where farm chemicals kill the essential soil microbes', says Dr Jones. 'When chemical use is added to intensive cultivation, which exposes and oxidizes the humus already in the soil it is easy to see why soil has become a huge net source rather than a net sink for the atmosphere.'

Unfortunately nothing is ever that easy, least of all soil science. Since Dr Jones provides no references to support any of this, I looked on the internet. The first article I turned up was by researchers in Wisconsin attempting to test the hypothesis that mycorrhizal colonization would be higher under organic dairy and grain farming systems, than under continuous corn receiving chemical fertilizer. They found precisely the opposite: 'This study found that mycorrhizal colonization was not suppressed but actually enhanced in the continuous corn system. These results were contrary to the initial expectations ... Lower mycorrhizal levels were observed in the organic managed systems.' The authors suggest that levels of fungi in the organic soils might have been lower because more tillage was used. They observe:

> The complexity of nature, especially that part of nature which lies underground, makes the study of the effect of farming systems on root biology a challenging proposition. The current knowledge of soil biology can be illustrated by the fact that soil ecologists cannot explain the feeding strategies of more than 90 per cent of soil biota.[57]

Organic Matters

However, the Wisconsin study is probably the exception rather than the rule. In November 2009 the Soil Association published its long awaited report on 'Soil Carbon and Organic Farming', by Gundula Azeez. This is a monumental piece of work, not so much on account of its main text, which is 130 pages long, but because of its 1288 references, covering 60 pages. Azeez is just as messianic

---

54   Jones, C (2008), *op cit. 50*.
55   Porteous and Smith (2008), *op cit. 50*.
56   Jones, Christine, *Soil and Carbon Credits*, 5 March 2006, http://www.soilcarboncredits. blogspot.com/
57   Brock, Caroline *et al* (n.d.), *A Comparison of Mycorrhizal Colonization in Corn in Continuous and Organic Systems*, University of Madison Center for Integrated Agricultural Systems, http://www. cias.wisc.edu/wicst/pubs/vam.htm

in her advocacy of organic carbon farming as any of the enthusiasts discussed above, but is more measured in her assessment of the potential benefits, and more diligent about providing evidence and sources for her conclusions. For anyone struggling to get their head round the complexities of soil carbon, this report is like a Baedeker, though one cannot be sure to what extent Azeez may or may not have selected her data to suit what was clearly her brief – to show that organic farming sequestrates significantly more carbon in the soil than chemical farming.

Azeez carries out a very useful analysis of 39 separate studies comparing soil carbon sequestration rates on organic and non-organic farms in Europe, the USA, Australia and New Zealand. In 32 of these, soil carbon levels were higher under an organic regime, and on average organic farming had 20 per cent higher soil carbon levels than non-organic farming, and 28 per cent in Europe. From this, Azeez projects that organic farming could sequestrate 560 kilos of carbon per hectare per year – a figure which is in line with IPCC and other mainstream projections, though far less than that anticipated by the likes of Yeomans and Jones. This would offset 23 per cent of the UK agricultural GHG emissions, or about 1.75 per cent of all our GHG emissions. Not that much really, considering it would involve a massive change in land use and diet.[58]

Azeez's discussion of the 39 studies, as well as other research, paints a clear picture of what factors help to make organic farming more effective at retaining soil carbon:

(i)   Converting arable land to grass;

(ii)   Use of legumes to bring in fertility;

(iii)   Mixed arable farming systems where arable is alternated with ley;

(iv)   Use of catch and cover crops;

(v)   Application of biomass in the form of compost, because much of the carbon is secured in stable form as humus;

(vi)   Application of manure in solid form rather than as slurry;

(vii)   Use of straw in manure (which contains sufficient nitrogen for healthy decomposition) compared to ploughing it in;

(viii) More biomass in the form of crop residues (eg longer straw) and weeds.

(ix)   Avoidance of chemical nitrogen, on the uncertain grounds that it inhibits benign microbial action, and discourages deep rooting;

(x)   Avoidance of some other chemicals such as herbicides and the wormer avermectin.

All of these, except (ix), are well supported in the scientific literature, though it is usually possible to find exceptions that demonstrate the contrary. Azeez also concludes that the main argument levelled against organic farming in respect of soil carbon – that it cannot easily operate minimum tillage systems which rely upon herbicides – carries little weight in the UK. In this she is supported

---

58   Azeez does postulate that there might be further gains if organic management methods were applied to permanent pasture, but concedes that these are unproven. Azeez, G (2009), *Soil Carbon and Organic Farming*, Soil Association, 2009, p 97.

by scientists at Rothampsted and at ADAS who estimate that the soil carbon benefits of minimum tillage are modest, precarious and may be completely offset by additional N_2O emissions.[59]

Most of the factors listed above are not exclusive to organic farming and many can be and sometimes are practised by non organic farmers, but they are either integral or heavily associated with organic farming. The first four are probably the most effective because they involve the roots, as opposed to straw, manure and compost, all of which are derived from the shoots (although (v) the use of compost, particularly associated with Biodynamic farming techniques, performs very impressively). The first three of these are all integral to grass farming, either pure livestock or mixed livestock and arable with leys, and to a large extent that is what Azeez is gunning for: 'Reversing the trend of intensification of dairy farming, reversing the expansion of pig and poultry farming, halting imports of intensively produced chicken and pork, and reinvesting in grass-based livestock systems on the large areas of existing farmland'.[60]

There is only one difficulty with this scenario, and that is a big one, whose presence loomed more and more ominously in the background as I read further into the report without its being acknowledged. One has to read over two thirds of the way through before there is any mention of yields, and yet these are crucial. The kind of ley-based mixed farming that Azeez (like myself) advocates has every possible advantage over chemical arable farming, except one: that, in Europe anyway, it has much lower yields. This is partly because the tonnage per hectare of organic grain production is barely two thirds that of chemical farming; and partly because if our non-organic cropland is converted to organic farming, then around a third of it will have to be converted into leys. This means that if grain production levels were to be maintained nationwide, more permanent grassland elsewhere would have to be ploughed up for arable, obliterating in a rather shorter time all of the soil carbon gains that could be made over 20 years by putting existing arable over to grass.[61] Some idea of the extent of extra cultivation required and the inroads made into permanent pasture can be seen by comparing Tables B and E in Chapter 9, which compare chemical and organic provision of precisely the same diets.

This is not what Azeez intends since she states that there will be 'carbon gains from the smaller area of arable land and the greater area of permanent pasture'.[62] This can only mean that grain production would be cut by around half, and that the UK might become much less self sufficient in food. Azeez does not give explicit details about what we could produce, apart from citing a report by Philip Jones of Reading University, estimating that the production (not necessarily the consumption) of white meat would fall by 70 per cent.[63]

Jones' report, which was commissioned by the Soil Association, anticipates that under organic agriculture, as well as this drastic cut in white meat production, self-sufficiency in grain would fall from around 100 per cent to 60 per cent, and in milk from around 90 per cent to 60 per cent. The recompense would

---

59   Goulding *et al* (2009), *op cit.* 7; A Bhogal *et al* (n.d.), 'The Effects of Reduced Tillage Practices and Organic Material additions on the Carbon Content of Arable Soils', Scientific Report for DEFRA Project SP0561, ADAS and Rothampsted.

60   Azeez, *op cit.* 58, p 30.

61   In a shorter time, because soil carbon is normally lost more quickly than it accumulates, Baritz *et al* (2004), *Task Group 5 Land Use Practices and SOM,* cited in Azeez, *ibid,* p 132.

62   Azeez, *ibid,* p 100.

63   Jones, Philip, with Crane, Richard (2009), *England and Wales under Organic Agriculture: How much Food could be Produced?* Executive Summary, Centre for Agricultural Strategy, University of Reading, http://www.apd.reading.ac.uk/AgriStrat/projects/org_exec.html

be a 68 per cent increase in beef production and a 55 per cent increase in lamb production, but this would only make up about half the loss of white meat. We would either have to reduce meat and dairy consumption (a solution that Azeez mentions, but not until page 126, and then only briefly) or alternatively import meat, milk or animal feed from elsewhere. Where might these come from? Azeez suggests 'there is significant potential to increase production in temperate regions: in Central and Eastern Europe, Caucasus and Central Asia agricultural production is now only 60 to 80 per cent of 1990 levels.' But an expansion of arable production in these regions might result in losses of soil carbon equivalent to the gains made in the UK.[64] As well as feeding off 'ghost acres', we would be emitting 'ghost carbon'.

The prospect of sacrificing the UK's food security for such modest gains in soil carbon is unthinkable, especially when these gains might be negated by additional cultivation abroad. The Soil Association's report is, in a sense, a disguised argument for reducing meat consumption. By comparison, the Permaculture scenario depicted in Table F in Chapter 9, which also turns arable over to ley farming, ought to make some soil carbon savings over the Chemical Livestock scenario in Table B without resorting to ghost carbon; but this is achieved by reducing permanent pasture, reducing beef and lamb consumption rather than dairy, and bolstering pork production with food waste (which Azeez advocates should be composted).

It is also possible that an organic ley farming system, boosted with a judicious amount of chemical nitrogen fertilizer, would achieve higher yields and economize sufficiently on land for the soil carbon so accrued to outweigh the carbon emissions generated by the manufacture of the fertilizer. That largely depends upon whether or not artificial nitrogen has an inhibiting effect upon soil carbon sequestration, a matter that is disputed.[65] But that option, of course, is not open to purist advocates of organic farming, who need most of all to find ways of increasing yields.

The above difficulties are all connected with changes in land use. There remain other improvements that can be made without taking up more land, for example using cover crops, employing compost or farmyard manure in preference to slurry, or possibly the adoption of minimum tillage techniques. However, with the possible exception of cover crops, it isn't uncertain whether any of these, even if carried out on a grand scale, would have the potential to alter the balance of soil carbon very noticeably without a simultaneous expansion of legume based leys.

I therefore incline towards the opinion of scientists at Rothampsted who conclude one of their reports with the words: 'We believe that there is too much emphasis on soil C sequestration and too little attention to major climate change threats such as land clearance for agriculture (food or biofuels) other forms of deforestation and wetland drainage'. A far more effective and immediate way to reduce global carbon emissions from agriculture would be to stop importing soya, rainforest beef and palm oil.

If soil carbon sequestration is to have any serious and rapid effect upon climate change then it has to function at an order of magnitude higher than the modest levels advanced by mainstream scientists, and achieve results like the 14.4 tonnes recorded in improved grassland in Colombia, or the 23 tonnes

---

64   Azeez cites Greenpeace, *Cool Farming*, 2008, p 15 for the figures on E Europe, etc. Greenpeace anticipates that arable agriculture will increase both through greater application of N fertilizer, and a 10–14 per cent increase in arable land area in Russia.

65   Azeez, *op cit. 58*, p 81.

that Christine Jones seems to think can be gained from bi-cropping grain with
perennials, or the rich black humus that Yeomans claims to be able to create
out of Australian dust. That is why, despite my scepticism, I have spent some
time describing their efforts. These people attract adherents because they claim
to have found what we all hope exists – the 'win-win' solution that combines
increased yield with ecological benefits. It's not impossible, that is what the
discovery of the properties of clover and lucerne gave us, with huge benefits
and no ill effects (except more nitrous oxide emissions). The impact of the Haber
Bosch process has been just as radical but not as benign. Biodynamic farmers
consistently outperform both organic and chemical farmers in the trials reported
by Azeez, and while this can be attributed to their use of compost, maybe it
also has something to do with their cow horn preparations and planting by the
moon. There is nothing to be lost by encouraging mavericks to experiment with
ways in which nature can store carbon whilst releasing in greater abundance the
nutrients that we need.

Credits for the Credulous?

While experiments in carbon farming are to be welcomed, I am sceptical about
the wisdom of extending this welcome to schemes for selling carbon credits.
There is no guarantee that these are any less likely to attract con artists and
sellers of indulgences than the other carbon offset schemes which have surfaced
in the last ten years – indeed judging by the Carbon Farmers of America, they
are already on the case. Soil carbon credits are handicapped by a number of
difficulties, not least the huge expense of monitoring various different types of
carbon down to a depth of at least a metre. But there is another flaw in all the
schemes mentioned at the beginning of the chapter: farmers are paid for the
amount of carbon they sequester in their soil; but logically they ought to be paid
for the amount of carbon they extract from the atmosphere – and that is by no
means the same thing.

To illustrate this, take Polyface Farm, which features in both Michael Pollan's
and Graham Harvey's books as an example of a well run and highly productive
grass farm – which indeed it is. Polyface produces annually 'twenty tons of beef,
fifteen tons of pork, ten thousand chickens, twelve hundred turkeys, a thousand
rabbits and four hundred thousand eggs'.[66] 'These are extraordinary levels of
production' says Harvey, in harmony with Pollan who comments: 'This seemed
to me a truly astonishing amount of food from 100 acres of grass'.

And so it would be, if it all did come from 100 acres of grass; but as any
farmer can tell at a glance, it is not just astonishing, it is impossible. I wrote to
Joel Salatin asking him what his feed inputs were and he was kind enough to
reply that the beef were completely fed on grass, the broilers got about 15 per
cent of their feed from pasture, the egg-layers a little more and the pigs about 25
per cent, though in a good mast season they can get nearly 100 per cent of their
feed from acorns. As one would expect, a sizable proportion of the feed comes
from other farms.[67]

With these feed imports comes fertility, to compensate for the fertility that
goes out with all the meat that is sold. The carbon and the nitrogen in the feed is
fed to the chickens who follow behind the cows, and who deposit it on the ground
in the form of manure; some of the carbon may be directly absorbed as organic

---

66   These are the figures given by Harvey. The figures Pollan gives are presumably from an earlier
year and lower. Pollan, *op cit*. 30, p 222; Harvey, *op cit*. 1, p 74.
67   Personal communication.

matter, while the nitrogen enhances grass growth and photosynthesis, which also contributes to soil carbon. Of all the carbon added to Salatin's pastures over the years, some will have come directly from the atmosphere; but a proportion will have come, directly or indirectly, from another farm in the form of soya, corn or whatever feed Salatin buys in.

However productive Polyface may be, it is in a sense only half a farm, and it doesn't help to analyse the carbon sequestration on one half, without knowing what is happening on the other. In the case of Polyface if the feed is bought from a responsible organic grower, it may well be that the carbon sequestration on the two farms added together is positive. But in another situation it could well be different. There are plenty of stock farmers who bump up the productivity and (perhaps unwittingly) the organic matter on their farm by buying in feed from a chemical grain farmer who has stripped the carbon content of his fields close to the bottom threshold. Stock farmers in Europe might be buying soybeans that have required the clearing of forest in Brazil. Unless carbon accreditation schemes are equipped with monitoring systems a lot more sophisticated than simply taking a soil sample once a year, they could end up rewarding stock farmers who are a small part of a farming cycle that is on balance adding little carbon to the soil, or even removing it.

When I put these concerns to Christine Jones she did find time to give a one sentence reply: 'building soil carbon has very little to do with biomass and everything to do with photosynthetic capacity and soil biology'. This is the crux of her dissatisfaction with the Roth C Model, and there is plenty of evidence that the carbon introduced into the soil through transfers of biomass does not stay there very long. But biomass brings with it fertility, and fertility enhances photosynthesis. And farmers know only too well that it is possible to strip a field of its fertility and transfer it to another.

One carbon farming website[68] has an article on a North American grass farmer called Martha Holdridge, whose pastures were tested and shown to have risen from 4.1 per cent organic matter in 2002 to 8.3 per cent in 2007, an impressive rise. The land had been managed without fertilizers, herbicides or pesticides since 1987, so why had organic matter increased so suddenly in the final five years? It couldn't possibly have increased at that rate over the entire 21 years. Martha Holdridge explained to me in an e-mail that they had cut hay on the acres in question for 10 or 12 years out of the first 15, and often half of the hay was hauled away to another farm. It is a basic fact of grass farming that if you take hay or silage off a field without adding nutrients you deplete it; and if you pasture animals on a field, especially if you are also feeding them hay or other feed, then you are likely to enrich it.[69] It is for this reason that (as some farmers have pointed out) a farmer who has been pasturing livestock for years and whose soil is already saturated with carbon won't get a cent from carbon accreditation schemes.

All of this threatens to make a nonsense of any carbon credit scheme that doesn't adopt a whole systems approach. Christine Jones' Australian scheme allows farmers 'to register up to four 20 hectare so-called Defined Sequestration Areas on any portion of their property'. If I was a stock farmer entering such a scheme, with say 80 hectares of established hay meadows and 80 hectares

---

68   Donovan, Peter (2008), *What Grass Farmers Have Known all Along – Research Shows Grass Sequesters Carbon*, June 2008, http://soilcarboncoalition.org/holdridge

69   Schnabel, R R *et al*, 'The Effects of Pasture Management Practice', in F Follet *et al, op cit*. 14. Skinner cites cases where high levels of hay removal turned pastures into net sources of carbon emissions. Skinner, R H (2008), 'High Biomass Removal Limits Carbon Sequestration Potential of Mature Temperate Pastures', *J Eviron Qual*, 37(4) 1319-26.

of pasture, I would register my meadows as sequestration areas, and leave the pasture out. Once accepted in the scheme, I would convert the registered meadows to pasture, and put the unregistered pasture over to hay. That way I would be paid for carting fertility and organic matter from the unregistered area to the registered one. After a few years I would swap everything over and get paid for carting it all back again. A regular switch over from arable to ley every ten years or so would achieve the same result.

A more sophisticated regulatory system could no doubt stop this kind of scam taking place on a single farm, but it would be harder to account for traffic of soil carbon throughout a region or a nation. Farming, and especially livestock farming, involves moving nutrients around, and there is little point in measuring carbon levels at one point without taking stock of what is happening everywhere else in the system. Where there is only voluntary adherence to an accreditation scheme, the farmers who have an interest in joining will be (a) those who manage to acquire the most manure, feed, hay or whatever other organic material is available – but who may not be adding anything to the collective wealth of organic matter; and (b) those who convert from arable to grass.

To an extent (but only to an extent) this is a zero sum game. A recent paper by a Dutch agronomist, G van der Burgt, takes a 'whole systems' approach to the prospects for carbon sequestration in the Netherlands. He and his colleagues conclude that there is scope for increasing the organic matter by 0.2 per cent in individual arable fields throughout the country but:

> At national level ... the situation is completely different. Compost, being an external source at farm level, is an internal source at national level. If the increase is to be realized with compost only, the need is 8.3 million tonnes per year. Total Dutch production at the time is 1.5 million tonnes per year, and this is already used in or outside agriculture for soil improvement. Other organic sources are not easily available.

As for grassland, the situation is similar: 'As mentioned before there are no substantial external sources of carbon – organic materials – available.' And in any case, with an average of 4.3 per cent organic matter, most Dutch grassland is already fairly high, no doubt owing to the vast quantities of biomass imported from abroad. Taking agriculture as a whole, the paper concluded that any increase in soil organic matter would be 'quantitatively very limited compared to the national CO_2 emission'.[70]

The situation is similar in Britain. Most existing organic materials already find their way into the soil, the exceptions being a small percentage of the sewage, most food waste, slaughterhouse waste, about 580,000 tonnes of poultry manure (burnt for energy), compostable paper, and whatever green waste is burnt needlessly or landfilled.[71] There is room for improvement in recycling these items, but by comparison to our greenhouse gas emissions, the soil carbon improvements derived from them would be negligible. Scientists from Rothampsted have argued that if more of the manure currently deposited on grazing land were incorporated into arable soils, a higher proportion of the carbon would be sequestered. However, the potential gains in soil carbon are small, little more than a quarter of those identified by the same authors for conversion of arable to woodland or to ley-rotation.[72]

70   Van der Burgt, G J *et al* (2008), *Dutch (Organic) Agriculture, Carbon Sequestration and Energy Production*, Fifth International Scientific Conference on Sustainable Farming Systems.
71   Bhogal, *op cit. 59.*
72   Smith, P *et al, op cit. 10.*

It is clear that rewarding any localized increase in soil carbon that has been recycled from a limited common pool is a waste of money. We should be focussing on ways of extracting carbon from the atmosphere through photosynthesis and root biomass; this fresh carbon is the only carbon worth paying for, and then only if it is not offset by a corresponding drop in photosynthetic carbon gain elsewhere. It will be a remarkably sophisticated monitoring system that can work all that out.

We need, anyway, to be increasing the biomass productivity of our land, and its biodiversity, by whatever sustainable ways can be found. If this can be achieved through organic and permaculture techniques such as greater use of perennials and legumes, better recycling of biomass, and rotational grazing systems such as at Polyface Farm, then increased soil organic carbon will result, and this will be of direct benefit to the farmer in other respects. The Soil Association advocates, and I agree, that rather than paying farmers for questionable amounts of sequestration, we should provide incentives for farming activities that provide multiple benefits, including carbon sequestration.[73]

We also need to be wary of the exaggerated claims for carbon farming from its proponents, because it is a distraction from the main issue. Carbon dioxide currently represents 77 per cent of greenhouse gases.[74] Two thirds of the historic gaseous build-up of CO_2 between 1850 and 1998 was caused by fossil fuels, while 19 per cent is estimated to have been caused by loss of carbon from soils.[75] In all, less than 15 per cent of greenhouse gases are a result of soil carbon loss. Any claim that we can put all the carbon we have dug up from the bowels of the earth back into its skin has to be viewed with the deepest suspicion.[76]

Such views are suspect, not only because they are likely to be inaccurate, but because they are an expression of the same environmental ideology that often lurks behind the overemphasis on methane. Claims that carbon sequestration in soil and trees can singlehandedly reverse climate change are a bid to make the biosphere responsible for remedying global warming problems that are primarily caused by fossil fuels – as can clearly be seen in the Carbon Farmers of America's promise that, for $175 dollars, they will turn your car into a carbon sink.

Repeatedly, advocates of carbon farming remind us how carbon sequestration will compensate for the damage cause by industry and transport. This example is from Ken Yeomans, who clearly shares his brother's antipathy for bicycles, as quoted by Graham Harvey: ' Don't believe the lobby groups. We won't need to decrease our standard of living. We won't need to abandon the family car. Nor will we need to live the austere life of monks or ride to work on push-bikes.'[77]

Actually, monks lived rather well off their vast herds of sheep; and I'm very happy with my push-bike. These cowboys need to get out of their four-wheel-drives and back on their horses.

73   Azeez, *op cit. 58*, p 138.
74   Baumert et al (2005), *Navigating the Numbers: Greenhouse Gas Data and Intenational Climate Policy*, World Resources Institute.
75   Lal, R (2003), *op cit*. 33; Azeez, *op cit*. 58, puts the figure at 10 per cent, p 15.
76   The same argument applies to the current enthusiasm for biochar.
77   Harvey, *op cit*. 1, p 63.

LAND USE CHANGE

15

THE GREAT DIVIDE

The symbolism of meat-eating is never neutral.
To himself, the meat-eater seems to be eating life.
To the vegetarian, he seems to be eating death.

Mary Midgely

When I have asked vegans what they think Britain would look like if it went vegan, I have several times received the reply: 'Oh! I've never really thought about that. What an interesting question.' If I give them a bit of time to think, they come up with: 'More trees; and more wildlife.'

If some vegans have never really thought about this matter, it is probably because they live in a city, and have never had to think about land use at all. Yet the strongest environmental argument for veganism is that grazing land could be better used for something else. A comparison of the vegan and livestock scenarios in Chapter 9 shows that the difference in arable land use is not enormous, but the vegan options release huge areas of grassland for other uses. While the inefficiency of meat is a good reason for eating less animal food, it is not a reason for stopping eating it altogether, and in one sense efficiency dictates that we *should* eat animals grazed on land which cannot be more productively used. Vegans are on safer ground when they argue not that grazing is inefficient, but that there are better things we could do with less fertile and accessible land, of which the two most obvious are wildlife conservation and forestry.

This was what tipped Edward Goldsmith towards the vegetarians (though it didn't persuade him to stop eating meat) when, as chairman, he summed up *The Ecologist* debate of 1976:

> Abandonment of meat-eating would undoubtedly free vast areas
> of marginal land, at present used for rough grazing, for forestry
> and wildlife conservation. This is indeed a very alluring prospect,
> especially in view of the terrible shortage of trees in this country and
> the equally unacceptable shortage of nature reserves … Our meat
> eaters tended to underestimate this. Seen from the conventional
> farmers' point of view, it simply meant a larger amount of land that
> was not suitable for farming that must, with the abandonment of meat
> eating, be taken out of production. I must say that here my sympathies
> lie with the vegetarians.[1]

Despite the power of this argument, thirty years on it is hard to find any vision of what a vegan countryside would look like, in the UK or anywhere else. There is nothing to this effect on the Vegan Society's website, the sections on 'land' being devoted solely to explaining how livestock is connected to existing

---

1   The Ecologist (1976), 'Should an Ecological Society be a Vegetarian One?', *The Ecologist* 6:10.

problems. Other vegan sources tell us that vegan farmers 'can harvest more per plot and thus spare some of today's cropland for nature', but that's usually about all the detail given.[2] When I issued a challenge to vegans to broach this subject in *The Land*, Jenny Hall volunteered that a vegan and organic Britain might have 7.2 million hectares of arable land, one million of fruits and berries, ten million of managed woodland, one million of managed wildlife conservation, and 2.8 million of 'wildland'.[3] That gives us an abstract grid, but it leaves the reader to to paint in the details. Will there be hedgerows in this liberated vegan countryside? Will there be grass? Will there be predators? Will there be rats?

After reading every tract in my possession and skimming through vegan websites, I tumbled upon this, from Paul Appleby, which is sensible enough to be worth quoting at some length:

> Opponents of vegetarianism often predict dire consequences for our countryside 'if everyone went vegetarian' or, even more calamitously, 'if everyone went vegan'. Such prognostications ignore the fact that the UK is not, alas, going to turn vegetarian, let alone vegan, overnight, and consequently that any changes to the landscape will be gradual, reflecting a decreasing dependence on food from animals. They also suppose the landscape is sacrosanct, and that any widespread change in diet or lifestyle that threatens the status quo should be viewed with suspicion, if not totally opposed …
>
> The British landscape has been shaped by human hands ever since Bronze Age man began clearing native deciduous woodland for timber and farming several thousand years ago … Had we inherited Lakeland as it was before its settlement by pastoral and agricultural people some 5,500 years ago, it would be a land of unpolluted lakes, teeming with fish and rich in waterfowl. The lake margins and valley flats would be covered by extensive tracts of fen woodland. Alder woods would flank the wild rivers. Ospreys would breed in trees around the shores. Forests, dominated by oak, would mount the valley sides.
>
> What we now have is a cultural landscape created by the interplay of terrain, wildlife and human use over the centuries. Would it be a disaster if, following the sad loss of many flocks [from foot and mouth disease] farmers decided not to restock? The ecosystems, eaten out of existence by the sheep, would slowly extend their refuges in the gills and crags to take over more land. It is customary to attack such embryonic woodland as 'scrub', but it is nonetheless rich in birds and insects and is high forest in the making. No one can predict with certainty what the landscape of a vegetarian or vegan Britain would look like. Perhaps we will never know. However, it seems certain that such a landscape would have more room for wildlife, not less, and that doom laden scenarios of cabbage monocultures bereft of animal life are plainly nonsensical.[4]

There is force in this argument, and I was touched by the writer's note of sad resignation: 'perhaps we will never know'. Farmers who complacently assert that they are 'guardians of the countryside' should pay attention, for all they are

---

2   Fox, Michael (1999), 'American Agriculture's Ethical Crossroads', in Tansey, Geoff and d'Silva, Joyce, *The Meat Business*, Earthscan, 1999.

3   Hall, Jenny (2008), 'Stockfree Britain', *The Land* 5.

4   Appleby, Paul (2004), *What If We Were All Vegan?*, http://www.ivu.org/oxveg/Articles/whatifallvegan.html

really guarding is their own narrow vision of what the countryside ought to be.

But Appleby's passage is unusual, and I have been unable to find any vegan literature that takes this further. However much vegans and animal rights activists might care for the welfare of domestic animals, they devote surprisingly little attention to the habits and habitat of wild ones. Could it be that people who feel uncomfortable about eating meat don't like to take too close a look at wild animals because they don't like what they see? Whatever the reason, no one has painted a detailed picture of what a vegan landscape would look like or how a vegan society would relate to wilderness.[5]

That is what I attempted when I embarked upon this chapter, to imagine what our agriculture and society might be like if we pursued a vegan ethic. Since I am an enthusiastic (though modest) meat-eater, vegan readers might object that I am hardly the best person to paint an unbiased picture, and I would agree. The scenario I envisage here of a world where humans find themselves increasingly separated from nature is an exaggerated one, and one that I find disturbing; I don't dispute that there could be other more benign ways of organizing a vegan agricultural economy. However, I subsequently found that a number of commentators had already taken a similar theme a good deal further than I had and developed vegan or semi-vegan visions of a future society that most of us would regard as madcap, or science fiction. I sincerely hope that they are; but there is some value in elucidating them here if they illustrate, in magnified form, consequences that might arise if an industrialized society drastically reduced meat consumption, particularly if it did so through the gradual spread of a vegan ethic.

The Fence

In the 1960s, the American biologist Robert Paine conducted an experiment involving the removal of a predator species from a seashore environment:

> When he removed the main predator, a certain species of starfish, from
> a population of fifteen observable species, things quickly changed.
> Within a year the area was occupied by only eight of the fifteen
> species. Numbers within the prey species boomed and in the resulting
> competition for space, reasoned Paine, those species that could move
> left the area; those that could not simply died out.[6]

Commenting on Paine's experiments, Allan Savory remarks: 'I witnessed a similar disruption in two much larger communities in Africa', namely the Luangwa valley in Northern Rhodesia (Zambia) and the lower Zambezi Valley in Southern Rhodesia, where he worked as a biologist:

> Both areas contained large wildlife populations – elephant, buffalo,
> zebra, more than a dozen antelope species, hippo, crocodiles and
> numerous other predators. Despite these numbers, the river banks
> were stable and well vegetated. People had lived in these areas since
> time immemorial in clusters of huts away from the main rivers,
> because of the mosquitoes and wet season flooding. Near their huts

---

5   Mark Fisher's vision of 'self-willed land' seems to be highly compatible with a vegan society that safeguarded large areas of land for trees; but I can find nothing stating that it is vegan, and it could also be compatible with a highly efficient factory farming system. Self-Willed Land http://www.self-willed-land.org.uk/

6   Paine, R T (1966), 'Food Web Complexity and Species Diversity', *American Nauralist*, vol 100 no 910: pp 65-75.

they kept gardens that they protected from elephants and other raiders by beating drums throughout much of the night or firing muzzle-loading guns to frighten them off. The people hunted and trapped animals throughout the year as well.

But the governments of both countries wanted to make these areas national parks. It would not do to have all this hunting going on, and all the drum beating, singing and general disturbance, so the government removed the people. Like Paine, we, in effect, removed the starfish. But in our case we put a different type of starfish back in. We replaced drum-beating, gun-firing, gardening and farming people with ecologists, naturalists, and tourists, under strict control to ensure they did not disturb the animals or vegetation.[7]

The result was a change in grazing behaviour by many of the animal populations, and a rapid deterioration in environmental quality. 'Within a few decades, miles of riverbank in both valleys were devoid of reeds, and most other vegetation. With nothing but the change of behaviour of one species these areas became terribly impoverished and are still deteriorating as I write.'

The ecologist or park manager faced with these problems, will therefore normally try to simulate the role of predator. Culling large animals near the top of the food chain is the easiest way of controlling what goes on in a wildlife park. Anything else can be very labour intensive. It is true that there is a growing avant garde amongst nature conservationists advocating that animal populations should be left to sort themselves out – as at Ostvaardersplassen Reserve in the Netherlands where ancient varieties of cattle are uncontrolled and are killing off trees by bark stripping, making the area more open.[8] There is much controversy about the wisdom of this approach, and some also about its disregard for animal welfare: in the absence of predators, many animals die a painful and lingering death, unless they are culled. And if they are culled, why not eat them? Culling is either hunting, or else it is a waste of good food.

These problems loom even larger for ecologists in a vegan society, since vegans, by definition, refuse to be predators. Vegans cannot cull – at least not with any degree of ease or consistency. And there is a further problem to be faced: what to do about poaching? Poachers present a problem for all managers of wilderness, but they present a more awkward one for vegan wildlife managers for at least two reasons. A vegan society cannot buy off miscreants with factory-produced meat; and if vegans cannot cull, the pressure to get rid of nuisance animals rises.

So how would wildlife parks function in a fully vegan society? New animals could be introduced, but only with difficulty could surplus animals be removed, by capturing and taking them somewhere else where they might cause the same problem. It is easy to imagine that certain populations might grow, quite quickly, to the point where they started causing damage, not just within the park, but outside it. How would the vegan park manager stop wild boar descending from the woods to dig up gardens, squirrels in their hundreds crawling over nut plantations or destroying timber trees, badgers rolling neighbouring wheatfields flat, or herds of hungry elephants stampeding through cropland?

There are a number of courses that a vegan wildlife manager can pursue. One is to introduce predators and hope that these – in conjunction with a dearth of food during cold or dry seasons – will keep the prey population in balance.

7   Savory, Allan with Butterfield, Jody (1999), *Holistic Management*, Island Press.
8   Fenton, James (2004), 'Wild Thoughts: A New Paradigm for the Uplands', *Ecos*, Vol 25:1, 2004.

This might work in some cases, but in others it might not. Quite a few pests – for example elephants, badgers, wild boar and kangaroos – don't have much in the way of predators and, like rabbits and rats, have long been controlled by the hand of the supreme predator, man. Whether that hand can be removed in such a way that animals do not cause intolerable damage to our agricultural interests is questionable.

A predator species can keep a prey species under check in relation to the available food supply. But if that food supply includes not only the wildlife park, but adjacent edible cropland, the predators will not stop the prey advancing into the cropland – indeed it is to the predators' advantage to let them advance. In the absence of any pest control by the farmers, it is not difficult to see deer, feral goats, rats and rabbits expanding their field of operation, possibly followed by wolves, bears or large cats, not to mention hyenas, badgers and other scavengers and opportunists who would avail themselves of whatever else they could find in people's dustbins and gardens. The bears and large cats would occasionally take out a human for good measure.

Of course, this would never happen on any scale, because the farmers wouldn't allow it, any more than the 18th century Berkshire farmer, quoted on p 121, tolerated the presence of his Lordship's deer. If the park managers failed to control pests, then the farmers would turn to poaching, not only to save the crops upon which their livelihoods depended, but also because they would reason that if wolves, bears, lynxes and foxes are allowed to predate, why aren't humans? Enforcement would be near impossible, no matter how many vegan policemen were deployed, while reinvoking the Black Act and making hunting a capital offence would presumably conflict with vegan ethics. In such a situation it is likely that poaching (and probably poisoning as well) would be quietly tolerated, in much the same way that the eating of sacred cows is tacitly accepted in India.

It is for this reason that the Vegetarian Society's 'Green Plan' drawn up by vegan Alan Long sensibly allows a measured amount of meat-eating:

> Domestic species of farm animals, at present travesties produced by domestic breeding, would be allowed to assume the feral state in these reserves … In such feral conditions, animals may have to be culled … Casualties could be used for meat for those who want it.[9]

One wonders, in passing, whether Long refers to cultivated cabbages and beans as 'travesties'. But for Peter Singer and more purist vegans the idea of 'harvesting' excessive populations of feral or wild animals is another form of 'speciesism'. His solution to the poaching problem, and to the problem of inedible pests, is to control their population through the use of drugs or the release of infertile males to reduce female fertility. I will come to that shortly. But what is most revealing about Singer's coverage of pests is the tiny proportion of his book which he devotes to them – just one page, compared with an entire chapter on factory farming and another chapter on vivisection. Pests, in Singer's view are a side issue: this is how he introduces the subject:

> It is possible to think of more unusual cases in which there is a genuine clash of interests. For instance, we need to grow a crop of vegetables and grain to feed ourselves; but these crops may be threatened by rabbits, mice, or other 'pests'.[10]

9   Long, Alan (1979), *The Green Plan, a Synopsis*, The Vegetarian Society.
10   Singer, Peter (1995), *Animal Liberation*, Pimlico, p 233.

Unusual? Rabbits, mice and other pests? Far more rodents have died as a result of traps, poisons or targeted anthropogenic disease, than have ever been killed in the laboratories he campaigns against. Singer seems blissfully ignorant about the perils of growing vegetables. Virtually every herbivore in the animal kingdom, from slug and carrot fly up to deer and wild boar, has long since sussed out that humans are more proficient at growing tasty food than nature is, and all do their utmost to partake of the feast. The smell of bacon may not awaken murderous feelings in the breast of vegetarian gardeners, but the sight of all their pea seedlings ripped out by pigeons often does. And nothing causes sleepless nights for conscience-stricken vegans so much as the sound of rats scuttling in the cavities of their walls.

Most vegans are currently protected from the ravages of pests through the discreet measures taken by the rest of society to keep them under control. Nonetheless a shift towards a vegan ethic has altered our relationship with wild animals. Fifty years ago a fox who dared approach a village in daylight would have been greeted by kids throwing stones and telling him to piss off, thereby establishing a mutually beneficial barrier between civilization and nature. Now we welcome nature with open arms, and when Reynard saunters down the street people point and say 'ooh look, a fox!' – with the consequence that foxes feel at liberty to take chickens in broad daylight. Savory complains that whereas baboons naturally run at the sight of people, in his neighbouring national parks they became so tame that 'they sat on cars or got into them and trashed everything and had to be destroyed as a nuisance'. Similarly:

> The remaining elephants in the park no longer fear humans. Although they are being culled at a high rate, they do not know that humans are doing it, as whole families are gunned down so that none lives to tell the tale. This deception is considered necessary as it is a national park and tourists require tame elephants. And they have become remarkably tame: their response to human scent is very different from what it was in the late 1950s when they were much wilder. Unfortunately tame elephants, or any other game interdependent with predators, are not natural and therefore lose their natural relationship with the plants in their community. Basically they linger too often and too long in the most favoured areas and thus overbrowse or overgraze.[11]

They are becoming like cows who hang around the gate waiting for feed. It is not only animal lovers who are responsible for this domestication of the wild, but also people who hunt for sport (as opposed to those who hunt for the table). As Ortega y Gasset noted, hungry hunters kill the first animal that comes along, and thus select for wildness – whereas sports hunters demand a challenge, ignore easy prey, and so select for domestication. When I worked as a beater on our local shoot, nothing we could do – shouting, firing guns, throwing stones, despatching dogs through the water – would persuade the ducks on the pond to take flight. Having observed what happened to those who flew on previous shoots, the survivors had sussed that their best tactic was to remain on the pond, because sports hunters don't shoot sitting ducks.

Singer's proposal of 'contraception for wild animals' is perhaps the best option for vegans who want to maintain wildness. However it is not necessarily as simple as it sounds. The feral camels in the Australian bush, descended from animals imported to carry bales of wool, are now 'the wildest camel herd in

---

11   Savory, *op cit.* 7.

the world', largely because nobody in the country which pioneered 120 foot transcontinental 'road-trains' has any use for ships of the desert any longer. There are already a million of them, and left to their own devices would double in number every ten years, so they are culled by marksmen from a helicopter. Interviewed on Radio 4, Tony Peacock of the University of Canberra commented:

> Nobody would like to do it with birth control more than I – but it's not possible. Our co-operative research centres studied birth control of wildlife for well over a decade and we spent probably in excess of AUS$80 million in research funds, but it just doesn't work well enough. You still have the issue of accessing the animals. Unless they are very easily accessible and very high value it's just impossible to do.[12]

Birth control of wild animals may become easier and more affordable in the future, but that prospect raises other issues. It sails worryingly close to genetic engineering, and would probably lead to domestication through a more sinister route. If we let scientists manipulate the fertility of wild animals they will inevitably start fiddling around with their genes, and before long, we risk finding large sections of the evolutionary process controlled by scientists – and in a vegan society, by scientists who might disapprove of speciesist activities such as predation. Left to their devices the lamb might indeed be lying down with the lion. Such an evolutionary outcome would presumably be acceptable to Singer, who not only coined the term speciesism, but is also a forthright advocate of genetic modification.

Singer omits to mention two other existing methods of pest control which, along with guns, are commonly used by farmers. The first of these is that adopted by the US Army in Vietnam – defoliation. Destroy the enemy's habitat, so there is nowhere for them to hide, and soon you outnumber them. Farmers nowadays do not spend hours on a tractor flailing hedges and fighting back the advance of woodland for the hell of it, but because the more cover you provide, the more pests you harbour. Plant peas or wheat in a small field half-surrounded by woodland and the chances are that pigeons will eat all your pea seed and badgers roll your wheat. Plant acre upon acre of them in the middle of an arable prairie, and there will never ever be enough pests to make any impact upon your crop.

One option for a vegan farming system beset by pest problems would be to get rid of all the pestiferous hedges which demand annual maintenance and which are no longer needed for enclosing cows and sheep. That is what habitually happens in areas where farmers have got rid of their livestock; or where livestock are kept indoors all year round, as they are in many areas of Europe. England's miles of hedgerows are unusual – a historical anomaly resulting from the fact that the countryside was enclosed before the invention of barbed wire. Permaculturist vegans would argue that the hedges and other biodiverse vegetation should be kept to provide a balance of predatory species – but that balance would be hard to maintain if the species that has been chief predator for several thousand years of evolution resigned from the position. The Campaign to Protect Rural England might protest at the disappearance of hedgerows, but from an environmental point of view removal could be justified if many more wildlife areas were created elsewhere on land formerly occupied by domestic animals or the crops grown to feed them. Efficient mechanized

---

12   Tony Peacock, on Radio 4 'Today', 12 August 2009.

vegan farmers, supplying food for millions, would be less inclined to share land with nature and more disposed to spare land elsewhere for it.[13]

The other recourse traditionally taken by farmers is to fence off land that is under siege from pests – in the UK, typically deer, badgers or rabbits. It can be an expensive option but in our hypothetical vegan society the farmers have a potential ally. If the wilderness areas are managed by conservationists who want to maintain a natural succession (rather than by scientists who want to engineer an artificial one) then they too will be keen to demarcate the zone between the wild and the human with as impermeable a fence as can be devised and afforded.

Fencing has often been a tempting prospect for wilderness creators on the grand scale (and true wilderness has to be on a grand scale). The palaeontologist Richard Leakey, when head of the Kenya Wildlife Service, proposed fences around the Maasai Mara Reserve and the Tsavo National Park, 'to keep animals away from people and people away from animals', but had to abandon the plan because it would have interrupted some animals' migration patterns.[14] Paul Tudor Jones, the billionaire managing director of Grumeti Eco-reserves, which has leased 340,000 acres to 'offer 54 guests 21st century service in a sumptuous bush-chic setting' wants to erect a fence along the Western edge of the Serengeti; and there are fences around National Parks in Zimbabwe and Namibia, which again interfere with animal migration routes.[15] In Botswana, where there are already thousands of miles of 'veterinary fences' designed to separate cattle from wild animals carrying tsetse and foot and mouth disease, the 240 mile long Makgadikgadi fence is:

> the first of a new breed of fence designed to reduce conflicts between people and wildlife. Specifically, it was built to stop lions from attacking cattle, to stop villagers from retaliating against the lions, and to protect the grazing land of the wildlife in the park from the cattle.[16]

In Scotland, the proposed 'Pleistocene Park' wilderness at Allandale, according to one visitor, is 'scarred by new fences' some of which are electrified and sport DayGlo notices stating 'Do Not Enter, Dangerous Wild Animals' – a matter which has come into conflict with Scotland's right to roam laws.[17]

Conservationists want to prevent humans wandering in and out of wilderness, because they might be poachers or because they might end up as prey. Farmers and villagers want to keep dangerous and nuisance animals inside wilderness areas. The easiest strategy for a totally vegan society would probably be to fence off the agricultural land from the wilderness on the same scale that the Australians erected their 'rabbit proof fence', or the Botswanans their miles of veterinary fencing. This fence would need to be tough enough to resist badger, wild boar or elephants, tall enough to prevent deer leaping over it, and well enough policed to deter poachers. It would probably look something like the

---

13   The concept of a choice between 'sharing land' and 'sparing land' has been developed by Tim Benson of the University of Leeds, from: Rhys Green et al, 'Farming and the Fate of Wild Nature', Science, 307, pp 550-5, 28 Jan 2005. See also Chapter 8, footnote 36.

14   Monbiot, George (1995), No Man's Land, Picador; Poole, R M (2006), 'Heartbreak on the Serengeti', National Geographic, Feb 2006.

15   Rogers, D (2006), Grumeti Reserves, Tanzania are the Ultimate Safari Destination. What are you Waiting for? October 2006, http://www.travelandleisure.com; Communicating the Environment Programme Factsheet 5, Musokotwane Environment Resource Centre for Southern Africa (n.d.) CEP Factsheet: Poaching, http://www.sardc.net/imercsa/Programs/CEP/Pubs/CEPFS/CEPFS02.htm

16   Flore, G (2006), 'Good Fences, Good Neighbors? Can Botswana Simply Cordon off the Conflicts Dividing Ecotourism, Cattle Farming, and the Interests of Conservation?' Natural History, June 2006.

17   Taylor, P (2008), 'Alladale's Wilderness – Seeing through the Fence', ECOS, 29:3/4.

fence around Glastonbury festival, sunk two feet into the ground. And it still wouldn't keep the pigeons in.

The contrast would be stark. Within the flat grade one and two agricultural areas, there would be a tendency to cultivate as much as possible, with only a small amount of land devoted to shelterbelts, woods, hedgerows, banks and other landscape features. Strips of grain and vegetables would be interspersed with strips of clover or lucerne, in a manner reminiscent of the open field landscape that can be viewed in the more denuded areas of Eastern Europe. Nut and fruit orchards would consist of close-planted, easy to pick, early maturing dwarf varieties, for in a society without animals, there would be no incentive to plant standards with a carpet of grass underneath them.

Whether the fence would surround the agricultural land or the wilderness would depend upon the relative size of each; though for the wilderness to function properly it would have to be large enough to support a viable population of predators. In fertile areas with high populations there would be some reservoirs of pseudo-wilderness fenced off; in outlying regions there might be islands of agriculture. But at some point between the two there would be the main continuous fence dividing Nature on one side from vegan agriculture on the other – the coastline where the ocean of wilderness washed against the shores of civilization.

If this is starting to sound like a science fiction film it is because the chances of it actually coming to pass are, mercifully, remote. The fence represents a logical conclusion of the vegan project, rather than its most imminent expression. I bring it up because it lies at the end of a path which some vegetarians and vegans are inviting us to take, and because it is the most graphic symbol of the rift between humanity and nature which I suspect would arise as a result of a refusal to eat meat.

Resigning from Nature

In his diatribe against vegan agriculture, the late Mark Purdey (best known for identifying a link between BSE and organophosphates) described how he started in agriculture 'under the great expanse of striated sky' of the fenlands, working 'thousands of bleak acres of vegetables':

> The farm workforce clearly felt estranged from what was once their indigenous native landscape. These labourers resented the fact that a mono-arable/vegetable system of farming had been installed two decades ago after a change in the land's ownership. This had left many of their former workmates jobless, whilst those remaining felt divorced from any aspect of management or relationship with their work.
>
> But I do remember admiring the stamina and reactionary insight of one fossilized, peewit eyed character called Reg Strawson. He'd vent forth his daily parable on the fate of this once-upon-a-time peaty, organic friable loam: 'See yon field now (being the tip of one big 190 acre amalgamated field) 'twas where I pastured the milch cows in former days. Now thee blessed bullocks have gone, the zoils' gone zour there. Stick zoil in with a beet fortch and it clatts like do clay. In May it's droughted up, by June its blowing over dyke yonder.'
>
> I, too, rapidly found myself unable to form any working relationship with this treeless prairiescape of sterile inorganic moondust. Disillusioned, I left my friends on the fens behind.

Upon arrival in the West Country, I quickly found my niche within the mixed, small farming landscape. Livestock pumped the economic heartbeat that enabled these smaller farms to survive. My first job was to muck out the yearlings' house and I remember experiencing an innate sense of wholeness the first time I watched a shower of dung being flail-fountained out of the back of the muckspreader; fertile fodder. All of the farms and their staff seemed vibrant with the ethereal relationship flowing between the soil, the crops, the livestock and the landscape.[18]

Purdy clouds his antivegan message by comparing animal farming to a worst case scenario of chemical arable agriculture. But much of what he says still stands. Under organic stockfree management the soil might have been in better heart, but would Reg Strawson's heart have been gladdened by a 190 acre field divided into strips of which one in three was clover green manure?

The contrast we observe between the wide arable prairies typical of some parts of Eastern England and the tree lined patchwork of the west is not the difference between organic and non-organic agriculture, but between modern arable farming and mixed livestock and arable farming. The difference extends, not just to the lie of the landscape, but also to the way humans relate to it. Most wheat, barley and potatoes are nowadays farmed without any need for the farmer to ever get down from the seat of his John Deere or his Dominator and touch the earth. But our meat is still farmed by men and women who handle their animals, who know how to straddle sheep and hoist cows on their feet when they are down, who trim their hooves and give them drenches and injections, who pull lambs in breach out of their mothers wombs' and give them their first breath, and who are used to coming home with their overalls covered in muck.

Moreover, in the pastoral west it is more likely to be women. K D M Snell has studied how in the 19th century, as arable farming in the east started to become increasingly differentiated from pastoral farming in the west, demand for male farm workers increased in the arable areas as demand for female farm workers declined. But in the pastoral west the situation was exactly the reverse: there was a rise in women's employment and a decline in men's employment.[19] The distinction is evident today. How often do you see a woman driving a 120 horse power tractor? Yet women stockkeepers are almost as common as men, and in the horse industry there is probably a female predominance. Women have traditionally been associated with dairy work and poultry production, and have a natural advantage in lambing, not only because they make better midwives, but also because they have smaller hands more adept at manipulating badly presented lambs. The work traditionally assigned to women in arable production – reaping, stooking corn, teddying, weeding, picking and gleaning – has either been mechanized or abandoned. Snell also remarks that 'a tight harvest time schedule' makes arable work less suited to women childminders, in comparison with the regular but interspersed demands of animal husbandry.[20]

A mixed farming system provides a more natural landscape than pure arable farming, is less mechanized, and gives humans greater contact with nature. Why should this be so? The answer is that mixed farming, like nature, is complex, whereas pure arable farming (whether it be for animals in feedlots or for vegans in cities) removes an entire order of creation from the system. Moreover it is the

18   Purdy, Mark (1999), *The (Vegan Ecological) Wasteland*, Weston A Price, http://www.westonaprice.org/farming/wasteland.html
19   Snell, K D M (1985), *Annals of the Labouring Poor*, Cambridge, p 45.
20   *Ibid.*, p 49.

order which is closest to humanity, which gallops and gives birth and suckles, which feels pain and anger and joy. Farmers talk to their animals and give names to them, perhaps not to all of them but almost always to some of them. What vegetable farmer ever gave a name to a cabbage?[21]

The wheel of Nature turns through the action of different species consuming each other. 'The whole of Nature' said Dean Inge, 'is a conjugation of the verb to eat, in the active and passive.' Every animal consumes living things, and in turn must be consumed, either by another living being, or by fire. Nature is cruel. By rejecting cruelty, by choosing not to eat nor to kill fauna, the vegan forces the greater part of the animal kingdom into exile from the human world, on the other side of the fence. Not only the prey must go, but also the bulk of the pests must somehow be persuaded to stay there. Within the vegan reconstruction of nature there is space only for pets, 'companion animals' who are compliant with the vegan norm. Less extreme vegans might include within that group well-behaved predators, such as domestic cats or foxes to cope with the birds and rodents, but there is no shortage of those who will not countenance such brutality. For example:

> The animals in our home are vegan like me. The dogs – Shim and Pluto
> – have been brought up on a commercialized 'complete' soya-protein
> food … I supervise them strictly when out walking and do not allow
> scavenging. They do of course sometimes chase rabbits or squirrels,
> but never catch or injure one … Because I cannot bear to have half
> alive mice or birds brought home and teased by my cats, the latter
> wear extra loud bells on their collars. This may prevent them being
> successful as hunters, although they still enjoy stalking and using their
> predatory instincts … My cats – my companions Brindle, Suzie and
> Miracle, follow the feeding regime advocated by the Vegan Society.[22]

This passage causes one to wonder, first of all, if vegan soya production can be as efficient as claimed if large amounts have to be fed to companion animals whose original purpose was to eat pests and unwanted meat, but whose purpose now is to provide vegans with a vestigial link with the animal kingdom which they have otherwise rejected from their lives; and secondly whether the author has ever given any thought to the difficulties of controlling mice and rats in a vegan society. There is even wackier stuff out on the internet: vegetarian ferret food, a vegan Rat Refuge, and a much replicated story about a lioness on the West Coast of the USA who refused to eat meat (though most accounts do not mention that she died of a virus infection shortly after making her first public appearance).

Let us assume, however, that on the other side of the fence wolves and pumas will not be weaned on to a soya diet, and the vegans will be content to let Nature continue in its normal way, red in tooth and claw. What business will the vegan have there? Unable to hunt, unable to cull, vegan folk may venture in there to look for berries and mushrooms, at the risk of meeting a large predator. They may come in to harvest timber but in a temperate clime, if it is a genuine wilderness, it won't carry very good timber. But they will not have a lot to do there because this is wilderness, Nature from which the human race has resigned. Mostly they will come in to look.

Nick Fiddes, in his study of the symbolism and semiotics of meat, cites a 1988 British television advertisement promoting a cereal bar called *Tracker*:

---

21   Stuart Andersen, who read a draft of this book, has informed me that he had a cabbage called Nigel.
22   Found on the internet, but no longer online.

> A man in hunting attire is shown tracking a deer through tranquil
> wooded scenery. He looks intelligent, strong, tenacious and purposeful.
> Presently the beast is squarely in his sights – when it is revealed that
> he aims to shoot his quarry with a camera, rather than with a gun. He
> then relaxes and enjoys his chewy cereal bar. The meaning is clear.
> This modern male prefers to appreciate nature's living beauty without
> having to injure it.[23]

Fiddes is on the side of the vegetarian camera-toting tracker. But to my mind this is a human who lives a dysfunctional existence. What he eats is entirely divorced from what he does. Unlike the trapper chewing on a lump of jerky cut a week previously from the carcase hung up back in his cabin, our modern tracker nibbles a candy bar bought 24 hours before when he got off the plane. He is a voyeur who peers at nature through the keyhole of his camera. Nature is spectacle, food a commodity, and he sees no need to connect the two. Has he ever walked through an oatfield, has he ever rogued wild oats, has he ever considered what a ticklish and painstaking endeavour it must be to remove those tight packed husks from every one of the grains that go into making his 50 pence cereal bar? It is unlikely. To the hunter and to the peasant, tracker man is an urban slob.

Modern urban man, vegan or otherwise, seems to have an insatiable desire to 'capture' the nature which he no longer is part of:

> Every spring in Yellowstone you are sure to see a large group of
> photographers standing around – or even sitting on lawn chairs
> – talking loudly right out some poor badger's birthing den … It's
> laughable to see photographers in a national park camouflaged from
> head-to-toe, sometimes including face paint, photographing a bull elk
> as he calmly grazes alongside the road – fully aware of their presence.[24]

'The unspeakable in full pursuit of the uneatable', as Oscar Wilde remarked about a different bunch of hunters. Bill McKibben, the author of *The End of Nature*, has proposed a moratorium on new wildlife photos on the grounds that there are plenty of photos already out there for newspapers and magazines to use.[25] Indeed there are, and there is something neurotic about our need to amass visual evidence not just of the wildlife we have seen, but of every event that happens to us, as though the record was more valuable than the experience.

The contest between those who live within nature and those who live outside it and wish to fetishize it is being played out in many parts of the world as the financial muscle of tourism engulfs indigenous livelihood strategies. On the edge of the Serengeti from which they have been expelled, the Maasai have been reduced to a half life, maintaining a bare minimum of cattle and parading themselves and their rituals before parties of tourists, who gawp at them as they if they were part of the African fauna, or toss them a shilling. In Uganda, the Batwa, who had been living for centuries in the montane forests of Southern Uganda, were evicted from their lands when these were declared a national park.[26] In Chad where the amount of land under protection increased from 0.1 per cent to 9.1 per cent of the country, there are now an estimated 600,000 conservation refugees, while the only other country which counts them, India,

---

23   Fiddes, N (1991), *Meat, A Natural Symbol*, Routledge, p 76.
24   Robertson, Jim (n.d.), *Empathy for Other Species is the Key to Ethical Wildlife Photography*, www.all-creatures.org/aw/art-empathy.html
25   *Ibid.*
26   Dowie, Mark (2005), 'Conservation Refugees', *Orion*, Nov/Dec 2005.

has 1.6 million.[27] In 2002, India evicted 100,000 tribal people from their lands in Assam, and in 2007, 3,000 villagers from Ghateya Madhya Pradesh fled as State Forestry officials tear-gassed, burned and bulldozed their village.[28] In Thailand the creation of well over a hundred national parks in the last 20 years, funded by the Global Environmental Facility, has resulted in dispossession of Karen and Hmong villagers.[29]

The list goes on. Globally, the amount of land under conservation protection has doubled since 1990, and now covers some 12 per cent of all the world's land. The number of conservation refugees is variously estimated at between 5 million and 14 million people. In 2004, at a meeting of the International Forum on Indigenous Mapping in Vancouver, all two hundred delegates signed a declaration stating that the 'activities of conservation organizations now represent the single biggest threat to the integrity of indigenous lands.'[30]

Similar issues have arisen in the debate surrounding a conflict taking place between hunters and greens in the developed world. The initial aim of the moratorium on whaling was (quite rightly) to reverse the damage to populations of great whales caused by industrial fishing fleets in the middle decades of the 20th century. But when, in the early 1990s, it was confirmed that populations of Minke whales and some other smaller species which were the target of traditional local fisheries were perfectly healthy, the anti-whaling lobby had to change its tune. The British Minister of Agriculture, John Gummer was forced to switch his argument from one of ecology to another of humane killing. 'The Whale' moved from being an endangered species (in fact about 75 species) to a symbol of endangered Nature at the mercy of Man, or as Arne Kalland has it, a totem:

> The whales have become models for us to emulate, and people do not eat their teachers and models, at least not in the Western urban world. Whalemeat has to become taboo, and eating it becomes a barbarous act close to cannibalism. Whales are taking on the characteristics of a totem – as found among the Aborigines and Indians in North America. But unlike traditional totemic societies ... and unlike the Hindus who in no way try to impose the prohibitions of killing and eating cows on the rest of mankind, whale protectionists try to make the prohibition universal. In their zeal they continue a form of Western cultural imperialism, initiated by Christian missionaries.[31]

There is nothing wrong with totems and taboos. White city dwellers who have never been whaling or even fishing in their life have just as much right to refuse whale meat as have overfed Indian babus to refuse cow meat. The licence given to the Japanese to hunt whales for transparently bogus 'scientific' reasons is on the same level of tacit hypocrisy as the derogation that allows Untouchables to eat beef. But it is not normally done for one religion or culture to impose its taboos on another, and here the whale-savers transgress. To the extent that they campaign against whaling on humane grounds, WWF, Greenpeace, Sea Shepherd and the like are no longer protectors of the environment, but have set themselves up as the world's ethical policemen. And as totem becomes fetish,

27   Ibid.
28   Ibid.; Amnesty International News Service (2007) India: Police Shootings and Forced Evictions Target Adivasi Indigenous Communities in Madhya Pradesh, 26 April 2007.
29   Lohmann, Larry (1999), Forest Cleansing: Racial Oppression in Scientific Nature Conservation, The Corner House.
30   Dowie, op cit. 26.
31   Kalland, Arne (1993), 'Management by Totemization: Whale Symbolism and the Anti-Whaling Campaign', Arctic, Vol 46 No 2, June 1993.

and fetish becomes commodity, they cash in on the process. Here is the view of a former Norwegian whaling skipper who, during the moratorium, reluctantly had to hire his boat out for whale watching trips:

> These so called 'researchers' and their WWF friends were the ones who stopped whaling for us. Now they are making money on adopted whales. They have robbed us of our livelihood, but make a profit on the whales themselves. They keep telling us that we do not own the whales, but they sell adoptions as if they own them.

The skipper keeps his harpoon mounted on the forecastle head during his whale watching excursions to let the spectators know that 'this is a whaling boat and I am a whaler! I am just waiting for the moratorium to end.'[32] Does he ever consider that the power harpoon which he sees as the symbol of his identity might be the very cause of his disempowerment? When the crew of the *Pequod* sailed by the wind, chased whales through the swell in tiny rowing boats armed with hand-held harpoons, and gave their lives to keep candles flickering in 19th century drawing rooms, they could hardly claim to be much more than pawns in the unending struggle of nature. The mainly Norwegian whaling factory ships of the early and mid 20th century put an end to that. 'The true graciousness of hunting' writes Roger Scruton, 'occurs when the species is controlled through the arduous pursuit of its individual members and so impresses upon us its real and eternal claim to our sympathy.'[33] Once technology grants domination, it imposes alienation; and the penalty, for alienated humanity, is to become a mere spectator of the world we were once part of.

If the vegan project were to spread outwards from the Anglo Saxon world, and taboos spread down the trophic hierarchy, from whale and elephant to fish and rabbit, millions of people living on the wild side of the fence would lose their livelihoods and find themselves faced with a triple option: of emigrating to soybean civilization; of becoming a tourist guide or vegan gamekeeper; or of disappearing into the woods, 'going native'. If ever we reach that state, the champions of a more natural human lifestyle will not be the environmental police, but the poachers.

Abolition of Nature

In his 2006 book, *The Revenge of Gaia*, James Lovelock maps out how he thinks Britain might look if we managed to stave off runaway global warming by what he considers to be the only solution – widespread use of nuclear energy.

Lovelock's nuclear-powered dystopia is divided into three sections. Most people would live in a dense compact metropolis 'of the kind now favoured by Richard Rogers', together with 'industries, ports, airports and roads'. The second part would be devoted to intensive farming, producing 'real food grown in soil'. However this real food might have to be reserved for 'the privileged' while 'the poor', Lovelock reckons, would have to make do on synthetic food made in factories from chemicals: 'tissue cultures of meats and vegetables and junk foods made from any convenient organism'.

The third area would be a wilderness, 'given entirely over to Gaia and left to evolve wholly without interference or management'. Lovelock proposes to spray the wilderness from time to time with nuclear waste, not just as a cheap

---

32   Ris, M (1993), 'Conflicting Cultural Values: Whale Tourism in Northern Norway', *Arctic*, Vol 46 No 2, June 1993.
33   Scruton, Roger (2000), *Animal Rights and Wrongs*, Metro, p 163.

way of disposing of it, but also to scare off any greenies inclined to 'go native' – but somehow I doubt this will be a successful deterrent. The more effective way of keeping out any of the poor who decide that they would rather eat wild boar than junk food will be to erect a massive fence around the wilderness area and patrol it with marksmen operating a shoot-to-kill policy. This shouldn't be difficult as any state dependent upon nuclear energy will have formidable security arrangements and a heavily armed police force at its disposal anyway. Security could also be enhanced by a vast network of CCTV cameras which will also, incidentally, allow the biodiversity and the wildlife to be gawped at in real time by urban citizens on their computer screens.

Now, imagine you are James Lovelock, hatching your vision of a civilization fenced off from nature, and are browsing through the various scenarios for feeding Britain given in Chapter 9, wondering which model best suits your schemes. It is not a hard choice: both vegan options concentrate all their husbandry into the arable sector and leave all the permanent and rough pasture obligingly empty of agricultural activity, ready to be fenced off for the biodiversity reserve. But the vegan/organic scenario is unnecessarily wasteful of land: why sow green manures and accept low yields when you can manufacture all the nitrogen you need with the aid of nuclear energy? There is no question about it, the chemical-vegan option is the one for Mr Lovelock.

There is, odd though it may seem, a carnivore variant on the theme of apartheid between humanity and the natural world. Paul Shepard's 'cynegetic' society, like Lovelock's, foresees eight billion humans fed on food cultured from 'bacteria, yeasts, protozoans and algae' and crammed into 'compact quarters in very large buildings ... built high into the air and far below ground'.[34] The 160,000 cities housing this concentration of humanity and microbes would be strung along a five-mile wide ribbon of development following the coastline of all continents and islands, while the interior would be kept as wilderness (Shepard was writing in 1973, before a rise in sea levels from global warming was viewed as a likely possibility).

But Shepard, unlike Lovelock, doesn't want to protect wilderness by keeping humans out – on the contrary, he encourages them to carry out hunting activities there on a part-time basis:

> From the age of 13 the adolescent youth would move into a series of increasingly extended and arduous expeditions for which childhood and juvenile skills had prepared him. All hunting would be done by groups of men, preceded and followed by ceremonial affirmation of the symbolic aspects of the hunt ... All hunting will be done only with hand weapons. There will be no machines, no guns carried even for emergencies, no ambulances, cameras or meat trucks, no lackeys to set up camp, no dogs or other domestic animals. It will be truly hazardous.

The hunting is a uniquely male activity, though women will spend as much time in the wilderness as men, participating in gathering expeditions, carrying meat back to the city, and butchering it (so perhaps there will be lackeys after all). Shepard's object is to:

> confront the division between man and the rest of nature, between ourselves as animals and humans, not by the destruction of nature

---

34   'Cynegetic' means related to hunting, and is derived from the Greek word for dog. The passages cited are from the final chapter of Shepard, Paul (1973), *The Tender Carnivore and the Sacred Game*, University of Georgia, 1998.

or by a dream of some return to the past, but by creating a new civilization … Nature would be separated from the works but not the lives of men, for men would live in both worlds.

Shepard, in contrast to Lovelock, does not view humans as a pariah life form to be banished from Gaia; instead the society he depicts is a schizoid one, where mankind's contact with animal nature is preserved by cleaving human nature into two spheres. The separation of 'work' and 'life', a rift that has appeared and widened as humanity has urbanized, becomes a geographical absolute. Shepard's enthusiasm for hunting is grounded in almost pathological hatred for 'the barnyard' and all things connected with farming, which he views as an 'ecological disease'. The 12,000 year history of co-evolution between mankind and the animal and vegetable kingdoms that created 1,000 breeds of cattle and 5,000 varieties of potato, he views as part of the process of 'industrialization of the earth's land and sea surface' which now subjects nature to a programme of chemical warfare and genetic manipulation. Better to let this 'debauched ecology' reach a confined apotheosis in the factory farming of microscopic organisms, and fence off the rest of nature as a place where humans (male ones) can exercise the vestiges of their prelapsarian hunting instincts.

Part of the instinctual legacy from our Pleistocene past that Shepard so values is our taste for meat, and one suspects that his global game park would encounter the same problems with poachers as its aristocratic and colonial predecessors – particularly since his education system is focussed around teaching youngsters the skills of hunting. Shepard rejects laboratory meat culture in favour of microbial stews only because he considers that it is not technologically feasible 'because growth-control processes soon deteriorate when only part of an organism is cultured'. But with recent advances in stem cell technology this seems less likely to be a barrier. If the controllers of Shepard's high rise settlements want to keep their part-time hunters satisfied they would be wise, like Lovelock, to include lab-cultured flesh in the diet.

The idea of growing meat in laboratory conditions has been around for some time. In 1932, Winston Churchill remarked '50 years hence…[we] shall escape the absurdity of growing a whole chicken in order to eat the breast or wing, by growing these parts separately under a suitable medium' and the idea has been developed by generations of science fiction writers, most recently Margaret Atwood, whose bio-engineered Chickie Nobs, in her novel *Oryx and Crake*, have eight breasts and no brains.

Recently there has been a spurt of research in this direction, initiated in 2001 by NASA, in an experiment designed to produce a source of fresh meat for space flights. Scientists chopped chunks of muscle about five to ten centimetres long from living goldfish, and immersed them in a vat of foetal bovine serum extracted from the blood of unborn cows. Within a week the chunks had grown 14 per cent in size. The NASA researchers claimed that their achievement held out the prospect of growing meat in industrial quantities from the muscle cell lines of various animals or fish. The gruesome method employed did not prevent project leader Morris Benjaminson claiming that 'this could save you having to slaughter animals for food'.[35]

In the years since then, lab-grown meat has shown signs of becoming a sunrise industry. The technology is similar to the stem cell techniques that have resulted in the growing of organs, such as human windpipes, outside the human

35   New Scientist (2002), 'Fish Fillets Grow in Tank', *New Scientist*, 20 March 2002 www.newscientist.com/article.ns?d=dn2066

body, under laboratory conditions. In June 2005 the magazine *Tissue Engineering* published what it claimed was 'the first peer-reviewed discussion of the prospects for industrial production of cultured meat.'[36] Two methods were described: growing cells either as flat sheets on thin membranes, or growing them on small three-dimensional beads. The challenge, said one of the authors Jason Matheny, who works with Benjaminson on a project called New Harvest, 'is getting the texture right. We have to figure out how to 'exercise' the muscle cells. For the right texture you have to stretch the tissue, like a live animal would.' However, once the difficulties are ironed out, he claims, 'cultured meat could appeal to people concerned about food safety, the environment and animal welfare'.

The scientists working on lab-grown meat can recognize an expanding market when they see one and the vegan/vegetarian market is tailored for their needs. Artificial meat, made, not from animal tissue, but from spun soya protein, has been around for several decades, and this in turn is a descendant of the 'nut cutlets' which gave vegetarians in the middle years of the 20th century the opportunity to sink their teeth into something at least analogous to flesh. Leafu (a paste extracted from inedible grasses or other plants), soya milk and soft margarine are all to a greater or lesser degree part of the same tendency. Processed protein and fat is a staple part of a great many vegans' diets. At a time when the organic sector of the green movement is campaigning for slow food, real meat and fresh local produce, the vegan/vegetarian camp has been nudging the industry in the very opposite direction: towards factory farming and factory food. Cultured muscle tissue is the dream product that lies at the end of this road.

The secret longing of some vegans for Chickie Nobs came out into the open in 2008 when Ingrid Newkirk, president of People for the Ethical Treatment of Animals (PETA) offered a $1 million prize to whoever can scale up stem cell techniques to grow edible animal tissue for a mass market. *The New York Times* reported 'near civil war' within PETA, with members leaving in protest. Jim Thomas, of the ETC group, picked up on the licentiousness inherent in allowing vegans to eat synthetic meat: 'Culturing exotic meats opens new markets: Anyone for lion? A panda burger? What about ethical human cannibalism?'[37]

But when PETA issued its challenge, veteran animal rights philosopher Peter Singer was not slow to voice his support:

> I always thought it would be a good thing, the same way that I think
> it's good that the abuse of horses for pulling loads has ended. … I think
> it would be good if the abuse of animals for raising them for meat were
> to end, because we had a technological solution to that. We had an
> alternative.[38]

Singer's support for lab-cultured meat is a logical extension of his vegan philosophy, which he states is based entirely on 'the principle of minimizing suffering'.[39] It is from this basic principle that he arrives at his definition and denunciation of speciesism:

> Just as human beings are speciesists in their readiness to cause pain to
> animals when they would not cause a similar pain to humans for the

36   University of Maryland (2005), *Paper Says Edible Meat Can be Grown in a Lab on Industrial Scale*, University of Maryland Newsdesk, July 6, 2005.

37   Thomas, Jim (2008), 'Flask Grown Flesh', *The Ecologist* July/Aug 2008.

38   Levine, K (2008), *Lab-Grown Meat a Reality, But Who Will Eat It?* National Public Radio, 20 May 2008, http://www.npr.org/templates/story/story.php?storyId=90235492

39   Singer, P (1975), *Animal Liberation*, Pimlico, 1995, Chapter 1.

same reason, so most human beings are speciesist in their readiness to kill other animals when they would not kill human beings.[40]

However, this philosophy was put to a closer test in 2009 when an academic philosopher from the USA, Adam Shriver, proposed the genetic engineering of 'pain-free' meat – an innovation that suggests that we could more easily engineer something like Chicki Nobs by dumbing down the live chicken, rather than building them up from cultured meat. Whereas Descartes postulated that animals were 'automata' who couldn't feel pain, in order to justify eating and mistreating them, Shriver proposes to manufacture automata for the same purpose. In an article published in the journal *Neuroethics*, he wrote:

> Though the vegetarian movement sparked by Peter Singer's book
> *Animal Liberation* has achieved some success, there is more animal
> suffering caused today due to factory farming than there was when
> the book was originally written … We may be very close to, if not
> already at, the point where we can genetically engineer factory-farmed
> livestock with a reduced or completely eliminated capacity to suffer.
> In as much as animal suffering is the principal concern that motivates
> the animal welfare movement, this development should be of central
> interest to its adherents.[41]

It seems that Shriver, a lifelong vegetarian, is serious and not just aiming to put Peter Singer on the spot, though that is what he does by steering Singer's ship so close to the rocks of carnivory. Since the ability to suffer is the crux of Singer's argument against killing animals, the factory farming of brain-dead chickens or pigs ought to be morally acceptable. But by the same token, unless you are a speciesist, the factory farming of brain-dead humans would be equally acceptable. Anyone for ethical cannibalism? Singer was quoted in the New Scientist as saying that 'his would be a moot objection if pain-free livestock could be engineered', which presumably means that he is not quite sure what he thinks.[42]

Are we witnessing the first signs of a convergence of interests between factory farming, veganism and genetic engineering? Singer's support for lab-cultured meat and ambivalence about pain-free factory farming are not only a reflection of his concern for the welfare of animals. In recent years he has increasingly focussed on biotechnologies, writing articles with titles such as 'Shopping at the Genetic Supermarket', in which he suggests that access to genetic enhancement and 'designer babies' should be allocated by the state rather than through a free market.[43] Singer is not a card-carrying transhumanist (the term for someone who favours replacing humans with a superior design of intelligent being through a combination of genetic engineering, cybernetics and nanotech). But he is listed on the World Transhumanist Association website as 'One of the most influential philosophers among transhumanists, and he is a defender of access to human enhancement'.[44]

40   *Ibid.*
41   Shriver, A (2009), 'Knocking Out Pain in Livestock: Can Technology Succeed Where Morality has Stalled?' *Neuroethics*, Vol 2 No3, November 2009.
42   Callaway, E (2009), 'Pain-free Animals Could Take Suffering out of Farming', *New Scientist*, 2 September 2009.
43   Singer, P (2003), 'Shopping at the Genetic Supermarket' in Song, S Y, *et al* (eds.), *Asian Bioethics in the 21st Century*, Tsukuba, pp 143-56.
44   World Transhumanist website http://www.transhumanism.org/index.php/WTA/more/hplusbioethics/

Singer's interest in biotechnology and human enhancement are not an inevitable outcome of his vegan and antispeciesist views, but the two are congenial bedfellows. On the other hand, transhumanists and others who take an interest in more extreme forms of bioengineering have a very obvious motive for adopting an anti-speciesist position. If genetic manipulation, stem cell technology, cosmetic surgery and robotics continue to advance at an exponential rate it may not be long before scientists have the ability to create semi-human life forms. The growing number of people who are enthusiastic about creating cyborgs, chimeras and similar beings need to be sure that humans do not claim the right to eat or kill them – and if this protection is afforded to cyborgs, then consistency dictates that it should be extended to all other sentient species. The brave new world of the transhumanist, unless it is to be red in tooth and claw, requires the veganization of human relations with nature – and to be fully consistent, the veganization of nature in its entirety.

Thus it is that an increasing number of transhumanists are becoming vegans, while some vegans, like Singer, are flirting with transhumanism.[45] The vegan transhuman tendency is also influenced by Buddhist ideas, emanating from sources such as the Buddhist Transhumanist Association and James Hughes, author of the book *Citizen Cyborg*.[46] But its principle ideologue is the Oxford transhumanist, David Pearce, who promotes his vision of a totally veganized nature under a variety of headings such as 'paradise engineering' and 'abolitionism'. Like most transhumanists, Pearce predicts that within the foreseeable future, the human race can be redesigned to eliminate both ageing and suffering:

> If you take suffering seriously, the only way to eradicate it is by
> biological reprogramming. In the short run this may involve superior
> drugs. In the long run the only realistic way to abolish suffering is
> through genetic engineering … With the right prosthesis one could
> have an enhanced body image, a greater responsiveness to a wider
> range of physical stimuli.[47]

But he doesn't stop at humans, for paradise engineering must spread throughout the whole of animal creation, and the first step in that project is to stop eating meat:

> Biotechnology will enable the human species cost-effectively to mass-
> produce edible cellular protein, and indeed all forms of food, of a
> flavour and texture indistinguishable from, or tastier than, the sanitized
> animal products we now eat. As our palates become satisfied by
> other means, the moral arguments for animal rights will start to seem
> overwhelmingly compelling. The Western(ized) planetary élite will
> finally start to award the sentient fellow creatures we torture and kill a
> moral status akin to human infants and toddlers. Veganism, though not
> in quite the contemporary sense, will become the global norm.[48]

This global vegan norm will encompass animals as well as humans. Some predators, particularly 'feline psychopaths', will have to be either reformed or eliminated:

---

45   The tendency can be observed by googling 'vegan transhumanism'.

46   Hughes, James (2004), *Citizen Cyborg*, Westview Press.

47   Pearce, David (2007), 'Mehr Rausch für Alle', *Vanity Fair* (Germany) 5 April 2007. http://www.hedweb.com/hedethic/vanity-fair.html

48   Pearce, David (1995), *The Hedonistic Imperative*, 1995 http://www.hedweb.com/hedethic/tabconhi.htm

Just as there is no need to recreate the natural habitat of smart, blond, handsome Nazi storm-troopers who can then prey on their natural victims ... likewise the practice of continuing to breed pre-programmed feline killing machines in homage to Nature is ethically untenable too.[49]

But other carnivores can be fed on cultured meat, while herbivore populations can be kept to a manageable size through drugs or bioengineering:

Essentially, in 50 years or so all that's left of 'Nature' is going to be our wildlife parks. What we will actually allow in our wildlife parks is debatable. Already in zoos we don't feed live animals to snakes or crocodiles, even though it would fulfil their 'natural' instincts ... Already the elephant population in a few parts of Africa has recovered sufficiently such that techniques such as depot-contraception have been introduced – in preference to cruel culling. Ultimately a whole ecosystem could be managed in the same way.[50]

Pearce's predictions may seem extreme, or even lunatic (like many in this chapter) but he is only a step or two further down the road than Singer – who is without doubt the most influential philosopher behind the spread of the vegan ethic. There is something to be said for people who pursue fashionable ideas to a fanciful but logical conclusion. Those of us who value the natural world, and more especially our relations with members of the animal kingdom, both wild and domestic, would do well to keep an eye on the vegan agenda, for it may not turn out to be quite as meek, disinterested and innocuous as it might seem:

We are what we eat, and by eating animals we help to ensure that we ourselves remain animals, participants in the food chain that momentarily we head before we too become flesh for worms. By declining to eat meat we abandon our status as predator, ostensibly to take on the more humble role of middle rank herbivore, but increasingly to assume the roles of manager and absentee landlord. As we detach ourselves from the natural world, it fades to a spectral image, glimpsed through the windscreen of a car or the screen of a computer, a world we can no longer be part of because we are too squeamish to partake of it. As a species we are slowly resigning from nature, and for those of us who lament this tragedy, there is at least one consolation: that for some time to come there will be poachers lurking in the woods, for the vegans and the wildlife managers will never catch them all.

---

49   *Ibid.*
50   Pearce (2007), *op cit. 47.*

16

THE STRUGGLE BETWEEN
LIGHT AND SHADE

Next in importance to the divine profusion of water, light and air,
may be reckoned the universal beneficence of grass.

J J Ingalls

The best friend on earth of man is the tree.

Frank Lloyd Wright

A strong argument for having fewer cattle and sheep in Britain is that there
would be more trees. This is because ruminants compete with trees by
forging an alliance with grass. Our natural environment is the field of a
perpetual struggle between light and shade, in which livestock are one of the
principal mediators. The other main mediators are fire (which favours grass) and
water (which favours trees).

Trees compete by outgrowing grass and starving it of sunlight. In order to
gain their height advantage they have to invest a proportion of their energy in
the necessary infrastructure – the trunk, boughs, branches and twigs that allow
them to spread their leaves over everything else.

Whereas trees compete vertically by shading out other species, grass in
temperate climes competes horizontally by crowding out other species; but to
be able to do this it enlists the service of ruminants and other herbivores. These
animals assist in two ways: they eat young tree saplings before they become
fully established, and they eat the grass. It is counter-intuitive that by eating both
kinds of plant, they should help the one and hinder the other, but there is a reason.
The saplings, in order to grow high, have to develop vulnerable infrastructure
above ground, whereas the grass, which has no need of a trunk, can keep all
its essential parts, namely its roots, safe beneath the soil, in the 'impregnable
fortress of its subterranean vitality' as J J Ingalls put it.[1] When a sapling is bitten
off, it has to grow its main stem all over again, and may not succeed. When grass
in temperate regions is eaten, it can put on more leaf growth immediately, and
tends to throw out additional shoots, a process known as tillering. The more grass
is eaten, within reason, the more it tillers, the denser the sward becomes, and the
harder it becomes for a tree seed to find a place in the turf to get established. That
is why mowing a lawn every few weeks keeps it healthy.

Because of this survival strategy, grassland as a whole is highly edible, but
only to ruminants and other fibre-eating herbivores; humans are only adapted
to eating fruits and seeds, which both trees and grass produce. However, we can

---

1   Ingalls, J J (1872), 'In Praise of Blue Grass', reprinted in USDA, *Grass*, Yearbook of Agriculture,
1948.

eat the animals that eat the grass, albeit at a low level of efficiency, and thus the whole of a field of grass is potentially edible.

A tree, by contrast, survives by growing to the stage where its trunk and boughs are too tough to eat. Except for the bark, its Achilles heel, a mature tree trunk is a triumph of inedibility, eaten by nothing until the tree sickens or dies. Up in the canopy there is often food, but not all creatures can get there to eat it. Birds, insects, squirrels, monkeys and fruit bats are welcome to dine on the flowers, nuts and fruit, and often that helps to spread the species. But humans have difficulty getting up there, and, more importantly for the tree, ruminants can't usually get up there either. A herd of tree climbing goats might polish off all the leaves of a tree in no time at all, and repeated attacks could leave an adolescent tree without any reserves to put into its infrastructure.[2] From time to time humans have found it advantageous to crop the leaves to feed to animals in winter in the form of dried faggots, but probably only when the small branches were useful for fuel as well. For the most part we have found it easier to make hay or silage.

The result is that landscapes dominated by grass normally offer humans more food than landscapes dominated by trees, both in regard to the accessibility of the seeds and the availability of meat. It is therefore not surprising that agricultural civilization evolved in the dryer areas of Europe and the Middle East, where trees were handicapped by a lack of water to support their superior bulk; and that early agriculturalists chose to improve and domesticate grass seed rather than tree seed, both because it was more accessible, and because all of the rest of the plant could be eaten by livestock.

These are the ecological dynamics that explain why our diet consists primarily of grass, and why a succession of commentators have come out with epigrams like 'the basis of human proliferation is not our own seed, but the seed of grasses' and 'grasses are the greatest single source of wealth in the world.'[3] The second greatest source, up until the discovery of fossil fuels, was woodland whose inedible infrastructure provided fuel, fibre and building material, but this remained secondary to the need to eat.

Grass's strategy of competing through edibility has proved spectacularly successful, not because herbivores have been particularly effective at keeping the tree population down, but because human predators have intervened on their side. There is a mutual dependence between predators, prey and fodder plants, whereas trees are have made themselves so unpalatable that they don't have many allies. As a result the frontier that separates forest from grassland has receded to the point that the planet's ecological equilibrium appears to be threatened, because trees provide other vital services, such as protecting water resources, preventing erosion, sequestering carbon, and mediating climate.

To redress this there are a number of courses humans can adopt. One is to stop the mostly wasteful procedure of growing grass seed to feed to herbivores who have evolved to eat fibre and to omnivores such as pigs who can survive on waste – a matter already discussed. A more extreme possibility is the vegan project of eliminating livestock altogether to create a system where the complexity of herbivores eating fibre is sacrificed in order to focus on ultra-efficient seed production. This would result in a drastic reduction of grassland and a huge increase in tree cover.

---

2   There are some interesting exceptions to this. The argan tree in Morocco is browsed by goats, but they relish its fruit. In dry *garrigue* landscape in the South of France, the leaves of scrubby holm oak are the staple diet of goats in the winter – but they hardly touch the leaves in the summer, when there are grasses and other foods around, allowing the holm oak to put on growth.

3   Agnes, Chase, *First Book of Grasses*; Evan Eisenberg, *Back to Eden*, cited in Harvey, Graham (2001), *The Forgiveness of Nature*, Jonathan Cape.

A third course of action, so far hardly mentioned in this book, is to enhance the edibility of trees, in order to bring them into the food economy. There is not much to be gained by harvesting them for animal fodder therefore any enhancement must be for seed.

Nuts

Nuts are to trees what grains of wheat or rice are to grass – the part which provides humans with essential nutrients. The rest of a tree is either edible only to animals, or inedible to everything. Some trees produce fruit, but in temperate climates fruit – healthy and full of vitamins though it may be – contains little protein, fat or energy, apart from sugar. Any proposal to replace grassland with trees, has an interest in demonstrating that as much nutrition can be derived from nuts as from grazing animals, if not more.

In the UK, the vegan case for expanding tree and nut production has been put forward by the Movement for Compassionate Living, pioneered by the late Kathleen Jannaway. Unfortunately their arguments have not been well substantiated; even a reviewer in the Vegan growers' magazine *Vohan News* took them to task because 'nowhere are we given a convincing argument that people in temperate zones could feed satisfactorily from trees'. Jannaway responded, reasonably enough: 'I do not claim that people in temperate zones could feed satisfactorily from trees alone, only that they could use them as a main source instead of animal products'.[4]

The permaculturist Patrick Whitefield, on the other hand, is keen to see nuts developed as a staple crop, an alternative for wheat and oats:

> Whether nuts are a practical proposition as a staple food in this country
> has to be an open question at this stage, though I am personally
> convinced that they will prove to be so. Much depends upon
> answering these questions:
>
> • How much food can they yield?
> • What's the quality of that food?
> • How easily can we grow them?
> • How easily can we breed improved varieties?[5]

No one seems to have a definitive answer to these questions. Wildly different yield figures are given – Jannaway gives 31 tonnes per hectare for walnuts while Whitefield cites 7.5 tonnes and the Agroforestry Research Trust in Devon claims to be getting about the equivalent of three tonnes per hectare from their best varieties. FAO estimates are even more erratic with China, France, Turkey and the USA all reporting average yields of around 2.5 tonnes per hectare for walnuts in their shells in 2007, while the Czech republic reports 6.6 tonnes, Azerbaijan 6.7, Pakistan 10 tonnes, Romania 22.6 tonnes and Slovenia 32 tonnes. Any advance on 32 tonnes? Yes, in 2004 Slovenia claimed to be producing 60 tonnes per hectare! Poor old Bulgaria, by contrast, only harvests 480 kilos per hectare, and Russia just 240 – while the UK doesn't produce any at all.[6] To put these figures into context, 32 tonnes is equivalent to the entire biomass that can be

---

4   Dave from Darlington (c. 2000) review of: Jannaway, Kathleen, *Abundamt Living in the Coming Age of the Tree*, Movement for Compassionate Living, 1991; and a reply from Jannaway, both in *Vohan News International*, 3.

5   Whitefield, Patrick (2004), *The Earthcare Manual*, Permanent Publications, p 216.

6   FAOSTAT: faostat.fao.org

extracted off a reasonably productive Sitka Spruce plantation[7] – the idea that an equivalent quantity of walnuts (plus all the rest of the biomass on the trees that bear them) can be produced on a hectare is not credible.

The yield of hazelnuts is more modest: the USA tops the league with an average yield of 2.8 tonnes per hectare, and otherwise only Croatia, France, Greece and Slovenia exceed two tonnes per hectare. As for chestnuts, Azerbaijan comes out on top with 10.6 tonnes per hectare, followed by China with 7.1 tonnes, Romania with 6 tonnes, Peru with 4.2 tonnes and Albania with 3.3 tonnes. Every other country produces less than 2.2 tonnes per hectare, which is pathetic, considering that chestnuts – which, nutritionally speaking, are the potato of the nut world – contain virtually no fat, less than a quarter as many calories as walnuts and less than a fifth as much protein.

The problem, for Britain, is that nut trees don't perform brilliantly in northern climes. The northern extremity of commercial chestnut production in western Europe is about 500 miles south of Paris. The English chestnut belt, across the weald of Hampshire, Sussex and Kent has been maintained for the value of its coppice wood. Walnuts are produced commercially in the Périgord, 250 miles south of Paris, not in Normandy or Brittany. 'The English have long cultivated walnuts with optimism, if not with success', observes Colin Tudge.[8] We know that 1826 was a good year for walnuts because William Cobbett, arriving at Blunsdon in Wiltshire in September of that year, remarked: 'I saw a clump, or rather a sort of orchard, of as fine walnuts as ever I beheld, and loaded with walnuts. Indeed I have seen great crops of walnuts all the way from London.' But we have no way of knowing how great the yield was; and almost in the next breath he proclaims: 'This is a cheese country'.[9] However accomplished 19th century farmers might have been at growing walnuts, it clearly didn't seem to them to be as rewarding or as reliable as keeping cows. Hazelnuts produce in a UK climate, but their yield is lower than that of chestnuts or walnuts, and is frequently zero if there are grey squirrels around. Even oak trees only produce a decent crop of acorns about one year in three, whereas further south in Europe they 'abound in most years, and pigs (and occasionally men) may depend on them.'[10] According to Oliver Rackham, that was why pannage, the grazing of pigs in woodlands, was never as widespread in Britain as it was on the continent.[11]

Whitefield pins his hopes on breeding new varieties that will fare better in the British Isles, and this is not unreasonable. Colin Tudge reports that 'breeders at Oxford University are now seeking to create varieties [of walnuts] that really can tolerate our decreasingly harsh, but increasingly fickle climate.'[12] But the most promising experiments in nut cultivation in the UK are being carried out by Martin Crawford of the Agroforestry Research Trust on land at Dartington in south Devon. Crawford has reported yields from his top varieties of trees equivalent to three tonnes of walnuts per hectare, or 1.35 tonnes of edible kernels. This is less than two thirds the weight of soybeans that can be obtained from a hectare, but with a 60 per cent fat content this quantity would yield more oil than a hectare of soybeans, though less protein. The yield for hulled hazelnuts (after Crawford succeeded in controlling squirrels through the use of live traps) was

---

7   16m^2 yield class, assuming 50 per cent of total biomass.
8   Tudge, C (2005), *The Secret Life of Trees: How they Live and Why they Matter*, Penguin Allen Lane, 2005, p 206.
9   Cobbet, William (1826), *Rural Rides*, Everyman, vol 2, 4 September 1826.
10   Rackham, O (1986), *The History of the Countryside*, Dent, p 122.
11   *Ibid.*, p 122.
12   Tudge, *op cit.8*.

about 1.1 tonnes.[13] The timber yield is negligible because these are low bushy trees, but productivity can be enhanced by feeding animals on the grass growing between the rows.

These yields, if they can be replicated on a wide scale, suggest that there could be a viable role for walnuts and hazelnuts grown in the South of England as a source of strong tasting oil and as an alternative for meat. 1.35 tonnes of walnuts probably has a higher nutritional value than the 1.6 tonnes of pork you could get by feeding a hectare of high yielding grain to pigs at a conversion ratio of 5:1. And this high pork yield cannot be achieved by organic cultivation, whereas Crawford's system is organic, with 30 per cent of the cropping area put over to fertility building plants, which can be grazed by animals.

The results for chestnuts are less convincing. Crawford gets the equivalent of five tonnes of chestnuts per hectare off his best varieties, which is a respectable yield compared to the international figures given above. The nutritional value of chestnuts is too low for them to provide a satisfactory substitute for meat, but with roughly twice as many calories as potatoes, and about the same protein content, chestnuts can serve as a staple. However, five tonnes per hectare is only one fifth the average UK yield for organic potatoes, which means that a hectare of potatoes produces over twice as much energy as chestnuts, and five times as much protein. Similarly a four tonnes per hectare yield of organic wheat would produce over 1.5 times as many calories and more than four times as much protein as chestnuts.[14]

My own reaction to the proposal to develop chestnuts as a staple is coloured by what I witnessed when for a few months I lived in the remnants of what must be the most northerly agricultural economy founded upon a tree-sourced staple in Western Europe. The mountains around the southern French town of St Pons (and a number of other areas in France, Spain and Italy) hosted, until the early 20th century, an agricultural economy whose main staple crop was chestnuts. The chestnut woods still cover the sides of the steep mountains up to the point where the beech take over, but they are by no means the only components of this once self-sufficient ecosystem. There are cherries in the narrow valley bottom, vines on the lower slopes, gardens and potato fields on flat land, and clearings for pasture and meadows on the less steep bits of the mountainside for sheep and goats, meat and milk. What more could anyone need?

On top of that, the ecosystem, even though it is semi-abandoned, remains very productive, bearing out what Whitefield says about the potential of trees. The quantity of chestnuts produced on these steep, schistous mountainsides sprinkled with a thin layer of acid soil is impressive, and there is an abundance of building wood, fencing materials and fuelwood. Overall productivity, as well as water capture and erosion prevention, is far greater than could be gained from felling the trees and turning the steep mountain slopes over to sheep pasture.

This once self-sufficient rural economy had its animal component, which was probably necessary since the protein to calorie ratio of chestnuts is only half that of potatoes. The meadows fed not only sheep and goats, but also, until the 1970s, mules who did the work.[15] The sheep mopped up the chestnuts that didn't get picked. They returned to the *bergerie* every evening and their manure ended up on the vines and the potatoes. The meadows also served the invaluable purpose

---

13   Martin Crawford, pers com March 2008 and November 2009.

14   Five tonnes per ha of chestnuts at 1700 kcals per kilo is 8.5 million kcals per ha; 25 tonnes potatoes, at 750 cals per kilo is 18.75 million kcals. Since the protein per kilo for both crops is roughly the same and the yield of potatoes is five times as high, you get five times as much protein per hectare with spuds.

15   I bought the last mule in the village from a retiring farmer.

of alleviating the woodland. Anyone who has ever lived in a forest knows what a wonderful thing a clearing is. Most humans find too many trees claustrophobic, and animals are the best way of keeping space clear.

The railway which ran along the valley bottom was, I suspect, the death of this rural economy. Before its arrival the inhabitants of this valley were 20 miles or so from the nearest land capable of supporting wheat in any volume, and even further from any land that could produce wheat profitably. The Canal du Midi was just as far away. The postman, so I was told, would leave to start his round on foot on Monday morning and got back on Saturday, having passed the night at various mountain farmsteads on his route. I've always fancied that job.

When the railway arrived it brought wheat and cash, and the rural economy collapsed. By the early 1980s, the chestnut woods were still vigorous, but untended, and suffering from chestnut blight. All over the hillsides the *secadous* – the stone huts where the nuts were smoked and dried – were falling to ruin. 'How happy we would be', Patrick Whitefield remarks, 'if we could create systems which combine the yield and self-reliance of the wildwood with the highly edible nature of the wheat field'. Well, how happy would we be? The magnificently sustainable chestnut economy around St Pons, which did just this, capitulated, I suspect, partly because people were fed up with eating chestnuts, and when wheat became more accessible they found it was a better staple, with a higher protein and calorie content. Nowadays all the local people eat bread; and when I lived there, in a household committed to self-sufficiency, we ate bread too, even though we had chestnuts to last us for eternity. Chestnuts (like salmon, which I have also had the questionable privilege of eating every day for six weeks) have too dominant a taste to be a satisfactory staple. Wheat and rice are bland, and go with anything.

But (as Jannaway remarked) that is not necessarily the point. The vegan mission is not to supplant wheat and potatoes, but to find a substitute for meat. If people in Britain ate less meat, or no meat at all, it is likely that they would welcome more nuts in their diet. Since meat production from grass is not very efficient, it is reasonable to suppose that growing nuts could be a more productive way of using pasture land than running sheep on it.

If the yields that Martin Crawford has obtained from his best nut trees can be replicated on average quality pasture land through a substantial area of southern England beyond the maritime palm tree belt, there will certainly be a good case for converting some grassland over to more efficient nut production. There will also be a case, in an organic system at least, for replacing animal feed grains grown on arable land with nut production for direct human consumption. But such a success story wouldn't seriously alter the balance of evidence in a debate about the relative merits of grass and trees. It wouldn't suggest that trees are inherently more productive than the grasses, because a hectare of organic wheat still produces more calories than a hectare of walnuts, and over twice as much protein. It would confirm the fact that farmed plants can produce more than uncultivated grass fed to animals, but that we know already.

What the spread of viable nut production into the UK might more convincingly indicate is the gradual advance of Mediterranean levels of biodiversity into our North Atlantic island – possibly because our climate is getting warmer, but hopefully because breeders have extended the range of cultivars. Whitefield, in a short passage in *The Living Landscape*, makes this observation:

> By the time agriculture arrived in north-west Europe the cereals
> [originating from the Middle East] had been improved by plant

breeding and the methods of cereal farming were well developed. It was an efficient and well known package. No one went to the trouble to invent a northern form of agriculture using the indigenous edible plants. If they had the landscape might look very different now, perhaps more like native woodland and less like an imitation of the south-west Asian steppe. Animal farming was part of the same package, and here too exotics reigned. It wasn't the native roe and red deer but the imported cattle, sheep and goats which made up the herds.[16]

We were colonized, and had we not been, we might have developed commercially viable varieties of … well the two indigenous plants that Whitefield mentions are 'hazelnuts and the starchy tubers of bulrushes'. It's interesting that the indigenous hazelnuts in Crawford's trial are not significantly less productive than his walnuts which originated in Eastern Europe, whereas bullrush tubers were a non-starter. Both were accessible plants that absorbed a generous amount of sunlight because they were in areas too wet for trees to grow, or on the edge of the forest canopy. As clearings expanded it's easy to see how bullrushes, reliant upon copious quantities of water, were outcompeted by wheat, barley and rye, whilst hazel hung on at the margins.

Why have arboreal incomers such as chestnut and walnut failed to establish themselves as successfully in Britain as wheat, barley, oats and rye? No doubt for the same reasons that they have failed to compete with cereals in most areas of southern Europe, namely that trees devote a significant part of their energy to creating infrastructure that makes the end product, the seed, difficult to harvest. To my knowledge, walnuts have never been a staple in western Europe, whereas chestnuts have, but only in mountainous areas that couldn't grow decent crops of cereals, or indeed walnuts. That is where chestnuts, the potato of the nut community, found their niche.

In the 21st century, several thousand years after the arrival of agriculture in the British Isles, the imperative is no longer to open up new areas to sunlight in order to generate prolific and accessible seed production, because we have cracked that with cereals. Now we need to sequester carbon in the natural environment, and to find ways of providing the concentrated nutrients that most of us crave, without overindulging in the extravagance of meat. Well-endowed nuts such as walnuts and hazelnuts look as though they might foot the bill, though there is some doubt whether chestnuts fall into that category. The ability to grow nuts in Britain as prolifically as they grow south of the Loire, and to substitute them for our grain fed meat production, would fix carbon and enhance the biodiversity of our landscape. For the consumer, it would increase the variety in the homegrown diet, and make the progression towards a default level of livestock more appetizing.

This doesn't mean that livestock are incompatible with nuts, though they are incidental, and they can be complementary. Like fruit orchards, nut plantations take time to establish, have to be kept clean and have to be supplied with the necessary nutrients. As is the case with orchards, these objectives can be achieved through the use of chemicals, or of machinery, or as was traditionally the case, through grazing and livestock management. A high yield of walnuts and hazelnuts suggests the need for high levels of fertility – what goes out must come in. Crawford states that he operates on the basis that 'about 30 per cent of the cropping area of an agricultural system needs to be devoted to other fertility-

16   Whitefield, Patrick (2009), *The Living Landscape – How to Read and Understand It*, Permanent Publications.

building plants to remain sustainable'. In this respect, his nut plantations are like any other kind of intensive organic grain cultivation – they need to be supplemented with an area of green nitrogen-fixing manure which may or may not be circulated through livestock, according to the preference of the farmer.

Tree Fetishism

'It is possible to get over-enthusiastic about trees', Patrick Whitefield warns at the beginning of the woodland chapter in his authoritative book on permaculture in temperate climes, *The Earth Care Manual*. 'They are not 'better' than other plants, nor are woody ecosystems 'better' than those without trees or shrubs'.

It is not difficult to imagine who he is thinking about. There is a fringe of the green movement which has managed to reduce the complexity of nature to the formula 'trees good, no trees bad'. This outlook is buttressed by the symbolic role played by trees in protests against road schemes and similar developments. If such people get a hold of an area of grassland, often the first thing they want to do with it is plant trees all over it. The fact that someone, a long time ago, went to a lot of trouble to get the trees out, and that generations of people have spent energy making sure that trees stayed out, is lost on them.

Years ago, on television, I saw a wildlife documentary about insects in an African wildlife park, most of it shot within 100 yards of a safari lodge. There was no point in humping the cameras any further into the bush. Every so often, in between shots of the mating rituals of beetles or the internal politics of the ant world, the camera would pan over to another load of tourists piling into their land rover for a trip to the lions and zebra. 'Do they realize', the commentator asked, 'that African wildlife is happening under their feet and on their doorstep?' It is like that with some of the tree-huggers: they don't seem to be aware of what is happening beneath their feet. They do not have much conception of grass either as a complex biological system, or as a crop, or as a way of life. For them it is just default vegetation, a place to walk on, to put things on, and to plant trees on.

However, it is not just the eco-warriors. The iconic status accorded to trees is coming to be accepted by a widening spectrum of the population. Planting trees is commonly viewed as an instant solution to environmental problems ranging from over-intensive farming to global warming, and we can expiate our carbon guilt by contributing to tree-planting programmes in distant lands. In the 1990s, John Major's Conservative Government announced in its rural White Paper that it wished to double the amount of woodland in England. Blair's Government stated, more cautiously, that it was 'determined' to oversee 'a significant increase in woodland cover across England', with a doubling of the area in some regions.[17] The latest advice in the Read Report is to increase woodland coverage by a million hectares.[18] Few people voice any objections – which is unusual for a land use change on such a scale, especially when we bear in mind that the country is currently busy trying to eradicate the coniferous legacy of the British Government's previous drive to double tree cover in the 1960s. Imagine the outcry if the Prime Minister announced that he wished to double the number of wheat fields, or horses, or houses or roads. Planting trees is more uncontentious than motherhood and apple pie.

17   Brown, P (2001), 'England's Woodlands Growing to 1,000-year Record Total', *The Guardian*, 22 November 2001. MAFF and DETR (now DEFRA) (2000), *Our Countryside: The Future*, Rural White Paper, HMSO.
18   Read, D J *et al* (eds) (2009), *Combating Climate Change: A Role for UK Forests*, The Stationery Office, Edinburgh, 2009.

The public have pitched into the tree planting boom with enthusiasm. Buying a small neglected woodland, or planting a new one, or establishing a community woodland are fast becoming national pastimes. The price of a small acreage of woodland, which a few years ago was about half that of pasture land, has shot up and would now be higher than pasture were it not for the fact that horseyculture has pushed the price of pasture up correspondingly.

The move to bring existing neglected woodlands back into a state of management is to be welcomed. An economy which rewards its urban élites and 'creative classes' handsomely, and doesn't pay its landworkers properly, has to rely on voluntary labour. If there are people with urban incomes who get more satisfaction from driving out to the countryside at weekends and hacking down brambles and rhododendron than they do from flying out to Indonesia for a fortnight in the rainforest, then that is well and good. It is a bit of a cock-eyed way of going about land management, but it is arguably better than letting land go derelict.

The planting of new areas of trees on grassland, however, is a different matter. It involves a reversal of land use, and often it is not at all clear that this is carried out with any coherent objective in mind. When villages and parishes were land-based communities there was a logic in the way they laid out land uses according to the topography, fertility and distance from the village, and much of this pattern still remains. The same concern is reflected by modern permaculture designers in their advice to 'zone' different land-based activities according to the amount of attention they require and other factors. It is unlikely that many private landowners take such matters into account when they decide to reverse several hundred years of history.

Take for example a certain village in Somerset where there are two areas called Broad Mead and Little Mead. These were the meadows which served the village when it was more communally operated before enclosure. In those days, copyholders, who rented small properties of ten acres or so, also received, with their property, the right to graze either 25 or 50 sheep on the commons. They would fold their animals on their lands for manure, bringing them down what are now hollow lanes, gouged into the hillside by the passage of hooves. Presumably they also had access to some of the meadow land, near the small river, otherwise they would have had problems feeding their stock over winter.

Little Mead is now the name of a housing estate. Broad Mead is divided into a number of fields, the majority of which still provide grazing or hay. But there is a move towards a change of land use. Already two of the fields have been planted out with amenity woodland. Since the profits to be made from farming are low, it is understandable that their owners should turn their field over to a crop which involves little maintenance, and provides a green and pleasant land use. Just two fields out of a dozen or so do not in themselves constitute a radical change, and they add texture and variety to the neighbouring landscape.

But the plantations show how far removed the dormitory villages of England are from any notion of collective or strategic land management. Centuries ago somebody went to a great deal of trouble to clear Broad Mead of trees, because the land was best suited for use as a meadow: flat, fertile, well watered and close to the village, a place where you could easily watch grazing animals or move to quickly if the weather looked like turning. Hay has been made there every year for centuries, hay which served to feed the villages' animals and subsequently to fertilize their fields. The meadows played an integral role in the economic life of the village; it is not clear what purpose the new woodland will serve other than supplying firewood and providing cover for foxes and badgers to launch attacks upon the villagers' chickens and wheelybins.

If there were a shortage of woodland in the area, or the settlement were marooned in the middle of a windswept prairie, then planting meadows with trees might make sense. But the village is already fringed with a belt of wooded hillside which for the most part nobody has looked after. The former commons, which had its cattle grids removed about 20 years ago, became a 'country park' and the sides of the hill are covered with scattered woodland which is neither grazed nor managed for timber, coppice or firewood, and which is now carpeted with an understorey of bracken and scrub. The hill opposite was logged for timber in the First World War, and has since become covered with neglected woodland consisting of about 50 per cent dense laurel which has killed off every other tree and under which nothing grows. Meanwhile large swathes of good quality arable land were put over to orchards in the middle of the 20th century, and a large fruit-packing complex was built in the centre of the village – then at the end of the century all the orchards were grubbed up and the factory sold off for housing. The village, like most in England, has lost its collective sense of what constitutes an integrated system of land use.

Its new plantations, which are mirrored in similar well-intentioned ventures in villages throughout much of England, raise a number of issues. First, the Small Woods Association, in the late 1990s, estimated that in Britain there were around 175,000 hectares of broadleaved woodlands under 10 acres in size which had had no management for over 30 years, and many are becoming invaded by rhododendron and laurel.[19] If we can't look after our existing woodland, how likely is it that anyone will look after the new ones we plant? The situation is improving as many small woodlands are currently being bought up and cared for by people with green aspirations. But anyone who wants to enhance the state of Britain's woodland should still consider taking over an existing derelict wood before thinking of planting a new one.

Second, trees do not require high quality land and some places are better for woodland than others. The most sensible places for woodland are hilly land, inaccessible spots, poor quality soils, wet soils, north facing slopes, and where windbreaks are needed in wide open plains. Woodland does not need much maintenance – and hardly ever does it need urgent attention – so it tends to be sited further from human habitation than those land uses requiring frequent attention. When I see decent quality pasture or potentially arable land covered with rows of tree-guards like a military cemetery, I am inclined to wonder whether the trees might have been better planted somewhere else.

The third matter is the bias against timber plantations, and especially conifers. There is an understandable reaction against the overplanting of Sitka Spruce in the 1950s and 1960s. But the pendulum has now swung violently in the other direction. Not only is almost everybody who plants trees in many parts of England planting native broadleaves, but often they plant them too wide to produce straight timber – unsurprisingly since the same grant is available whether or not they space them densely enough for timber production. In some places there is a virtual pogrom against conifers. Dorset, until 2009, had a policy to banish conifers from its Area of Outstanding Natural Beauty, which occupies about half the county, and replace them with broadleaves.[20] 'Sitka can actually

19   Nichol, Lucy *et al* (2000), *Planning for Sustainable Woodlands*, publised by Chapter 7 and the National Small Woods Association, 2000, p 2.

20   Dorset AONB (2004), *A Framework for the Future of the Dorset AONB 2004 to 2009*, Dorset AONB Partnership. The new plan does not appear to be so prejudiced against conifers: Dorset AONB Management Plan 2009-14, http://www.dorsetaonb.org.uk/partnership/dorset-aonb-partnership/33.html

be classed as a destructive weed' states one tree-planting charity.[21] According to a Forestry Commission regional officer I interviewed, the number of conifer plantings in his region is 'very small' and he confided that he found it 'depressing' that there was 'such a pronounced move away from productive woodland'.[22]

In our Somerset village, until recently the only mature wood of any size being managed was a stand of Douglas fir and larch, planted by a firm called Economic Forestry as a tax scam in 1960, and bought by 'Happy Valley' community in 1994. Since then it has been progressively thinned, two areas have been replanted and about a hectare of invasive cherry laurel has been removed. The conifers have been looked after rather better than the five acres of broadleaves in the same wood, because larch and Douglas are first class building timbers; they provide the walls, floors and roofs of the community's buildings, the posts for its fencing and surplus timber is planked up on a sawmill and sold locally. The broadleaf trees, by contrast, aside from any ash which has escaped canker, are not so obviously useful for anything beyond beanpoles and firewood.

All those people who are insouciantly planting native broadleaves perhaps need to think a bit about who will look after their woodland in the decades to come. 'The wood that stays is the wood that pays' is a foresters' motto, and it is a main conclusion to be drawn from the historical information provided in Oliver Rackham's *The History of the Countryside*. 'The wood that pays' doesn't necessarily mean a wood that makes a financial profit, but a wood that meets the need of the rural economy. The landscape of Britain was crafted according to people's needs – though not necessarily everybody's needs. When foresters planted all those Sitka Spruce, it was to meet the needs of industry, but now they have reached their maturity in a global economy where it is not clear that there is any need for them. Woodland today is often planted according to the whims of people whose material livelihoods are more or less unrelated to the rural economy, so if these plantations meet the needs of future generations, it will be more by luck than by design.

This undiscriminating approach towards trees and tree-planting is shared by people of all sorts of persuasions. But it is prevalent amongst vegans and not uncommon amongst adherents of permaculture, and we can locate some of its intellectual roots in both the permaculture and the vegan movements. These movements are more influential than they might appear, and tend to influence the mainstream with their ideas through a sort of intellectual osmosis long before they are eventually accepted into the mainstream in their own right. The influence of vegans and vegetarians can be detected in an overall decline in meat consumption, and a shift to a more balanced diet; similarly, the current public interest in trees probably owes something to the permaculture approach. Both these trends are very welcome – it is when they become obsessive or absolutist that I worry.

The Gentlemen's Agreement

Permaculture, or permanent agriculture, was first formulated by Bill Mollison and David Holgrem in Australia in the 1970s, and until recently most of its important texts have been written with tropical countries in mind. It is important to bear in mind that the discussion in this section focuses primarily on the UK and to a lesser extent on other countries enjoying temperate climates.

---

21   Tree Spirit (2006), 'Maes y Mynach – Sitka Control', *Tree Spirit*, Autumn 2006.
22   Fairlie, S (2004), 'In Praise of Douglas Fir', *Chapter 7 News*, 15, Summer 2004.

Patrick Whitefield's *The Earth Care Manual* is a long awaited assessment of the value of the permaculture approach to farmers and growers in temperate zones. The fact that I use it to question certain aspects of the UK permaculture movement does not diminish my respect for Whitefield's work, or for his book, which is full of shrewd advice, and which I use as a teaching aid. I don't know of any other book in the UK which covers the same issues in anywhere near as much detail. There is not a lot to be gained from subjecting to scrutiny a work which is not the best in its field, which *The Earth Care Manual* undoubtedly is.

Permaculture means different things to different people', says Whitefield in the opening paragraph of his first chapter, 'but at root it means taking natural ecosystems as the model for our own human habitats. Natural ecosystems are, almost by definition, sustainable, and if we can understand the way they work we can use that understanding to make our own lives more sustainable.'[23]

You might think at this point that any sensible vegan would decide: 'Permaculture is not for me'. All natural ecosystems have animals that eat the vegetation as well as one another, and are integral to the ecological balance, so an agricultural economy which eschewed animals would seem to be inherently inconsistent with permaculture. But bizarrely a great many vegans are interested in permaculture, and a considerable number of permaculturists are vegan or vegetarian. I have just keyed the words 'vegan permaculture' into Google and have got 516,000 results; when I key in 'livestock permaculture', I get 78,200.

You might also think that the author of this opening sentence would be reluctant to expunge the animal kingdom from his vision, and here you would be right. If you read all of Whitefield's book you will see that he shares broadly similar opinions about the relative value of animals and grass and trees to those I express here. He warns prospective pioneers about the danger of buying land for tree-planting which is obviously more suitable left as grass.[24] He is aware of the value of livestock as a means of providing nutrients and moving them around the farm.[25] And although, like many people, he accepts what I argue is an exaggerated account of the inefficiency of livestock production, he is neither a vegan nor a vegetarian.

It is not the main body of Whitefield's book that I have difficulty with, but the headlines, and therefore the emphasis. The titles of his five chapters on land use give an indication of a slant: 'Gardens', 'Fruit, Nuts and Poultry', 'Farms and Food Links', 'Woodland' and 'Biodiversity'. Animals (with the exception of poultry), arable crops and grassland are all lumped in under 'Farms and Food Links', and animals are allocated six pages to themselves, compared to nine pages for nuts and 42 pages for woodland.

Particularly noticeable is the scant attention paid to grass. It's not that Whitefield doesn't have sensible things to say about grass – he does. But he doesn't say very much, when you consider how much grass there is in Britain, how well it grows, and that grass is the most likely ecosystem that permaculture pioneers, when they buy a bit of land, are going to have to deal with. It is also usually the vegetation they know least about – 'of all things most common, grasses are the least known.'[26] The word 'grass' doesn't appear in the index, nor do the words 'meadow', 'pasture' and 'hay'. By contrast, 'trees' have 22 entries, and 'woodland' has 51 entries. 'Squirrels' have three entries, whereas cattle have only two.

23   Whitefield (2004), *op cit.* 5, p 3.
24   *Ibid.*, p 306.
25   *Ibid.*, p 261.
26   J C Mohler, cited in the frontispiece of Harvey, Graham (2002), *The Forgiveness of Nature*, Vintage.

Partly this is inadequate indexing: there are two pages entitled 'Grassland with Trees', and meadows and pasture pop up elsewhere in the book. There is a thoughtful assessment of foggage, the wintering of cattle and sheep outside on grass. But I couldn't find anything at all on making hay, even though it is by far the most biodiverse crop that any British farmer is ever likely to harvest at one go. Grassland is dealt with incidentally in the book – as though the author doesn't want to advertise its benefits – whereas tree cover is accorded pride of place, and the ecological functions and resource benefits of a woodland ecosystem are explained in detail.

This bias is uncharacteristic of Whitefield (whose latest book, *The Living Landscape*, devotes almost as much space to grassland and moorland as it does to woodland, and rarely mentions permaculture), but it is prevalent throughout the UK permaculture movement.[27] To focus on one example, a land resettlement project I know of involving 21 acres of land in the West Country, of which more than a third is pasture and the rest arable, commissioned from separate experts both an ecological survey and a permaculture design for the land. The ecological survey identifies three different types of grassland on the site, including an area of 'species rich MG5c' which contains 'seven species considered as indicators of species rich lowland meadows.' Its authors conclude that 'the grasslands should be kept as grassland' and managed by grazing livestock and cutting for hay, specifying a low stocking rate of 0.5 livestock units per hectare.

The permaculture survey of the same land, on the other hand, is almost blind to the grass. A list of resources to be found on the land cites 'timber and poles from the hedges and border with the river, freshwater from the spring, clay rich earth, good range of terrain suitable for a range of activities, good existing fences' – there is not a word about the main crop actually growing on the land at the time, namely somewhere in the region of 75 tonnes of grass a year. The word 'grass' occurs just twice in nine pages of text, whereas the ecological survey uses the word 'grass' 28 times in a document of the same length. On the other hand the word 'tree' occurs 30 times in the permaculture design, although the only trees are in the hedgerows and along the banks of the river, and the word 'mushroom' occurs 11 times.

There is a wide ideological gulf between these two approaches which I would like to see bridged. I don't believe it is a given that all 'the grasslands should be kept as grassland', certainly not the less species rich zones – there is no shortage of average quality grassland in the West Country. On the other hand the livestock and grassland are part of a complex and diverse biosystem which for centuries was the preferred choice of people whose carbon footprint was far lower than ours, so it seems daft not to give it due consideration.

There are plenty of similar examples, where, for example, permaculturists take over pasture land, and concentrate on planting trees or creating ponds in small areas whilst abandoning the grazing so that the grass becomes weedy, overstood and a nightmare to manage. One permaculture group I know, with about 20 acres of grassland, currently rents much of it out to a local sheep farmer, who not only grazes it but removes some of the fertility in the form of hay, because the community has a policy of not keeping livestock for meat.

Why do so many permaculturists in the UK shun the virtues of grass and focus on trees, forest gardens and horticulture as a means of producing food? In some cases it may be because they only have access to a small amount of land; I know of several instances where conversion of an acre or two of pasture to a

---

27   Whitefield (2009), *op cit.* 5.

mixture of raised beds, forest garden, orchard and coppice provides a far more productive and biodiverse environment than would keeping a few sheep or a couple of horses on it.

A bias towards trees may also stem from permaculture's preference for perennial crops. There are several good reasons for favouring perennials over annuals: they require less tillage, they sequester carbon, you don't need to put aside a proportion of the harvest for seed and they don't suffer the diseases that annuals are prone to when planted on the same ground over a number of years. Growing perennials, in Whitefield's words, is 'a direct imitation of nature. Annuals are rare in natural ecosytems. They normally need bare soil to get established, and are quick repair specialists, able to move in quickly and heal the wound. But they are soon succeeded by perennial plants.'

Perennials also often play a main role in another key permaculture concept, stacking – growing two or more crops at different height levels with a view to obtaining a higher total yield, even though the yield from each individual species would be lower than if grown as a monoculture.

There are, naturally, some disadvantages. Perennials take longer to improve through conventional plant-breeding methods – a fact that Whitefield suggests explains why prehistoric farmers focussed on annuals like wheat and barley. They also have to devote a proportion of their energy into maintaining and expanding their inedible infrastructure. On the other hand, Whitefield observes:

> Although we can't harvest all the food produced by a perennial crop,
> it's in a position to produce much more in total than an annual. This
> is because it starts out at the beginning of each growing season as a
> full-sized mature plant, with the energy resources to start growing at
> maximum rate as soon as the temperature and other conditions are right.

All of the above is textbook knowledge for followers of permaculture. What some seem to forget is that permanent grass is an entire ecosystem of perennial species (with its own internal stacking system) which doesn't have any above ground infrastructure to maintain. Its infrastructure is entirely in the roots. It can therefore spring into life sooner than most other perennials – to quote J J Ingalls again, it 'withdraws into the impregnable fortress of its subterranean vitality, and emerges upon the first solicitation of spring'.[28] In the south west of the UK it never completely retreats, but carries on growing at a very diminished rate throughout most of the winter. According to Whitefield 'pasture grasses carry out 60 per cent of their annual photosynthesis before ash trees come into leaf, so the two can be combined in a stack which is more productive overall than either planted alone'. Similarly there will already be a crop of grass in an orchard by the time the apple leaves are fully developed.

Besides being 100 per cent edible to animals, grass has numerous other advantages that one would have thought would commend it to permaculturists: it is highly biodiverse and resilient, it creates organic matter in the soil, it introduces nitrogen and improves fertility, its fertility can be moved easily from one place to another with the aid of animals, it can be cut for mulch, it opens up the ground to sunlight, it can be walked on or driven on when mown or grazed, it provides the easiest surface for picking up windfalls or shaken fruit, and it is good for playing football on.

But while it is hard to see why many permaculturists accord such a low status to grass, it is easy to see why vegans are not interested. Perennial grass

---

is absolutely useless as food for someone who doesn't eat animal produce. If a vegan wants to grow anything edible on non-arable land, it has to be trees or shrubs. I have a memory of Ken Fern, as he was showing me around his remarkable landscape of edible trees bushes and shrubs in Cornwall, kicking at the waves of grass licking around the base of his beloved plants, muttering 'horrible stuff, horrible!', whereas I was thinking that from a cow's point of view it looked really tasty.

In the absence of any comprehensive land use strategy of their own, it is understandable that many vegans should gravitate towards a theory promoting the view that perennials such as trees and shrubs can be highly productive. This is not true of all vegans. The stockfree grower's magazine *Growing Green International*, is wary of adopting the permaculture label, and promotes 'vegan-organic' agriculture. But the sheer volume of vegans at the radical end of the UK spectrum seeking an agricultural expression for their dietary beliefs has, I fear, skewed permaculture in a graminophobic direction. If this is the case, the alliance between the vegan movement and the permaculture movement is potentially helpful to veganism, but doesn't offer anything to permaculture – the one is parasitic upon the other.

Perhaps this is happening because everybody is too polite to discuss the matter properly. There sometimes seems to be a gentleman's agreement within the permaculture movement that differences between meat-eating and vegan permaculturists should be glossed over. Nobody, to my knowledge has attempted to examine this fault line publicly. There have been a few spats in the letters column of the UK-based *Permaculture* magazine, but the editor, Maddy Harland, tells me that a few years ago she (understandably) got bored with publishing vegan and anti-vegan rants.

This polite reticence is in evidence in a 60 page booklet called *Permaculture, A Beginner's Guide*, written by Graham Burnett. Burnett is an active campaigner against meat-eating: he has written another pamphlet entitled *Well Fed – Not an Animal Dead!*[29] But his guide to permaculture breathes not a word of his vegan beliefs, not at least until we get to page 50, where there is a mention of the excessive amount of land devoted to animal raising and the entirely reasonable suggestion that 'if we are to reduce our environmental impact we will all need to think about life-styles which are less dependent upon animal products and the inputs these entail'. In the spirit of politeness which pervades the public debate (though not private conversations), he adds, 'not everybody might be ready or want to make such changes'. Other than that, there is not a single mention of domestic animals in his entire book, except for ducks kept to eat slugs. A plan drawn for a farm in a cool temperate zone shows areas for orchards, cereals grown in clover, vegetable crops, forest garden, woodland etc., but no pasture and not an animal in sight.

There is nothing wrong with Burnett's brand of vegan permaculture and there is no reason why this farm could not function successfully and contribute to a society which is less dependent upon animal products. Burnett's book would be fine if it was called *Vegan Permaculture: A Beginners Guide*. I don't think he is meaning to deceive, I think he is just being polite. But as it stands the book gives the beginner a distorted picture of what permaculture is or could be.

Whitefield, a meat-eater, does provide a two page discussion on page 258 of his book, headed 'Omnivorous or Vegan?'. He phrases this discussion within the

---

29   Burnett, Graham (2000), *Permaculture, A Beginner's Guide*, Land and Liberty; Burnett, Graham (n.d.), *Well Fed – Not an Animal Dead!*, Spiralseed.

context of organic farming, which he has just soundly endorsed, and prefaces the main part of the discussion with these words:

> The practical question is 'how important is the keeping of animals to sustainable farming? On the one hand, we can get a far higher yield of food from the land by eating plants rather than animals. On the other hand, animals have an important part to play in maintaining soil fertility.

Whitefield goes on to explain the principles of fertility building through ley farming, and examines experimental variations such as stockless rotations, and bi-cropping (sowing grain in a sward of perennial clover, which can be mown for fodder or grazed after the harvest). He concludes:

> The total output of grain from a wholly organic agriculture would be less than at present, partly because of lower yields per hectare, and partly because a proportion of the arable land must be down to ley at any time. We would have plenty of grain for ourselves, but not much to spare for feeding animals. So production of pigs and poultry would have to be reduced. But grass-eating ruminants – sheep, cattle, goats – would probably have a central place.

I could hardly agree with him more. It's all there in a sense, yet it's tucked away when it ought to occupy a central place in his book, and in the permaculture canon. One thing I found curious about Whitefield's discussion of the delicate meat/vegan issue is that he never mentions the word permaculture. 'How important is the keeping of animals to sustainable farming?' he asks. In a permaculture handbook you would have though that the crucial question was 'how important is the keeping of animals to permaculture?'Is he being polite to permaculturist vegans? Does he consider that ley farming, though sustainable, is not part of permaculture? Or is this just an accidental choice of words that I am attaching too much importance to?

That doesn't really matter. What does matter is that nobody has yet found a way to grow continuous high yields of grain crops organically other than in conjunction with grass and legumes – which may not strictly require animal husbandry, but certainly lend themselves to it. Grain is our primary staple, and however successful breeders may become at developing food from trees, it is going to remain the staple for the vast majority of people in the world for some time to come. Staple crops, by definition, lie at the heart of any agricultural system, and it is very difficult to see how permaculture, unless it espouses agrochemicals, can afford to neglect the grass and legumes that are so essential for the organic production of grain, and so propitious for the rearing of animals. If it does neglect grass and grain – horn and corn as they used to call it – permaculture is likely to remain on the fringes of agriculture, producing secondary and niche commodities such as vegetables, mushrooms and berries.

Permaculture is a contraction of the words 'permanent' and 'agriculture'. In 1948, the US Department of Agriculture published as its *Yearbook of Agriculture* a volume entitled simply *Grass*.[30] It is a veritable encyclopaedia, 892 pages long and containing over 120 contributions about grass farming from dozens of authors. It was the outcome of widespread concern at the time to reverse the destructive farming processes resulting from the 'Great Plow-up' of the 1920s that had led to the dustbowls of the 1930s. The first section of the book was headed 'Permanent

---

30   USDA (1948), *op cit*. 1.

Agriculture', and its opening lines were 'Our goal is permanency in agriculture'.
It continued:

> Permanency in agriculture is a goal to be sought always, by all
> people everywhere. So, in the wake of war and in the glow of our
> unprecedented production, this country looks to the future and
> considers again the land and its management – this time as never
> before in terms of grass. For around grass, farmers can organize
> general crop production so as to promote efficient practices that lead to
> permanence in agriculture.

The opening chapter by P V Cardon, and indeed the entire book, put the case
for a sustainable agriculture, founded on mixed farming, with grass providing
protection against erosion and restoring the organic matter and the nitrogen to
the soil that had been depleted by excessive cultivation of demanding crops.
Grassland was viewed as 'inseparably linked to livestock production'. Grasses
and legumes, Cardon writes, 'are dominant in a flexible pattern designed to
conserve the land and its productivity, but at the same time keep it adjustable to
emergency needs'.

Grass, being adaptable, resilient, nutritive – 'the forgiveness of nature' –
would heal the wounds that had been inflicted upon the Mid West by years
of extractive agriculture. And so it did to some extent, until in the early 1970s,
when President Nixon's Secretary of State for Agriculture, Earl Butz, famous for
urging farmers to 'get big, or get out', launched the United States into another
fossil fuelled 'plow-up', which still supplies the world with factory farmed meat
and soft drinks.[31] Now, with a need to reduce fossil fuel use and increase soil
carbon, grass farming is emerging from its subterranean fortress and coming
back on the offensive. One day soon, I hope, some bright fellow may pick up
a copy of the USDA Yearbook for 1948, take in its message about permanent
agriculture, and feel inspired to produce the *Permaculture Book of Grass*.[32]

The End of Natural History

Patrick Whitefield, in the first paragraph of *The Earthcare Manual*, speaks of 'a
wild woodland, which is the natural vegetation of this country', a remark which
might have passed unnoticed a few years ago, but is now open to challenge.

Whitefield makes no mention of Frans Vera's reappraisal of how the ecology
of deciduous woodlands developed in prehistorical times and throughout the
Middle Ages (though he does allude to it in a later book).[33] In the few years since
the publication of *Grazing Ecology and Forest History*, Vera's thesis has been taken
seriously by most ecologists; but it does not yet seem to have been noticed by
many in the permaculture world.[34] This is surprising, because Vera's conclusions
lend weight to a permaculture approach.

The traditional view of Europe, after the retreat of the last Ice Age, is of a
continent covered by belts of thick closed forest, which were gradually cleared by

---

31   For a good account of Earl Butz and the rise of King Corn, see Pollan, Michael (2006), *The Omnivore's Dilemma*, Bloomsbury.

32   The copy at my disposal of *Grass*, is now held at the British Library's supply centre at Boston Spa, after starting out life in the Science Museum. From what I can tell, it was last lent out in 1971. Prior to that, in 1966 and before that 1952. Sadly neglected.

33   Vera, Frans (2000), *Grazing Ecology and Forest History*, CABI, p 347. Whitefield does briefly allude to Vera's thesis in his later book.

34   I am, however, grateful to Mark Fisher, a permaculture practitioner and theorist from the North of England, for alerting me to Vera's work.

humans as they established agricultural settlements. This was an unchallenged orthodoxy which dominated in the latter half of the 19th century and persisted throughout the 20th century. Part of the evidence for this was the observation that if an area of pasture is fenced off from animals and left to 'do its own thing', scrub begins to appear, then sun-loving trees such as oak and hazel, until finally these are shaded out by a canopy of trees such as lime or beech which require less light to regenerate. These shade-tolerant trees are regarded as the climax vegetation – the biological equivalent of 'the end of history' – which can only be reversed by a catastrophe such as fire, hurricane, climate change or the arrival of man.

Vera argues that this climax was not what normally happened in many lowland areas in Europe. Before farmers arrived with their cows, sheep and pigs, Europe was populated by large wild herbivores: the aurochs (wild ox), the tarpan (a kind of horse), the European bison, red deer, beavers and wild boar. These animals, which between them ate grass, leaves, seedlings and bark, applied constant pressure on the tree population with the result that, far from being blanketed with a thick mass of closed forest, many areas of Europe consisted of savannah type landscape, more like what can be seen in parts of the New Forest, in Extremadura in Spain, or in Serengeti National Park, all of which are grazed landscapes.

Much of the evidence cited by earlier ecologists for the existence of closed forest relies on fossil pollen counts showing a high proportion of tree pollen and comparatively little grass pollen, which, at first sight, does suggest a predominance of trees in the landscape. Vera suggests that this is consistent with a savannah type landscape, because much of the grass gets eaten before it goes to seed. Modern pollen counts taken from grazed or mown fields next to woodland show a similar spectrum. Moreover, the fossil pollen counts show very large amounts of hazel pollen, which are inconsistent with closed forest because hazel has a hard time surviving in heavily shaded woodland, and only flowers and seeds at the woodland edge.

Vera's hypothesis is that the open savannah type woodland followed a cycle along the following lines:

> In an area of grassland, grazed by herbivores, thorny shrubs which herbivores shun will begin to appear. Light loving pioneer species of tree, particularly oak, hazel, and wild fruit trees, whose seedlings get eaten when they appear in open grass, grow safely in the protection of this scrub.
>
> When the light loving trees become mature, they create a grove with a canopy which is too dark for their own seedlings, and so the seedlings of species such as beech, lime and elm are at an advantage. Eventually the shade loving trees dominate the oaks and hazels, killing them off in the centre of the grove.

So far this succession is consistent with the climax theory, but this is where Vera's hypothesis diverges. The orthodox view is that when trees die in the centre of the woodland grove, creating gaps, more beech and lime seedlings grow in their place. However Vera postulates that because it is too shady within these groves for the thorny bushes to grow successfully, herbivores come to eat the grass which colonizes the gaps almost immediately, and then either eat or trample the beech and lime seedlings, maintaining a grassy clearing within the woodland. The clearing may grow because the exposed trees around its edge are more susceptible to windblow.

Meanwhile, the grove of trees is expanding, because the thorny scrub is spreading out into the grassland, allowing more oak and hazel to grow at the periphery of the grove. So the woodland is an ever-expanding American donut (not as climax theorists postulate an English doughnut) invading the grassland at its periphery, while fresh grass is hollowing out the centre. Eventually the grassland in the centre becomes sufficiently large and well lit to support the emergence of thorny shrubs which mark the beginning of a new grove of trees.

The process is aided by various other participants in the ecosystem: jays, nuthatches and squirrels may bury acorns and hazel nuts, assisting their propagation; wild boar snuffle in the sward making holes where scrubby bushes can establish themselves; beavers create large areas of grassland in the middle of woods by damming streams and creating flooded areas which kill off trees; bark eating herbivores such as bison may ring bark trees when the area of grass becomes limited; and predators will control the number of herbivores when they become excessive.

The result is a shifting mosaic (this word crops up repeatedly in texts on the subject) of shade-tolerant and light loving ecosystems, supporting high levels of biodiversity.

On its own this would just be an interesting aetiological theory, of secondary significance to the modern agriculturalist. But Vera completes his hypothesis by arguing that humans continued to maintain this silvo-pastoral landscape on common grazing lands up until the 19th century. Typically these lightly wooded areas provided grazing for cows, sheep and horses; pannage consisting of mast (acorns, beech nuts, hazel nuts etc), roots and insects for pigs; faggots of brushwood for firing bread ovens and glass furnaces; coppice wood for fencing or charcoal; and standard trees, of which the most valued was oak, for house and ship building.

These commons were managed according to regulations which were designed to maintain the desired balance of woodland and pasture (which of course might vary from one location to another). The population of grazing animals was limited by stints – the allocation of a certain number of livestock to each commoner. Sheep in particular were restricted in number because they are more destructive of seedlings and saplings than cattle.[35] Newly coppiced woodland had to be fenced off from livestock until it was sufficiently high to deter browsing. When oak trees were felled others had to planted, and it was common to plant a thorn bush in the same hole as the oak seedling. In some areas fenced nurseries of hawthorn and blackthorn were maintained for this purpose.

Vera's interpretation is mainly based on Northern European evidence (and hence previously unavailable in the English language), but much of his historical material matches the history of English woodlands portrayed by Oliver Rackham. However, Rackham (writing ten years earlier) does not share Vera's view of prehistoric tree cover : he attributes the sizeable pollen presence of hazel (bizarrely in my humble view) to its being a 'canopy tree', rather than an understorey tree germinating and pollinating in the light at the edge of a woodland.[36] After Vera's thesis appeared, Rackham, in his 2006 book *Woodland*, criticized Vera's assessment of prehistoric pollen counts, and suggested that the Vera model, while it may have applied in parts of Europe, was followed less closely in the wetter British Isles, where prehistoric grazing animals died out much earlier. That view would place the British Isles as an isolated dark damp

---

35   Vera, Frans (2000), *op cit.* 33.
36   Rackham, *Woodlands*, Collins, 2006, p 68.

fringe at the edge of a more mottled European patchwork. George Peterken, another prominent expert on British woodland, is also sceptical, but postulates:

> Large herbivores can generate clearings, which allow more large herbivores to thrive, but, equally, heavy tree cover could militate against large herbivores and thus reduce populations. These two possible feedbacks could generate two stable states, open wooded pasture and closed high forest, the probability of which might vary by site type.[37]

But Rackham's and Vera's views converge once humans enter onto the scene. Rackham's *History of the Countryside* devotes one chapter to 'Woodland' and another to 'Wood Pasture – Wooded Commons, Parks and Wooded Forests'. He suggests that 'much of the "woodland" of Norman England was really "wood pasture"' and notes that oak 'is an ideal wood pasture tree; the oaks of old England (like the cypresses of old Crete) are in part the result of cattle, deer and goats eating their more edible competitors'. He estimates that in the late middle ages, besides wooded commons, there were around 3200 private parks in England, whose purpose was not so much hunting as the 'prosaic supply of venison, other meat, wood and timber', and which bequeathed us the breed of White Park cattle. In the 180 or so wooded forests which technically belonged to the crown, such as the New Forest, 'it was the commoners and landowners who did most of the grazing and woodcutting'.

As populations grew, and an expanding market offered better opportunities to exploit the commons for profit, pressure upon them increased; in some commons, but by no means all, regulations were undermined or ignored and the quality of the pasture and the woodland deteriorated. This deterioration supplied arguments for those who advocated enclosure of the commons, but combating overgrazing was not their only priority. Vera's evidence shows that another of their aims was to move to what they regarded as more efficient monoculture.

The first assault on the English commons, in the 14th–17th centuries, came from monocultural sheep farmers, who could derive more profit by fencing land and putting it over to the production of wool than they could from mixed farming. Later enclosure in England involved the draining of highly biodiverse fenland for arable production, and finally the enclosure of common fields and waste for improved arable farming. But the evidence from the continent supplied by Vera suggests that much of the pressure for enclosure in the 19th century came from the forestry industry.

'Agriculture has existed for 10,000 years,' says Vera, 'forestry for only 200 years', which is another way of saying that until two centuries ago agriculture and forestry were the same activity. The move towards forestry as a separate, monocultural activity occurred as a result of the industrial revolution. The arrival of coal, steam power and steel resulted in a radical change in the kind of timber the market required.

When Europe was reliant upon handtools, timber of relatively small girth was more economical for many purposes, because of the effort of chopping or sawing it up. Faggots of small branches were more convenient for cooking wood; roundwood and split coppice wood was more convenient for fencing; tool handles were cut straight from the hedgerow. Before coal, brushwood such as thorn and gorse was required for flash firing (bringing ovens and furnaces quickly to a high temperature) and coppice wood was more practical for making charcoal. Cruck architecture relied on curved oak trunks for the frame of a

---

37   Peterken, G (2001), review of Vera (*op cit. 33*) in *British Wildlife*, Feb 2001, pp 225-6.

house, while wattle and daub walls and thatched roofs could be laid on wiggly coppicewood rails. Shipbuilding required a large number of curved oakwood members to provide bows, struts and bracing. By a happy coincidence, the broad, spreading oak trees of the open forest, ideal for cruck frames and shipbuilding, were also the oak trees which produced the most abundant mast for pigs.

With the advent of mechanization the market required completely different products. The arrival of sawmills meant that timber trunks had to be straight to pass through the mill. Factory-made roofing tiles and stud walls required straight sawn members. Teak and steel substituted for oak in shipbuilding. The need for charcoal and brushwood was replaced by a need for pit props. Purchasers of firewood demanded cordwood rather than faggots. Meanwhile, large oak trees supplying mast became redundant because pigs could now be fed from rows of potatoes. To operate efficiently, mechanization required the rectilineal regimentation of the landscape.

The main aim of modern scientific forestry is to produce straight trees of reasonable sized girth, suitable for the sawmill. Trees like this cannot be grown in a savannah type landscape, because the abundance of light makes them grow bushy, like the mast oaks that grow in park land. They have to be grown in the artificial equivalent of a closed forest, namely a plantation, where they stay straight because they are competing for light. Like the supposed climax of a beechwood closed forest, a plantation is a monoculture. And like a closed forest, a forestry plantation can only be created if grazing animals are excluded, to prevent them destroying seedlings and saplings, and barking trees.

Besides, the foresters argued, there is nothing for animals to eat in a properly managed forest. The place for grazing animals is a field of grass, preferably ryegrass. The common lands of non-Mediterranean Europe had to be changed from a kaleidoscope of vegetation rotating through the spectrum from light to dark – not unlike the Chinese board game *Go* – to a chessboard whose monotone squares were either grass or plantation: 'Certain areas of land were designated as pasture; others for the provision of wood. Foresters had been insisting on this division for some time.'[38]

The diet of animals suffered accordingly. Whereas 'the natural foodstuff of the ancestor of our domestic cattle was soft, bushy, leafy material, the lower branches of trees, sedges, herbs and grasses', once fields were enclosed with hedges it became imperative to stop livestock over browsing them – so thorn became the primary hedgerow tree.[39] The plant which in the silvo-pastoral mosaic had served as armour for opportunist oak seedlings, was employed to fortify entire plantations and improved pastures from the ravages of haphazard grazers, condemning cattle to a monotonous diet of 'improved' grass. Now, when cattle escape from their fields, they make straight for the herbs and forbs on the other side of the hedgerow.

Rackham reports on how enclosure affected English woodland:

> An even sadder story is that of the Forest of Dean, which has a rich
> history of pastures, coppices, and outsize timber trees, of deer and wild
> swine, roadside trenches and industries going back to the Romans …
> Nearly all this heritage has been effaced. Dean is now blanketed with
> plantations of uniform, poorly grown oaks whose later replacement, in
> part by conifers, is hardly to be regretted.[40]

38   Vera, *op cit*. 33, p 174.
39   Crawford, M and Marsh, D (1991), *The Driving Force*, Mandarin.
40   Rackham, Oliver, 1986, *The History of the Countryside*, Dent, p 147.

Epping Forest suffered a similar fate. The Act which in 1878 saved it from enclosure requires the Conservators of the Forest to:

> protect the timber and other trees, pollards, shrubs, underwood,
> heather, gorse, turf and herbage growing on the Forest ...
> Unfortunately the early Conservators pursued their duty of protecting
> timber trees with more enthusiasm than their duty of protecting the
> other, more historic and precarious features of the Forest. They took
> a dislike to pollards, thought to be the 'maimed relics of neglect', and
> promptly terminated the woodcutting rights. They disapproved of
> hornbeams and bogs, did nothing to prevent trees from overrunning
> the heather and gorse, and demolished the medieval New Lodge to
> save the cost of repair.[41]

However, the pressure for enclosure coming from the forestry profession was probably less pronounced in the UK than it was on the continent, partly because such a high proportion had already been enclosed by the agricultural sector, and partly because much of Britain's timber was coming from abroad. Whilst France and Germany were developing systems of scientific forestry on their home ground, British experts were off enclosing forests in Burma and press-ganging the local farmers into a forestry rotation compatible with slash and burn cultivation known as *taung ya*.[42] After the First World War, when the pressure for a strategic timber reserve in Britain resulted in the planting of millions of acres of conifers, it was mainly on land that had already previously been taken over for sheep.

Vera suggests that 19th century natural historians were contaminated by a bias against grazing animals that emanated from the forestry profession. Grazing animals were the enemy of trees, and vice versa: therefore there could be no prehistoric landscape where grazing animals and trees had coexisted in equilibrium. Closed forest was the natural state of wilderness, grazing animals were opportunists who found a temporary niche after fire or some other disturbance, and had subsequently been replaced by the great disturber, man. Nature, in their view, like forestry, aspired towards monoculture.

A compelling feature of Vera's theory is that his vision of our past natural history is inherently permacultural, if by that term we mean rich, anarchic and biodiverse. The view of European ecology as a progression tending towards a monocultural blanket of beech trees, disturbed only by fire or catastrophe, lends itself to a hierarchical interpretation, at odds with permaculture's perception of nature as a web of complex relationships. By placing grass on an equal footing with trees, and assigning herbivores as catalysts in a Manichaean struggle between the forces of light and dark, Vera allows us to perceive nature as a spectrum of different species, none of which can claim precedence in a linear progression, because all play their role in myriad interwoven cycles of emergence and decay. The shifting frontier between grass and woodland provides a maximum of 'edge' – a collective term for areas, like the sea shore, where rival ecosystems overlap and throw up high levels of biodiversity.

I'm not in a position to judge how accurate Vera's prehistorical analysis is, but his thesis bears the hallmark of a theory whose time has come. It is almost as though the conservation establishment has been hoping for something like this for years. Given that it overturns some entrenched orthodoxies, it has received recognition (though not universal acceptance) remarkably quickly, and

---

41   *Ibid.*
42   Bryant, Raymond (1994), 'The Rise and Fall of Taungya Forestry', *The Ecologist*, 24:1.

even those who share some of Rackham's and Peterken's misgivings recognize that Vera raises important questions about how temperate natural ecosystems work. His theory has generated a debate within UK nature conservation circles, aired in the pages of *Ecos* (the magazine of the British Association of Nature Conservationists) between advocates of the 'English approach' and the 'Dutch approach', here compared by James Fenton:

> As part of a recent conference at Lancaster University, we went on a field trip to the Pennines where staff of English Nature proudly showed us an experiment in the 'wilding' of the eastern flanks of Ingleborough. To them being 'wild' meant removing all grazing and planting some trees. Next day we were back at the University to hear inspirational thinking from Frans Vera about returning wild nature to Holland – at the Ostvarrdersplassen – and we heard even grander plans to create large-scale wildlife corridors from there to Germany and France.
>
> The essence of these Dutch schemes is the reintroduction of wild herbivores. Being wild in Holland does not mean excluding grazing, but the introduction of a range of large herbivores, in this case wild cattle, horses and red deer, and seeing what happens. These animals, of course, have a major impact on the vegetation pattern, the only constraint on their numbers being the amount of forage available in winter.[43]

The debate does not just revolve around the preference given to either grazing animals or trees; a further issue is the extent to which humans should interfere in areas devoted to wildness. At Ostvarrdersplassen the policy is 'seeing what happens': the population of grazing animals is not controlled, trees may be killed by bark-stripping and 'the vegetation and species mosaic that develops is completely unpredictable.'[44] The area is fenced off and the public excluded on the grounds that the animals could be dangerous.[45]

In situations where wild herbivores are allowed to multiply without human intervention, it could be that over time most trees will eventually be killed, in which case the resulting vegetation might not look that different from conventional sheep pasture. This has led some conservationists to suggest that the heavily grazed uplands of Britain may not be so far removed from their natural wilderness state as formerly thought:

> Maybe, in much of upland Scotland at least, we have had our Oostvarrdersplassens all along – large tracts of land with significant numbers of indigenous herbivores, resulting in a relatively natural vegetation pattern. Maybe we have them throughout upland Britain, the only difference being that sheep have replaced the red deer. It is a common observation in Scotland that if sheep are taken off a hill, red deer come in.[46]

It is ironical that the government should finally bow to pressure to reduce subsidies for grazing animals on hill farms, just at the point when the conservationists' pendulum may be starting to swing in the opposite direction. But by no means everybody welcomes the changes brought about by Vera's

---

43   Fenton, James (2004), 'Wild Thoughts: A New Paradigm for the Uplands', *Ecos*, Vol 25:1.

44   Kirby, K *et al* (2004), 'Fresh Woods and Pastures New', *Ecos*, Vol 25:1, photo caption.

45   Harris, Neil (2006), 'Ecosystem Effects of Wild Herbivores: Lessons from Holland', *Ecos* 27 :3/4,.

46   Fenton, J (2004), 'Wild Thoughts Followed Up', *Ecos*, Vol 25:1.

hypothesis. Mark Fisher, a UK permaculturist and tree enthusiast, observes:

> It was only a matter of time before the theories of Frans Vera would
> end up in a justification of agriculture in nature conservation. I
> have no problem with his challenge to a primeval blanket of climax
> woodland across Europe because instinct suggests that it is too simple
> an explanation, and it belies evidence from other countries showing a
> variety of grassland, scrub and open and closed woodland. But if we
> go with Vera's theory we have to examine ... the stage at which wild
> herbivores exerted their influence, and their population size. Would it
> have been equivalent to the 10 million cattle, 30 to 40 million sheep and
> untold rabbits and horses that we have now?[47]

Almost certainly not. Reintroduced herbivores are rarely subject to the pressures
that they would have to face in a truly wild situation. Even when wilderness
is the objective, animal welfare legislation intervenes, as Peter Taylor, another
contributor to *Ecos*, notes : 'They have a right to slow death by natural diseases,
death by combat or predation by big cats. Even the laudable effort of the Dutch at
Ostvarrdersplassen failed to get the first of these, despite getting the second, and
is a far cry from bringing back the northern European lion.'[48] The manufacture of
wilderness, its seems, is not for the squeamish.

And Fisher takes up the point about the missing carnivores: 'In a natural
system that has plants, herbivores and carnivores, [there] would be more
woodland of various types compared to the area of open grassed spaces, even
in most of our upland areas.'[49] His recipe for what he calls 'self-willed land'
(meaning that it does its own thing) involves removing livestock to improve
upon the 'diminutive woodland coverage of today ... along with a landscape
wholly fashioned by people. I think wild nature can do better and I am willing
to give it a chance.'

The values expressed in this last sentence imply that it may not be wildness
for its own sake that Mark Fisher seeks but the trees that he hopes it will bring
about. But fencing out herbivores in order to compensate for an absence of
carnivores sounds more like estate management than wilderness; and so too
does the establishment of predator-free deer and cattle parks, linked by wildlife
corridors. For all the talk of 'seeing what happens' and 'self-willed land', the
option of doing nothing doesn't appear to exist – short of rebuilding Hadrian's
Wall somewhere to the north of Stirling and evacuating the Highlands and
Islands of the remainder of their inhabitants. Human beings, much as they might
like to, cannot abdicate from decisions about how land should be managed;
and the crucial decision, the fulcrum upon which every other scale within
the ecosystem pivots, is the level of livestock grazing – whether it should be
unlimited, controlled by predators, culled, husbanded or forbidden.

At one end of this scale we will get the heavily grazed landscape that is
obtained by running a national flock of 45 million sheep, or the equivalent
amount of wild grazing animals. At the other extreme we have wildwood. In
between lies the Vera-esque blend, something close to the landscape in the New
Forest. The vast majority of the public have never heard of Frans Vera, and
are not preoccupied with what constitutes genuine wilderness or indigenous

47   Fisher, Mark (2004), 'Self Willed Land: Can Nature Ever Be Free?' and 'Wild Follow Up' in
Ecos 25:1.
48   Taylor, P (2004), 'To Wild or Not to Wild: The Perils of "Either-Or"', *Ecos*, Vol 25:1.
49   Fisher (2004), *op cit.* 47 (quote slightly cut).

habitat, but they do like to experience a landscape that feels wild and is varied. People are fed up with seeing their natural world divided into a chessboard of monocultures, which in the UK's less populated areas often consists of heavily grazed sheep pasture and dense Sitka spruce. That is why the subsidies available for both these activities have been lowered in recent years; why the New Forest, the UK's largest silvo-pastoral commons, attracts nine million visitors a year; why treeplanting policy has done a U-turn and focussed on native broadleaves; and why there is a revival of interest in coppicing and woodland crafts, and in the maintenance of heathland and wildflower meadows.

What is needed is a philosophy of land use which pulls all these activities and aspirations together and shows how they can be made productive and viable in the event that we find ourselves having to feed and support 60 million people on quite a small island. Permaculture is an obvious candidate for this role, but there may be a need to reassess the view that 'wild woodland is the natural vegetation of this country' – not because we can be certain that it wasn't, but because 'it is too simple an explanation'. A permacultural approach, post Vera, will not be one that favours trees on the grounds that they have a superior indigenous pedigree; it will be one that juggles with the dynamic between light and shade to produce landscapes that are rich, biodiverse and convivial for humans.

Towards a Permaculture Livestock Economy

Political economy has hitherto insisted chiefly upon division.
We proclaim integration.

Peter Kropotkin, *Fields, Factories and Workshops*

In this last chapter I want to map out some of the features of a post-industrial rural economy which is the diametrical opposite of the urban dystopias pictured *in extremis* in Chapter 15. Its defining characteristic is that it is in a state of 'energy descent': most of its energy is derived from biomass, muscle power and other renewable sources such as wind, water and solar, and there is not very much of it, though enough, let us say, to manufacture, maintain and operate serviceable computers and a reduced amount of machinery – roughly the levels available during Cuba's 'special period'.

To a greater or lesser extent most of its other features are a natural consequence of this one condition. This is a society which can only afford default levels of livestock, which values human labour and animal power highly because energy is expensive, which is broadly self sufficient in its food and fibre without being xenophobic about imports, which has more people living and working in the countryside, and where a larger area is covered by trees.

Mostly I focus on the rural economy and rural land use, because this is a book about livestock, but that does not mean there are no towns or cities. How large they are will depend upon two things: how much energy there is available to support the demands of industry and an urban lifestyle; and how many people wish to live there. The rural landscape and way of life described here will appeal to some, but not to others.

Nor should it be assumed that this is an entirely carnivorous society. There will be vegan and vegetarian households, farms and communities – there might be a vegan county just as there are Amish counties in the USA. Conceivably the entire country could be split between corn and horn, rather as it was in mediaeval times between the central 'champion' country of open fields, where livestock became progressively harder to maintain as population grew, and the less fertile 'woodland' counties on the fringes where there was a mosaic of trees and grass, and people kept animals and practised convertible husbandry.

Doubling Tree Cover

In Chapter 9 I outlined how a Permaculture Livestock rural economy could function without excessive dependence on the lands and resources of other countries. In that scenario the 7.9 million hectares required for staple food production through arable cultivation and leys are non-negotiable – we cannot do without them. The 3.5 million hectares of permanent pasture can only be allocated to another use if we relinquish some of the dairy and beef which provides a significant amount of the high quality protein available in the diet, thereby propelling us in the direction of the Vegan Permaculture scenario. However, the 11 million hectares of less productive land, about half the entire country, is negotiable. The food it produces – a bit of lamb and game – constitutes little more than one per cent of the national food budget, so we could if we wanted, dispense with that meat altogether, and allocate the land entirely for uses other than the production of food.

The range of choices we can make about how this land is used correspond to two spectrums of land management raised by Patrick Whitefield in the first paragraph of his *Earthcare Manual* – how woody do we want it, and how wild? The matter of woodiness boils down to a single question: do we allow herbivores in or fence them out? The matter of wildness is dictated by the intensity of human management. Figure 4 shows how different land uses are positioned across these two axes. In an abbreviated form, 'wild and woody' means natural woodland; 'domestic and woody' means short rotation coppice; 'wild and grassy' means wild animals; and 'domestic and grassy' means cows and sheep. However, there is an area in the middle occupied by a range of silvo-pastoral habitats exhibiting the kind of mosaic landscape favoured by Vera, and found typically in remaining scraps of commons and a few big ones such as the New Forest.

Figure 4. Wildness and Woodiness in the UK Landscape

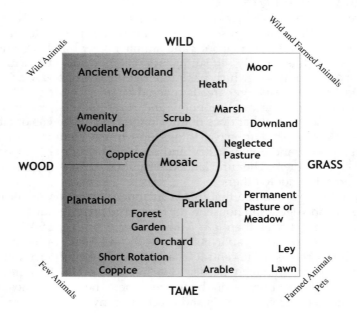

The Livestock Permaculture model allocates six million hectares, nearly a quarter of the UK's countryside, to woodland. The main reasons for this decision are that this hinterland does not currently produce much food, nor does it need to, while trees supply fibre and energy, the second of which will be in short supply. The UK, with just 11 per cent of its land under trees, has one of the lowest levels of tree cover in Europe.

It used to be even lower: at the end of the 19th century trees covered barely five per cent of the country. In Scotland this may have been a result of clearances, while in populated areas it was a result of the availability of coal. Landowners and farmers preferred to have coal trucked in for their labourers than to give up land for firewood.[1] A combination of coal, sheep and high concentration of landownership is a disaster for tree cover, and some areas such as Southern Lancashire were almost completely denuded.

When we hear the government talking about increasing the area of woodland, warning bells ought to ring, since it would not be the first time they've tried it. In the 20th century the area of woodland doubled, not to provide fuel, but as a strategic reserve of timber, pulpwood, and pit props. There was some logic in this programme. High quality softwood timber is hard to grow in Britain (because it grows too fast) and it seemed to make economic sense to grow low value pulpwood trees which are relatively expensive to transport, and import high quality timber, whose value better justifies the expense of shipping. But the result was many hundred thousand hectares of dense conifer monoculture, disliked by nature conservationists and the public alike. On top of that it did little to improve the rural economy of the uplands – few things provide less in the form of human livelihoods than pulpwood plantations. Timber is trucked from the hills of Scotland and Northern England down to pulp mills in the Midlands, close to the market, depriving thousands of upland inhabitants of a way of making a living from their land. Scotland has 19 per cent more woodland than England and nearly three times as many conifers – yet its forestry industry employs only three quarters as many people as England's.[2]

If we are going to plant more trees, we therefore need to establish a number of matters. What are these trees for? Do we want 'wild' semi-natural woodland or more intensively managed plantations or short rotation coppice? How much woodland do we want? And where are these trees going to go? I'll try to answer these questions in some sort of order.

Aside from the amenity and wildlife value, there are two main practical uses for trees: the supply of timber and the supply of firewood. Let's deal with timber first. The UK currently uses about 46 million m³ of timber, including panel board and paper and pulp (PPP) but less than 10 million tonnes of it is grown here.[3] Because of its climate, the UK cannot produce some of the high quality timber it imports, but it could in theory produce most of the PPP, and say half of the timber currently imported, which (net of exports) comes to about 40 million m³.

However, about 35 million m³ of this would be PPP, most of it from softwood trees, typically fast-growing Sitka Spruce. It is unlikely that we would want to replant the vast acreages of Sitka Spruce which we are currently keen to pull up – most people prefer sheep. And a sustainable society, forced to steward scarce resources, could quite easily increase recycling rates and reduce the amount of PPP currently squandered on junk mail, newspapers full of supplements which people never read and shelving units which are so tacky they get skipped after a few years' use. If we were intent upon making the country self-sufficient in timber, 23 million m³ (half the amount we currently consume) might be a sensible level of home grown timber production to aim for.

1   See for example Jeffries, Richard (1898), 'Wiltshire Labourers' in *The Toilers of the Field*, Longmans Green and Co, 2nd edition, 1898.

2   Forestry Commission (2009a), *Woodland Area, Planting and Restocking*, 11 June 2009, http://www.forestry.gov.uk/pdf/area09.pdf/$file/area09.pdf; Forestry Commission (1999); Forestry Commission (2009b), Forestry Statistics : Employment and Business, http://www.forestry.gov.uk/website/forstats2009.nsf/0/55E4B468BA06C76D8025734E00376F34

3   Or about 56 million if you include the pulp that goes for recycling. Forestry Commission (2008), *Forestry Statistics 2008*, 'Trade'; and *Forestry Facts and Figures 2008*.

This reduced amount is less than we currently grow. The annual increment in UK forests in 1999 from timber quality woodland alone was about 21 million cubic metres, which means that every hectare grew on average about 10 m^3 of timber per year. This, as the Forestry Commission delicately acknowledges, is 'believed to be significantly underexploited'.[4] We currently harvest only 48 per cent of this annual production, which suggests that many of the trees we planted in the 1960s are getting bigger and bigger and if nothing alters will continue to do so, until they start falling over and dying.[5]

The quality timber production only represents 2.2 million hectares, so there is another 800,000 hectares or so of amenity woodland, producing timber that is too wiggly or otherwise unsuitable. As we noted above, pre-industrial construction methods can cope with timber of all sizes and shapes, whereas sawn timber requires straight logs from trees that grow in dense woodland or plantations. However, pulpwood and some other more modern technologies can make do with any shapes and sizes, though they require certain economies of scale. If, as Prince Charles once claimed, conifer plantations are 'nothing more than industrial cellulose factories', then there is no intrinsic need for straight plantation grown trees – though it no doubt remains more efficient to grow and harvest them as such.[6] While there will always be a demand for sawlogs, there is now much more scope than there was 50 years ago for using timber from biodiverse semi-natural woodlands, at both ends of the technological spectrum.

The other main product is forestry biomass, otherwise known as firewood, which comes in a variety of shapes and forms that can be seen ranged down the dark side of <u>Fig 4.</u> Short rotation coppice (SRC), at the tame or intensive end of the spectrum, is most frequently of willow, and under good conditions can attain yields of ten tonnes, or even 15, of oven dry wood per hectare per year on suitable land.[7] Currently, in the UK it is mainly being tested on an industrial basis, designed to benefit from economies of scale (though there is a tendency for small scale permaculture settlers to plant a patch on their plot). Planting has so far taken place mainly on flat, relatively high quality farmland, congenial to machinery and close to a processing plant (although these have sometimes had problems securing planning permission). The corporate bias of research and development in the UK tends to steer bio-energy production towards large-scale centralized schemes based mainly on biodiesel, bioethanol and willow coppice, even though on-farm processing of biomass for local consumption ought to be more efficient, because of the reduced distribution costs.[8]

4   Forestry Commission (2004), *State of England's Woodlands, 2004*, http://statistics.defra.gov.uk/esg/evaluation/forestry/4.pdf
5   The Forestry Commission does not have any figures for annual increment after 1999, but timber production has declined slightly since 1999 (personal communication from Sheila Ward of the Forestry Commission, 20 April 2009). Figure from *UK Indicators of Sustainable Forestry,* 'D Timber and Other Products', figures for 1995-2000, Table D1, p 64. Formerly online but no longer available.
6   Prince Charles cited in Tompkins, S (1989), *Forestry in Crisis*, Christopher Helms.
7   RCEP (2004), *Biomass as a Renewable Energy Resource*, Royal Commission on Environmental Pollution, 2.14; Tubby, I and Armstrong, A (2002), *Establishment and Management of Short Rotation Coppice: Practice Note*, www.forestry.gov.uk/pdf/fcpn7.pdf/$FILE/fcpn7.pdf
8   I have been unable to locate any study comparing the efficiency of on-farm biomass energy for local consumption with centralized biomass energy production. There are dozens of papers highlighting the costs of transporting bulky biomass over long distances. E.g. 'Some practical problems are associated with the use of biomass material (sawdust, wood chips or agricultural residues) as fuel. Those problems are mainly related to the high bulk volume, which results in high transportation costs and requires large storage capacities', European Biomass Industry Association (n.d.), *Densification,* http://www.eubia.org/111.0.html
    However, all the studies I have seen look for solutions through logistics, processing the feedstock, or a similar centralized approach, rather than comparing the performance of local production and consumption. This one from Tennessee is typical:

Traditional coppice and mature woodland are lower yielding, but they have other advantages. They can be planted and managed on poor land deficient in nutrients, on steep slopes, and in areas too small for the machinery associated with industrial SRC to be used efficiently. A proportion of the energy of a fully grown tree is invested in its permanent infrastructure – its trunk, roots and branches – which provides both timber and a carbon sink. Full grown trees may not produce so much renewable energy as SRC, but they store more carbon, allowing us to emit a bit of greenhouse gas with impunity.

Mature woodland also ranges in wildness between monocrop plantations and mixed species woodland. Besides firewood, a well managed mixed woodland supplies timber, a wide range of coppice wood, game, mushrooms, pig grazing, foliage etc – all of which will be hard to find in an SRC willow plantation. And finally, in terms of the 'amenity' provided, biodiverse woodland wins hands down over SRC. According to DEFRA more than 300 million day visits to woodland suggest that 'woodlands are the most popular land-based destination for day visits'.[9] Few people would choose to take a walk through a field of fully grown short-term rotation willow coppice; even a wheatfield, which you can at least look over, and which changes in hue and texture over the seasons, is preferable to an impenetrable eight foot tall curtain of monocultural biomass.

There is a place for SRC, though overemphasis could easily turn swathes of the countryside into a biomass factory. One likely place is on the 876 hectares allocated for bioenergy in the Livestock Permaculture scenario in Chapter 9, where either SRC or Miscanthus is likely to perform better than bio-ethanol crops or rape for biodiesel. More remote pasture will not get such high returns from industrial scale SRC because the land is less fertile, more exposed and precipitous, and because inputs and outputs have to be transported further. The best option for biomass on the rougher kind of pasture land, which is the land we can most easily afford, is likely to be conventional semi-natural or plantation forestry.

However, in their enthusiasm for SRC, scientists and environmentalists drawing up carbon budgets tend to be snobbishly dismissive of the value of conventional woodland. It may be a viable option for Scandinavia, they say, but in the UK there are too many people and not enough trees.[10] The Royal Commission for Environmental Pollution (RCEP) states that 'at best 3.1 million oven-dried tonnes of wood-derived fuel could be made available' from Britain's 2.74 million hectares of woodland.[11] This equals about 3.75 million tonnes of air-dried firewood – about 1.4 tonnes per hectare or from two to three cubic metres

'Biomass facilities will always be faced with the dilemma that their feedstock cost will be highly location specific. Using supply curves that are based on aggregate farmgate price information and uniform transportation costs can be misleading. Such curves may obscure opportune locations where biomass supplies can be had inexpensively while at the same time overestimating the total number of facilities that can be supported by feedstock under a specific price. Capturing the geographic complexity of potential biomass supplies is a necessity and one for which GIS is well-suited.' Graham, R L et al, 'The Effect of Location and Facility Demand on the Marginal Cost of Delivered Wood Chips from Energy Crops: A Case Study of the State of Tennessee', from the Proceedings, Second Biomass Conference of the Americas: Energy, Environment, Agriculture, and Industry, pp 1324-3. http://bioenergy.ornl.gov/papers/bioam95/graham2.html

However for a paper concluding that the 'Development of economic farm-scale technologies' for deriving energy from straw 'appears to be feasible and necessary to reduce the costs of straw collection and transportation', see: Banowetz, G M et al (2008), 'Assessment of Straw Biomass Feedstock Resources in the Pacific Northwest', Biomass and Bioenergy, 32, 629-34.

9   DEFRA, The State of England's Woodlands, https://statistics.defra.gov.uk/esg/evaluation/forestry/4.pdf

10   Monbiot, George (2006), Heat: How to Stop the Planet Burning, Allen Lane, p 118.

11   RCEP, op cit.7, 2.76.

per hectare, depending whether it is soft wood or hardwood.[12] This seems rather pessimistic, even if we allow that some of the woodland is currently poorly managed, since typical yields range from four to six cubic metres per hectare for oak to over 20 cubic metres per hectare for some softwoods.[13] Estimates in other environmental reports vary considerably with the highest being 13 million tonnes of oven dried wood available throughout the country – more than four times as much as the RCEP estimate.[14] The RCEP is perhaps discounting the wood used for the timber and pulp industries, but there is no reason not to take this into account because it can nearly all be burnt when it comes to the end of its life and this will become progressively easier to implement as restrictions on dumping timber products in landfill come into effect.

The British Trust for Conservation Volunteers (BTCV) probably have more experience of harvesting firewood than the scientists who draw up these documents. Their woodland handbook tells us 'there is an old rule of thumb that an acre of woodland can produce a cord of wood per year indefinitely'.[15] Cords are piles of firewood, which are a convenient dimension for forestry workers (four foot, by four foot, by eight foot), but awkward for scientists because they are in imperial measurements and full of airgaps. On top of that green wood weighs more than air-dried wood, while hardwoods such as oak and ash weigh more than conifers or fast growing broadleaves such as willow. Another complication is that firewood figures tend to be given in tonnes, whereas timber production is usually measured by volume.

A cord weighs from 1.3 to 2.5 tonnes if it is green, and perhaps one tonne air dried if it is spruce, or 1.8 tonnes if it is oak. A cord per acre of mixed timber might therefore contain in the region of 1.4 tonnes of air dried firewood, the equivalent of 3.5 tonnes per hectare.[16] To allow for inefficiencies, I'm inclined to lower the yield to three tonnes per hectare. BTCV also give figures for air dried firewood from long rotation coppice, ranging from 2.5 tonnes per hectare for hardwood, to six tonnes per hectare for willow and poplar.[17] Again three tonnes per hectare seems a conservative compromise, yet this figure is over twice as high as RCEP's estimate of 1.4 tonnes per hectare.

A cord is about 2.4 cubic metres of solid timber, so BTCV's cord per acre is the equivalent of six cubic metres per hectare of mixed wood.[18] This is about a

12   I assume air dried wood to contain 20 per cent moisture. A cubic metre of softwood contains 400 kg of drymatter or 500 kg of air dried timber; hardwood about 700 kg of air-dried timber. Pieter Kofman, *Firewood*, Coford Connects, http://www.coford.ie/iopen24/firewood-p-966590.html

13   See, for example, tables at Agroforestry Forum, The Macaulay Land Use Research Institute: http://www.macaulay.ac.uk/agfor_toolbox/trees.html#swelow

14   Read, D J *et al*,2009, *Combating Climate Change: A Role for UK Forests, The Stationery Office*. Their figure for 13.1 tonnes is derived from Whittaker, C and Murphy, R *Assessment of UK Biomass Resources*, AtlannTic Alliance (*sic*) 2009.

15   BTCV (2002), *Woodlands* BTCV Handbook http://handbooks.btcv.org.uk/handbooks/content/section/3767.

16   BCTV *ibid*; Kuhns, M and Schmidt, T, *Heating with Wood: Species Characteristics and Volumes* Utah State University Forestry Extension; http://extension.usu.edu/forestry/HomeTown/General_HeatingWithWood.htm

17   BTCV, *ibid*. http://handbooks.btcv.org.uk/handbooks/content/section/3754; Barnsley Council report that the 12,000 hectares of woodland it has available produce 5 tonnes per hectare of new growth per year from which it aims to get 3.75 tonnes of firewood per hectare. New energy Focus (2008), *Red Tape Hinders Barnsley's Biomass Fuel Conversion Efforts*, New Energy Focus: Bioenergy and Waste News, 28 July 2008, http://www.newenergyfocus.com/do/ecco.py/view_item?listid=1&listcatid=87&listitemid=415

18   The solid timber in a cord is about 85 cu feet or about 2.4 cubic metres: Iowa Department of Natural Resources Forestry, *Forestry Definitions*, http://www.iowadnr.gov/forestry/definitions.html#cord. For measurements in respect of Sitka spruce in Scotland, see Scroome, S and Taylor, H, Power Point Presentation http://www.forestry.gov.uk/pdf/SamCroomeFEITraining.pdf/$FILE/SamCroomeFEITraining.pdf

third as much timber as can be produced in high-yielding conifer plantations, and is 60 per cent of the average annual increment on timber quality woodland. One would expect more than a cord of firewood per year off an acre of Sitka Spruce plantation. However a cord of seasoned spruce only weighs about a tonne, so I still view it safer to stick to the figure of three tonnes per hectare, which is the equivalent of about 6.6 cubic metres of spruce or about 3.8 cubic metres of oak.[19] Three tonnes of mixed timber might therefore be in the region of five cubic metres.

On top of this somewhat theoretical figure, there are all the scraps of firewood which do not make it onto the account book, but which those of us who go wooding are familiar with. There are the heaps of sneddings and brash which from a commercial forester's point of view do not class as timber. There are the loppings of urban trees and bushes. There is the very considerable amount of small gauge firewood that grows in Britain's hedgerows of which there are estimated by various sources to be anything between 200,000 and 500,000 miles long, occupying upwards of 67,000 hectares.[20] Much of this ends up flailed to shreds and left to lie in the banks and ditches; the 'long acre' which, since it is no longer grazed, is now the most fertile land in Britain. However, in the 1980s, hedgerow trees were the source of one-fifth of all home-grown hardwood marketed in Britain.[21]

Another neglected source of firewood is the scrub which grows on commons and poor pastures. The advantage of this kind of fuel, particularly gorse and blackthorn, is that it burns at a high temperature, providing the kind of flash fire ideal for bread ovens and kilns (apple prunings or vine prunings when well dried will serve the same purpose). An acre of gorse in Bedfordshire in the 18th century was worth about five weeks' wages.[22] Nowadays gorse presents a management problem, described here by the Royal Society for the Protection of Birds: 'Old and degenerate gorse is relatively poor for wildlife. Meanwhile, the accumulation of plant debris increases soil fertility, aiding colonisation by, for example, bracken. The accumulated dead material also presents an increased fire risk.' This is not a big risk in Britain, although burning overstood heather sets fire to peaty ground; but in Mediterranean areas, the fact that people no longer collect faggots of scrubwood for their bread ovens, along with a decline in grazing, is viewed as a main cause of forest fires.[23] RSPB advise land managers to 'Cut the old gorse and burn the arisings and litter in a series of fires across the restoration to encourage seed germination.'[24] If energy becomes scarce, they will be advising their employees to 'cut the old gorse and deliver to the village bakery' – that is if the bakery workers haven't already harvested it.

We are looking at a substantial supply of biomass fuel, even if we need to let a proportion of it remain on the ground to assist biodiversity. The next matter to consider is: how much of this woodland do we want? The UK officially has one of the lowest levels of tree cover in Europe at around 11 per cent – about 2.8 million hectares out of a total of 24.5 million hectares. On top of this there are

---

19 A cubic metre of seasoned spruce weighs about 450 to 470 kilos. A cubic metre of oak weighs about 780 kg, see Kuhns and Schmidt, *op cit. 16*, and Scroome, *ibid*.

20 BTCV state that hedges in Great Britain total about 251,000 miles (404,000 km) occupying about 167,000 acres (67,000 hectares), 1998 estimates. BTCV (2003), *Hedging*, BTCV Handbook, Chapter 2, http://handbooks.btcv.org.uk/handbooks/content/chapter/66

21 *Ibid*.

22 Neeson, J M (1993), *Commoners: Common Right, Enclosure and Social Change in England, 1700-1820*, Cambridge University Press, p 176.

23 Atherden, Margaret, *Upland Britain, a Natural History*, Manchester University Press, 1992, p 99.

24 RSPB, *Gorse Management Techniques*, http://www.rspb.org.uk/ourwork/conservation/advice/gorse/techniques.asp

the 872,000 hectares classified as 'other agricultural land including wood land', and all the trees in gardens, streets, parks, and hedgerows. It seems reasonable to assume that total tree cover might be nearer 3.5 million hectares, or about 14 per cent of the entire land area. When the Tory government advocated a doubling of the amount of woodland, both for the environmental benefits and as a carbon sink, few people appeared to disagree with that aim.[25] Does this mean a doubling of the 11.6 per cent forestry coverage or the 14 per cent tree cover? Let us go somewhere in between and propose an increase in tree cover to a quarter of the country. This would not be particularly drastic: we would still have a lower proportion of our land under trees than France (27.9 per cent), the EU (40 per cent) or the entire world (29 per cent).[26]

Twenty-five per cent of the UK's land area is a little over six million hectares. Allowing for a firewood harvest of three tonnes or five cubic metres per hectare per year this would produce 30 million cubic metres. In theory (since there is an increment of ten cubic metres per year currently achieved on 2.1 million hectares of timber quality land), this would still leave in the region of at least 10 million cubic metres for timber (which is what we harvest at the moment), or up to 20 million cubic metres if we opted for more fast growing plantations – most of which would also become available as fuelwood when it reached the end of its economic life.

The next question is: how much fuel do we need? The average three bedroomed home, occupied, typically, by a family with two children, requires about eight tonnes (roughly 13 cubic metres) of air dried wood per year to be self-sufficient in heat to the currently acceptable standard.[27] At this rate with 25 per cent of the country wooded and producing 18 million cubic metres, we could currently heat up to 2.25 million homes, housing nine million people.

However, eco-builders are now constructing houses which are so well insulated and passively solar heated that they can maintain a constant temperature of 20 degrees without any extra heat source at all. This is highly commendable, but a home without a hearth or a living flame is a sorry affair (and possibly dependent upon some other source of energy, like flying off to the Gambia in December, for spiritual nourishment). It seems reasonable to assume that a family of four, in a well insulated home, could heat themselves and maintain their hearth happily on the three tonnes of firewood per year which traditionally could be obtained off a hectare of well managed wood.[28]

At this rate our six million hectares of woodland would heat six million family sized homes, or 24 million people – or 40 per cent of the population – provided they were living in reasonably well insulated homes (or alternatively not very demanding about the degree of heat they required); and providing they lived fairly close to the source of the wood. If a vegan society decided that it wanted to convert as much as possible of our grazing land to natural woodland, it could probably heat most of the population on this basis.

Since residential heating currently accounts for 17 per cent of the UK's total greenhouse gas emissions, this would represent a reduction of 6.8 per cent in the UK's emissions, of which about 2.5 per cent would be attributable to fuel

---

25   MAFF and DETR (now DEFRA) (2000), *Our Countryside: The Future*, Rural White Paper, HMSO.

26   DEFRA, *op cit.*, 9.

27   Or 17 tonnes of spruce. BTCV, *op cit.15*; Scroome, *op cit 18*.

28   After I wrote this, I was commissioned to examine a number of proposals for eco-homes at the prospective Lammas project in Pembrokeshire, and noted that nearly all of these families anticipated an annual wood usage of 3 tonnes – though some required an extra tonne for cooking.

wood and the rest to insulation.[29] In addition, the creation of about three million hectares of new forest would sequester a considerable volume of CO_2 per year, though the amount would depend on how much was taken for firewood and how much harvested for long lasting timber products. The Read Report gives estimates of the CO_2 savings achieved by different kinds of woodland, through a combination of sequestration and fuel use. The average for productive and multipurpose woodland is around 12.5 tonnes of CO_2 per hectare per year, which suggests that three million extra hectares of woodland would represent a reduction of around 6 per cent of our current GHG emissions.[30]

It has taken a lot of figures to arrive at these calculations, but there is more to life than calories and carbon budgets. Firewood, the old adage tells us, is the fuel which heats you twice – which is another way of saying that people who do physical work get less cold than people who sit at a computer and turn on the heat with a flick of the switch. Scientists who leap to explain how most of the heat from an open fire goes up the chimney, rarely account for the fact that a fire with a living flame offers a spiritual warmth that cannot be obtained from central heating. This is evident from the fact that people in homes and pubs frequently light open fires even when they have central heating, that people who have no fireplace often buy a fire with an imitation living flame, and that people at parties and barbecues do not congregate outside warming their hands around a bank of electric storage heaters. There is a sizeable minority of people who find the fug of central heating debilitating and prefer the thermodynamics of a cool house where, when human energy runs low, the hearth springs into life, quickening the flagging brain with its flames and warming the blood with its coals. SRC on the other hand is likely to be delivered to the consumer, either via an electricity generating station, or else minced up and moulded into pellets. The difference between gas or wood-pellet fuelled central heating and an open fire or wood burner is like the difference between a spoonful of sugar and a shot of malt whiskey – same calories, different effect.

Restoring the Mosaic

The advantages of doubling the size of our woodland are enormous, so much so that some might be tempted to enlarge its area still further by eliminating sheep entirely, and no doubt many vegans would be keen to do so. However, before committing ourselves to any further woodland expansion, it would be well to assess how biomass forestry compares to animal production – which at present, for the sort of rough and permanent pasture we are talking about, mostly means sheep. A hectare of average land in these categories produces about three tonnes of dried firewood, or alternatively perhaps 100 kilos of meat. The calorific value of both meat and dried wood is roughly 3,000 kcal per kilo, so 28 times as much energy is produced by the woodland as by the sheep. However, sheep calories are eaten, while wood calories are burnt: a kilo of wood will keep you warm for an hour or two, while a kilo of mutton will keep your inner fire burning for a day or more.

The other convenient way to compare their value is by price. At the time of writing, upland sheep might produce approximately £350 worth of meat per

29   BERR (2007), *Estimates of Heat Use in the United Kingdom*, http://www.berr.gov.uk/files/file43843.pdf; DECC (n.d.) *UK Emissions of Carbon Dioxide, Methane and Nitrous Oxide by National Communication Source Category*, http://www.decc.gov.uk/en/content/cms/statistics/climate_change/co2_meth_n20/co2_meth_n20.aspx

30   Read, D J *et al, op cit.14.*

hectare wholesale.[31] A hectare of woodland might produce about £350 worth of fuel delivered to the door.[32] The lamb is worth a lot more retail – but so is the firewood if you stick it in fertilizer bags and sell it at the local garage. The woodland frequently has no export value (sawlogs are often worth no more than firewood). It has some amenity value, whereas the more intensive sheep pasture has less – but if there were more woodland and less sheep pasture, the values would be reversed. The woodland has more value as a carbon sink. Grazed land is necessary if you want to erect wind generators. The sheep also produce around five kilos of wool per year, which may have little value at the moment, but would have if the price of oil rose significantly. The woodland may provide some secondary products, such as mushrooms, game, pig grazing etc.

There seems to be little difference in value between pastureland and woodland used for fuel. But there is one other crucial difference. £350 worth of wholesale sheep weighs about 100 kilos, and can easily be carried to Manchester, London, or even the South of France where it can be sold for a profit. £350 worth of firewood weighs about three tonnes, and has to be consumed locally. Since the areas we are talking about tend to be depopulated (a fact which is largely due to enclosure for sheep) there is no market for all this potential firewood (which explains why the preferred land use activity is still sheep). The clearances created a vicious circle which it is hard to get out of.

The other economic problem for fuelwood is that it can't compete with the cheapness of fossil fuels. This partly comes back to the transport costs due to its weight, but also because it is awkward. Firewood from semi-natural landscapes varies from the huge and immovable to the small and fiddly and is not amenable to mechanization; a machine can cope so much more easily with a SRC monoculture on flat land. These problems may help to explain why the Royal Commission on Environmental Pollution gives such a low estimate of the amount of available firewood. They write:

> Forestry products are not suitable for all modes of biomass conversion. The dispersed nature of the supplies makes it unlikely that they will be used for large scale energy production. The fuel is often insufficiently homogenous for small-scale plants without considerable processing to increase the density and uniformity and reduce moisture content.[33]

and again:

> If forests are located in remote areas, there may not be access for harvesting machinery or transportation and it may be uneconomic or unattractive to invest in building roads … The long lead time, uncertainty of supply and a lack of expertise in harvesting methods all detract from the value of forest materials as a long term source of fuel compared to energy crops, for example, that are more controllable.

Corporate scientists and economists tend to think in terms of trucking resources to centralized processing depots, and 'homogenizing' them in a way that makes them easy to pack neatly on a lorry and pass through the machines. That is why, when they think of biomass, they think first of all of SRC or some such monoculture, frequently planted on flat arable land.

---

31   The gross returns vary between about £200 for hill flocks and organic upland flocks, to £500 for non-organic upland flocks. Nix, John (2007), *Farm Management Handbook*, Imperial College, London and Lampkin *et al* (2007), *Farm Management Pocketbook*, University of Wales, Aberystwyth.

32   £70 per load of about half a cord.

33   RCEP, *op cit. 7*, p 23.

In Port Talbot, South Wales, a company called Prenergy Power is planning to construct a wood-fuelled power station generating 2.8 billion kilowatt hours of electricity from three million tonnes of wood chips – about a kilowatt hour per kilo of wood. The wood chips to power it are to be shipped in from a 'variety of overseas sources' (the Ukraine, Canada and the USA have all been cited as possible sources). The environmental statement for this project, written by the Scottish firm Sinclair Knight Merz (SKM) (who usually provide technical support for new airports, roads and other nonrenewable energy projects), witters on about the absence of bats, great crested newts and badgers on the 21 hectare site.[34] Their absence is no great surprise since the Port Talbot steelworks, as viewed at night from the M4, are the closest thing in the UK to Dante's Inferno. It tells us nothing about what will happen on the million or so hectares – about half the area of Wales – that will provide the fuel, because it doesn't even know where they will be.[35] There is no attempt to examine what energy costs might be involved in planting, growing, felling, extracting and processing the timber, turning it into wood chips, transporting them to the coast, shipping them to Britain in 45,000 tonne Panamax boats, re-equipping the docks to unload them, constructing the 24 new buildings and silos at the £400 million plant (life expectancy 25 years), and constructing a new 250,000 volt connection to the national grid – not to mention all the ancillary stuff such as the journeys to work of the lumberjacks, dockers, power station operatives and electric meter measurers, or the computers used to provide environmental statements and calculate electricity bills. All this expense is required in order to turn three square metres of land in one country into an hour of a one bar electric fire in another, or alternatively a five mile journey into work in an electric car.

DEFRA, however, has done some work on these issues. A report entitled *Biomass: Carbon Sink or Carbon Sinner* concludes that when SRC chips are used to generate electricity they give off between 15 per cent and 65 per cent of the GHG emissions of a gas power station per unit of energy, and the emissions from clean waste wood chips are broadly similar. The report states that 'transporting fuels over long distances … can reduce the emissions savings made by the same fuel by between 15 and 50 per cent compared to best practice'.[36]

The permaculturist, on the other hand, thinks first of all of adapting to the resources nature provides, even if they are a bit awkward or fiddly. Sixty years ago the laurel planted for pheasant cover at the aforementioned Happy Valley was regularly cut down by local landless people who bundled it into faggots and took it home to burn. I do not approve of the economic system which gave the landlord ownership of all the fine timber, and left only the sneddings for the poor, but there is a lot to be said for a land management system which finds an economic use for its produce and wastes little. Nowadays laurel, like rhododendron, causes a problem because it grows out of control, suffocating mature trees and shading out all competition. When it is cut down, it is usually by volunteers or subsidized workers who do not take the timber back to their centrally heated homes, and who, in the case of volunteers, receive nothing more for their pains than the pleasure of physical work in the countryside. The same is true for substantial quantities of the timber and other biomass

34   Sinclair Knight Merz (2006), *Port Talbot Renewable Energy Plant: Non Technical Summary* http://www.prenergypower.com/nontechnicalsummary.pdf

35   This figure is based on the figure of three tonnes of dry wood per hectare per year that I used for estimates of UK firewood production. SKM provide no figures of land take.

36   DEFRA (2009), *Biomass: Carbon Sink or Carbon Sinner*, Summary report, 2009, http://www.environment-agency.gov.uk/static/documents/Biomass__carbon_sink_or_carbon_sinner_summary_report.pdf

in the countryside, which are cut down as part of a subsidized or voluntary management programme, and then left in a heap, or burnt.

In a situation where sources of zero carbon energy became scarce and valued, the fiddliness would be less of an economic constraint – but the transport costs over a given distance would increase. There are two methods for lowering transport costs: one is to make charcoal on site, which lowers the weight of the firewood so that it burns hotter and more efficiently, but is of lower gross calorific value. Charcoal-burning is a semi-industrial activity which requires all the timber to be of one size for a successful burn, and cannot cope too well with the 'fiddliness' of biodiverse waste wood; but when combined with other coppice crafts a very high percentage of the timber produce is used.

The other way of lowering transport costs is to bring people closer to the resources instead of bringing the resources to the people. The RCEP observes:

> In those areas close to forests the benefits of using an existing local
> resource for energy production are clear ... The rural location of most
> forests makes them ideally placed primarily (but not exclusively)
> to serve rural communities. There is an opportunity to link biomass
> energy policy with rural regeneration and fuel poverty strategies. The
> economic returns on rural schemes may be lower than in urban areas
> due to the lower heat demand density in rural areas.[37]

In other words, the more people who live in the countryside close to forests, the more people there will be who can benefit from the resource. Why bring Burnham Wood to Dunsinane? Why not move people to Burnham?

This is what has happened at Happy Valley, where some of the laurel, together with other thinnings, are now used for heating and cooking by the dozen or so people who live on 16 hectares of woodland. Everybody's heating, cooking and hot water is fuelled by firewood, but their low impact homes have an insulation standard which will probably soon be made illegal (there is less pressure to insulate when you have a surplus of firewood). On top of that the community manufactures and sells timber which it planks up on a rack-bench saw driven by a wood-powered steam engine, which is sometimes powered with laurel sneddings.

Permaculture designers have developed a system of 'zoning' whereby different land-uses are cited, concentrically, nearer or further from the centre of human activity – the farmhouse, village or town – according to the amount of attention required, and hence the amount of transport to and fro. Typically kitchen gardens and what the French call *basse cour* – the poultry and pigs that eat domestic and garden waste – are sited closest to the residence in zone 1; orchards, larger scale animal housing and maincrop vegetables a little further out in zone 2; main crops, pasture and in Whitefield's view 'small intensively managed belts of woodland' further out still in zone 3. Both Whitefield and Burnett place the bulk of woodland, which requires little maintenance, further out still, along with rough grazing in Whitefield's case and 'forage and collecting wild food' in Burnett's vegan scenario.

The concept of concentric agricultural zoning is common sense, and has been around a lot longer than permaculture. Cressey Dymock elaborated a zoning model for fenland improvement in 1651, but the concentric 'Thünen rings', proposed by the 19th century economist Heinrich von Thünen, are rather better

---

37   RCEP, *op cit.*7, pp 23-4.

known.[38] Von Thünen's zoning arrangement, which is focussed around larger settlements or markets, is virtually identical to the permaculture design, except in one detail. Market gardening and dairy production are again sited close to the centre of human activity grazing in zone 1, and grazing and wilderness at the extremities in zones 4 and 5. But in the middle belt, whereas permaculture theorists place woodland production outside the arable maincrop area, Thünen places it inside, in his zone 2.

This was because in early 19th century Germany, wood was the main fuel and there were no railways or internal combustion engines; firewood is much heavier than grain, in relation to its value, so the cost of bringing it to market from a distance would be greater. There is an issue here that manure is even heavier than wood in relationship to its value, and the von Thünen arrangement possibly places too much fertility (from humans and dairy cows) next to the market gardens, and not enough (only that from grazing livestock for meat and from rotation of pastures) contiguous to the arable maincrops. Nonetheless, the transport implications of firewood use are important. Collecting firewood on a haphazard daily basis, for those who have the time, is a pleasant and time-honoured occupation. In a heavily wood-based economy there certainly is a case for siting woodland close to human residences – or vice versa. Perhaps the woodland plantations whose siting I queried on page 240 do have a place close to our Somerset village centre after all.

The ascendancy of sheep over trees in our uplands over the last few hundred years has lately been assisted by the availability of coal, but the rise to dominance was well under way before the industrial revolution. 'Your sheep that were wont to be so meek and tame and so small eaters', Thomas More wrote in 1516, 'now, as I hear say, be become so great devourers, and so wild, that they eat up and swallow down the very men themselves.'[39] More was writing when the international wool trade was at its apex, and like all international commodities, wool, and later mutton, were an agency of enclosure and expropriation. In a countryside emptied of people, there was less need for folds and fences and firewood, no need for woodland, and all the more room for sheep.

Now that the sheep scenario is generally agreed to be rather overplayed, there is a good argument for reintroducing woodland, for biomass as well as timber, and that necessitates reintroducing people, because biomass has to be sited near where people live. That doesn't mean imposing a blanket of woodland where once there was a green desert of grass or heather. There are a number of reasons for keeping a sizable population of sheep in the country; in particular to bring surplus phosphate and nitrogen from outlying areas onto arable land, and to supply wool when the cost of importing fossil fuel dependent plastic fleeces from China rises. The objective will be to restore the balance of timber and grass which humans, and animals, and indeed the full biodiversity of nature require to flourish in any environment where there is sufficient water and warmth for trees and grass to grow.

The need to bring humans back close to their biomass is not the only reason for dispersing human settlements more evenly around the country, though it is probably the primary reason. I shall examine the need for ruralization more generally later. Any system of organic agriculture which prioritizes mixed farming rotations will be inclined to spread itself more widely around the country

38   Grove, Richard (1981) 'Cressey Dymock and the Draining of the Fens: an Early Agricultural Model', *The Geographical Journal*, Vol 147 No 1, Mar 1981 pp 27-37. Johann Heinrich von Thünen, Wikipedia, http://en.wikipedia.org/wiki/Johann_Heinrich_von_Thünen
39   More, Thomas (1516), *Utopia*, Everyman, 1994.

than the UK's population is at present. Up to half of the high grade arable farms in the east of the country will be required at any one time for fertility building through legume-based leys, steering them back to the kind of mixed livestock and arable farming propagated by Coke and his like. The corollary of this is that more arable land will be required in the west of the country.

Much of this arable land already exists and has been conveniently building up or retaining fertility as pasture for the last hundred years or more, except where farmers have been plundering it for silage. In 2006 I visited Andy Trim, a vegetable producer in Herefordshire who was having trouble acquiring permission for a residence on his holding. One of the arguments advanced by the planners was that the area was traditionally a pastoral landscape. But, as Andy pointed out, most of the barns in the locality that the planners had allowed to be converted to luxury dwellings were threshing barns, with wide central doors known as sails, because they were used to funnel the wind inside for drying and winnowing. He showed me a 19th century tithe map identifying which fields were arable and which were pasture, and over a third of them were arable. Andy's view was that not all the fields would have been cultivated at any one time, but they would have been folded into a convertible husbandry rotation.

When urged to diversify, the fossil fuel dependent barley barons of East Anglia habitually squeal that their 'four horse clay' is unsuitable for livestock, and that yields of corn in the west will never match those they achieve in the east. But their clay is no heavier than some I know of on dairy farms in Somerset. And the fact that yields of wheat from the UK's arable lands are higher than in any other country in the world except Belgium does not prevent us from importing grain and soya beans from low yielding lands in the Americas, so perhaps we should not be over concerned. It is also worth noticing that the country with the highest yield in oats – a remarkable 7.3 tonnes per hectare – is Ireland; that hilly Switzerland and boggy Ireland both have higher barley yields than the UK; and that the only grain in which the UK excels above every other country is rye, so perhaps we should grow more of that.[40]

There is no question that many pastoral areas like the Welsh borders or Devon can grow arable crops, because they have done so before. But there are rougher, less populated areas where nothing other than perhaps lazybed potatoes have been grown for as long as anyone can remember. It would be foolish to imagine that grain cultivation can or should be imposed upon every scrap of moorland, heath or peat bog, and it would be environmentally undesirable to do so. But there are few regions which do not have some areas appropriate for arable; even the extremities of Scotland hosts crofts, and there is nothing wrong with potatoes. Moreover, wherever there are livestock there is the opportunity to garner and concentrate fertility from the wider environment, and an overpopulated island striving for self-reliance can ill afford to waste the opportunity that livestock present to improve poor quality soils.

To some people 'improvement' in regard to land is a dirty word. It is the forerunner of that most contentious of modern terms, 'development'. Improvement is the last thing that conservationists want to see happening to precious remnants of biodiverse meadow or heathland. And those of us who bear the scars of enclosure on our ideological coat of arms remember that our ancestors were thrown off their land and herded into cities by those who considered it their God-given right, for the purposes of improvement, to depopulate fens and steal commons.

---

40   FAOSTAT, 2007 figures.

But just because improvement was the proclaimed objective of land-grabbers, that doesn't mean it is necessarily a bad thing. Admittedly, many nowadays would agree that the draining of the fens went too far and would like to see areas of fenland reclaimed. But the same people will also fight to preserve the hedges that were planted for enclosure and improvement of land in the 18th century. Many of the improvements of the agricultural revolution would have been welcomed by commoners and there is some evidence for the introduction of new crops and rotations on common fields.[41] However, commons were managed collectively, with social as well as economic objectives, and collective agreement often takes time. Improving landlords – those who were not simply using improvement as a cloak for avarice – were too impatient to let the democratic process take its course.

In fact, in the UK, agricultural improvement did not often extend very far from the centres of population. We do not find in the uplands many improvements on the scale of the terraces that capture eroding soil and conserve water on the rocky slopes of the Mediterranean, or in many parts of South East Asia. The Scottish Highlands were in no way improved by enclosure, but were degraded by the clearance of the population and their replacement by sheep. However, there is no reason why some of this wide expanse of semi-abandoned pasture land cannot be improved. Kenneth Mellanby writes:

> There is general agreement that much of our grass could be vastly
> more productive. It is common experience that, with care, the stocking
> density possible on many fields can be doubled. However the area with
> the greatest potential for improvement is some of the rough grazing.
> High in the Pennines, or in parts of Scotland we find occasional
> paddocks producing excellent crops of grass and hay. These have been
> modified by the farmer from something very like the surrounding
> land, covered with such poor grazing that each sheep needs some
> ten or more acres for its keep. It is not easy, or cheap, to effect this
> improvement, which needs much lime, basic slag and other fertilizer.
> Also it will rapidly revert to something even inferior (at least to the
> botanist) than it was before treatment if it is not well managed. But
> there are probably several million acres of rough grazing which could,
> at a cost, be ten times as productive as they are today. This could
> greatly enhance our ability to raise sheep and cattle.[42]

Or indeed to grow potatoes or rye. There is one peculiar historical serendipity which could greatly assist the improvement of some of our poor mountain land. Improvement requires a number of problems to be overcome: exposure to the wind and the elements, lack of fertility, drainage and acidity of the soil. Of these, exposure can often take the longest to rectify. It is therefore something of a bonus that the Forestry Commission and corporate tax-scam merchants, in the middle years of the 20th century, planted huge swathes of conifer plantations which, when they are in the right place, provide excellent shelterbelts.

A few years ago, David Gillen bought 30 acres of clear cut Sitka Spruce plantation on a mountainside in the Welsh borders, for which he paid the princely sum of £300 per acre. The firm which felled the trees left a generous shelterbelt standing all around the site, presumably at the instigation of the Forestry Commission who deserve considerable praise for this foresight. This

41   See, for example, Yelling, J A (1977), *Common Field and Enclosure in England 1450-1850*,
Macmillan; Beckett, J V (1990), *The Agricultural Revolution*, Basil Blackwell.
42   Mellanby, K (1975), *Can Britain Feed Itself*, Merlin, p 29.

meant that Gillen could move on to the site in a caravan and set up home in a sheltered glade which would have been a windswept moor if the trees had never been planted. He bought in pigs, plus a few chickens, and scratched a living selling pork and smoked bacon at farmers' markets. The pigs keep the land clear, whilst allowing natural regeneration of broadleaf trees. The stream provides potential for hydro-power and the remaining woodland provides firewood and construction timber.

By bringing in feed for his pigs, Gillen is gradually enhancing the fertility of the soil. With liming he could plant crops and within some years, when the tree stumps have rotted, he should have a potential arable holding that will have been improved beyond measure from its original condition of mediocre pasture. Gillen has toyed with the idea of weaving and staking lines of brash along the contours of the steeper slopes, and planting willow and other trees within them – the idea being that these would serve as barricades to retain the soil disturbed by heavy duty pig rooting, eventually forming terraces bound in place by living willow walls. Unfortunately it is perhaps one of those nice ideas that don't work in practice; it would require the compliance of the pigs (not something one should normally rely on) who might root too much, or not enough.

In Scotland, The Forest Village project is advancing a more formal proposal for the settlement of conifer plantations by new land-based communities. The timber would be felled gradually and processed on site for sale in the local economy, rather than shipped off for pulpwood. David Blair, the project's spokesman, emphasises:

> This doesn't have to be in the middle of nowhere, with few amenities. Many of Scotland's rural towns and villages back onto forests which currently provide few benefits to the local community. As well as managing the local forest and providing timber, affordable houses, fuelwood, crafts, and employment opportunities, the forest village community would bring in more people to support the local school, shop, post office and so on. For a forest village community to be viable, many of the basic needs of its members must be provided from the forest that surrounds it. That means that its housing, water and much of its food and power should be produced locally and sustainably.
>
> Sitka is a perfectly acceptable building wood and a resource which should not be wasted – it has a role to play in the construction of affordable sustainable housing. Judicious removal of mature conifers will provide space not only for housing, but for more durable timber species, such as European larch, Douglas fir and Scots pine.[43]

In 2005, these proposals attracted interest because of the plummeting price of softwood on the world market, and the Scottish Minister for Forestry and Rural Development requested the Forestry Commission to produce a report into the possibility of establishing forest crofts along these lines. However, in 2006 timber prices rallied, and in 2008 and 2009 the Scottish Government spent most of its energy trying to force through the sale of Forestry Commission land to private companies, rather than exploring ways of using the resource more sustainably. A bid from the Kilfinan Project in Argyll to buy 452 hectares for a community forest and forest crofts was turned down because the Forestry Commission required

---

43   Blair, David (2006), 'Forest Villages: How to Make Use of Sitka Spruce', *The Land* issue 1, Winter/Spring 2006; Blair, David (2005), Forest Villages: A Proposal for Sustainable Forestry, 2005; http://www.dunbeag.org.uk/images/upload/files/dunbeagdocs_pdf_33.pdf; Blair, David (2009), 'Transition to Rural Resilience', *Reforesting Scotland* issue 40.

full market price, while the Big Lottery Fund would not finance full market price acquisitions.[44] However in 2010, the money to buy 125 hectares was raised by public subscription with help from Highands and Islands Enterprise.

Kilfinan Forest Village aims to derive its most substantial income from adding value to the produce, and sees a future in improvement of the woodland; but the emphasis on 'localizing the source of essential resources' – food, housing, power and water – places it firmly within the sphere of permaculture. It would be a wonderful irony if these much-maligned conifer forests turned out to be the serendipitous seedbed for the spread of polyculture.

It is tempting to view the establishment of clearings and settlements within plantations as analogous to Vera's groves, hollowed out by the passage of time and the incursions of grazing. Similarly, the landscapes of formerly arable East Anglia and of formerly pastoral Herefordshire would be 'Vera-ized' as they became peppered with woodlands, fields, pastures, and orchards. As in Vera's model, the texture of this landscape would be in large part maintained by livestock. Only livestock can engineer the balance that any society seeks between the realm of light and the realm of shade on any scale beyond the arable. Even in a full-blooded fossil fuel economy, JCBs, timber harvesters and other wheeled monsters are fighting a losing battle with nature unless they enlist the help of quadrupeds. Where livestock are allowed to roam they bring grass, and where they are excluded trees grow, and it is a relatively effortless matter for humans to calibrate their performance to our will or whim.

The other task that animals perform more efficiently than machines is to move nutrients from where they are not needed to where they are required. That (together with the addition or removal of water) is largely what land improvement consists of. We employ plants to extract nitrogen from the atmosphere into the land, and we employ animals to move nitrogen and other nutrients from one bit of land to another. More importantly perhaps, plants cannot extract phosphorus from the atmosphere, so the role of animals in importing surplus phophorous from outlying areas could be crucial.

A fossil-fuel dependent economy not only uses livestock to move nutrients to the wrong place, creating a pollution problem, but also, through its requirement for cut-price factory farmed meat, undermines the economics of the grazing animals that we could be using to move nutrients to the right place. When lamb is cheap, wool worth nothing, and mutton unheard of, hill farmers can afford to do little more than leave sheep on the moor and see how many get through the year without dying. Ranging sheep in this manner is a labour-saving method of farming developed by English enclosers, which has spread to the conveniently (or deliberately) depopulated colonies. It was not the method employed in pre-enclosure days when sheep were folded overnight on cropland; and it is not the method employed in many parts of Europe where sheep are herded by shepherds, and brought back to the farm at night to pay their rent in manure. Occitan shepherds call 'Ven, véçi, ven!' ('come, come this way!') and their sheep follow; English sheep farmers drive their panicking flocks into the corner of a field with dogs or quad bikes. Was this what Thomas More was thinking of when he noted that our once tame sheep had grown wild?

In a ruralized agricultural economy, where less meat is produced, the value of farm animals is enhanced, and so, therefore, is the amount of attention that can be devoted to them. We might see a return to the folding of flocks at night, and even their shepherding by day, which would allow more care to be devoted to the

---

44   Blair, David (2009), *ibid.*

welfare of animals, and reduce or eliminate losses from predators. It would bring surplus nutrients, generated from the rocks, the atmosphere and the passage of wildlife, down the ancient hollow lanes leading from the hill pastures, to deposit them on arable land where they serve a human purpose.

A second advantage of shepherding is that it allows more watchful management of pasture land. The indolent air of the farmer leaning on his gate with a straw in his mouth belies the fact that he is busy observing his animals – and that is what professional shepherds do as a matter of course. Any shepherd, goatherd or cowherd, working for a number of years in the same pastures, is likely to have a much deeper understanding of what is happening to that particular area of land than can be gained from any number of ecological surveys.

Finally, on unfenced land, a shepherd can target the flocks' grazing patterns more effectively than can be achieved by leaving stock to wander at will. This is not simply a matter of steering the flock or herd away from vulnerable areas such as river banks; it can involve a sophisticated rotation of grazing areas based on knowledge and understanding of local conditions. Whether or not we agree with the theories of Voisin and Savory about pulse grazing, the advice of pastoral authorities always seems to be the same.[45] Allowing flocks promiscuous access to pastures tends to weaken the grass because animals will always head for the tender bite; better to allow flocks intensive access to restricted areas in a timed sequence, allowing grass to grow back. In pastures too wild and extensive for a network of fences, this can only be achieved by shepherding.

When I lived in the South of France, in the 1970s, at first light on certain days in early June, I would be slowly awakened from my slumber by a distant river of sound flowing into my dreams. It rose in a seamless crescendo – imagine Ravel's Bolero played by a gamelan orchestra – until it reached a gentle climax and then equally imperceptibly faded away until it could be heard no more.

It was the sound of *transhumance*, a thousand or more sheep being driven up to the summer pastures on the plateau, and it remains the sweetest noise I have ever heard. The sheep were folded in our village at the foot of the escarpment overnight and at four in the morning, when it was cool, decked in their brightly coloured crests and pom-poms and tinkling or clunking bells whose size and tuning denoted rank, they would start the six kilometre climb up to the mountain pass.

In the early 1970s there were about eight flocks which made the ascent, and the mayor of the village valiantly enlarged the enclosure which was rented out to the shepherds to accommodate their growing flocks overnight. But by the mid 1980s there were only three or four left, and although *transhumance* from the plains to the mountains still takes place, it is now carried out in a single day in lorries.

Graham Harvey has performed a welcome service in reminding us that *transhumance* was once a feature of the British (particularly the Celtic) agricultural economy, though like so many peasant practices it appears to have been abandoned in this country well before it fell into decline in mainland Europe. The shielings, as summer pastures were known both in England and Scotland, endured on the Isle of Lewis until the 1950s, but in most other areas they fell into disuse in the 19th century. They were much loved, as indeed *transhumances* are everywhere – they are after all a summer holiday. In Britain, where, as in the Alps, dairy cows for cheese and butter were more commonly summer pastured than sheep, the shielings were often the preserve of young women. Harvey quotes from a number of stirring reminiscences, this being one recorded in Ireland in 1943:

45   Voisin, André (1959), *Grass Productivity*, Island Press, 1988, cites abundant sources from the 17th century onwards.

I often heard my mother talking about the fine girls who were her companions in the shieling. There was in her company a fine woman from Feann-a-Bhuidhe above there called Úna Mícheáil Síle ... When she and my mother were in the shieling in their youth they were always close friends ... I heard her say that she was as strong as a stag. There was no man in the glen as strong and as active as she ... The summer pasturing continued in this district until my mother was twenty years old or more. That was about 90 years ago. I heard her say that the last summer was the most beautiful and pleasurable they ever had. They were a band of young spirited girls with little to trouble them.[46]

It is heartening to imagine that this admirable tradition might return as Britain re-ruralizes, though young dairymaids of the future might find their serenity troubled by tourists and hill-walkers. It would provide the British market with fine cheeses of the quality that is so fiercely guarded by the Swiss manufacturers of summer-pasture Gruyère. And it would bring to pastures that have been left to the ravages of unguarded and subsidized sheep, the watchful eye of the cowherd and the grazing habits of the cow.

Even without a revival of the shielings, we can anticipate a return to dairy cattle in upland areas, both to supply the local population with milk, and to make money from the export of cheese, which packs more value from less grass into a kilo than does lamb, and supplies pigs with whey. As well as assisting in the transport of nutrients, an increase in the number of cattle would be welcome in landscapes punished by an excess of sheep, and might assist in the control of weeds such as bracken.

While regions regarded as remote will see an intensification of agriculture and forestry commensurate with their repopulation, in areas which are already very intensively cultivated we may see some land taken out of intensive production. This is already happening, though not necessarily for the best reasons or in the best spot, as patches of land are turned over for forestry, or returned to the status of 'wildflower meadow'.

An obvious candidate is fenland, which could well allow a proportion of its fertile arable land to revert to waterland. Drained by Dutch engineers and Scottish prisoners of war in the 17th century for agriculture, much of the land is now kept dry by the use of diesel or electric pumps. A return to fenland is already being carried out by the National Trust at Wicken Fen in Cambridgeshire, where the aim is to rewild 3,700 hectares of farmland and return it to fenland. Currently some of the reclaimed area is being grazed by semi-wild Konik horses, though as far as I know the National Trust is not, like the Dutch wetland managers, retailing pony-burgers. The scheme will sequester carbon in the peaty soils that have been stripped bare by years of arable farming; and it will serve as a 'green lung' for residents experiencing spiritual suffocation in the new expanse of suburban sprawl planned for the Cambridge area. It would be good to think that Wicken Fen could become the haunt once again of some endangered species of human such as eel catchers and fowlers, who 350 years ago employed guerrilla tactics against the engineers destroying their homeland and livelihood.[47]

Throughout lowland England we might see some areas of intensively farmed land or pasture returned to something like a common. It is here that we

---

46 O'Dubhthaigh, Niall (n.d.), 'Summer Pasture in Donegal', translated by Danachair, C, *Folk Life*, Vol 22, 1983-4, pp 42-54, cited in Harvey, Graham (2001)The Forgiveness of Nature: The Story of Grass, Jonathan Cape.

47 Colston, Adrian (2004), *Wicken Fen – Realizing the Vision*, ECOS 25:3/4.

would be most likely to find the classic silvo-pastoral landscape described by Vera, and that indeed is what many remaining commons look like today. The degree of nutrient removal from these areas can be calibrated by the extent to which livestock are returned from the common to their farmyard or onto arable land at night. This is a mutually beneficial arrangement, not unlike that enjoyed by Mr and Mrs Spratt, both for farmers who require the fat of the land for their fields, and nature conservationists who want to see the ecosytems under their guardianship kept lean. The removal of surplus nutrients through grazing and folding often has a benign effect upon the maintenance of natural landscapes and biodiversity – though not if it results in the degrading of the flora and the spread of aggressive unpalatable weeds through overgrazing. In areas where grazing livestock on hillsides and folding them on arable was carried out over centuries, there does not seem to have been any great damage, and certainly nothing irreparable. But excessive nutrient loss has to be guarded against, and there are clearly limits to the amount of nutrients that can be sustainably removed from any given environment. One way of addressing such limits might be to rotate land, over very long time scales, between grazing, forestry and wildlife – another dynamic which echoes Vera's kaleidoscopic model of natural progression.

There is unfortunately a limit to the amount of land that can be given over to such extensive use. In our land budget, there are nearly five million hectares scheduled for wildlife and rough grazing – about one fifth of the country. But the bulk of this land is currently found in remote areas. Much of the land in lowland areas will have to be farmed productively, and can only be taken out of intensive production to the extent that land in more remote areas is improved and production increased. But there is also scope for intensifying some land use in lowland areas – for example by reviving water meadows – whilst letting other land nearby revert to common or conservation land.

The livestock management practices described in the last few pages, some more extreme than others, are the sort of changes which might occur if the current counter-urban trend in Britain blossoms into a rural renaissance. They do not, of course, preclude vegans either as consumers or as farmers: stockless rotations, green manures, nut production, forest gardens, mulching and so on will have a lot to contribute to the texture of the arable landscape. But it is hard to see how those who shun the use of animals can have any intimate involvement in the management of those parts of the country which are neither arable or woodland – that is to say the proportion of the country which we opt to keep under grass.

Low-Carbon Livestock

Over the course of this chapter I have been gradually covering the ground necessary to piece together a vision of what Britain might look like under a permaculture system of land use which is self-reliant, organic, low carbon and ruralized. There is a balance of land uses – principally arable, pastoral, woodland and wildlife – with more emphasis on woodland than at present. However, it is not as if these areas are in large blocks. In order to maximize connectivity – particularly in a society no longer dependent upon cheap fossil fuels – functions need to be dispersed around the country, more like the colours in a Jackson Pollock painting than in a Mondrian. With a ruralized population dispersed around the country, the pattern of land use is likely to be more fine-grained than it is now, and fractal, in the sense that similar arrangements are likely to be reflected at different orders of magnitude. Just as farmers place the land uses which require the most constant attention close to the farmhouse, so we can expect villages, districts and regions to follow the same logic. A kitchen garden is normally sited near the kitchen; likewise, instead of growing half the country's vegetables in Lincolnshire and wholesaling them in London, most market gardens will be located either inside or on the periphery of settlements. There will no doubt still be some bias between the grain producing east of the country and the pastoral west, but the balance will be closer to that of the mixed farm economy which existed 250 years ago – though with higher yields and more woodland.

Within this context there will be fewer domestic animals than there are now, though that may not be visually apparent, since it will mainly be the concentrate-fed and factory-housed animals that we will dispense with, and perhaps also because, with fewer stock, there will be less danger of the land being poached if they are left out all winter. But the main consequence of a reduction in the number of animals will be that they will have more value and so more attention will be lavished on them. Currently, although prices are very volatile, the margins on livestock are ridiculously small, sometimes negative. Nix in 2007 listed average gross margins as £14.69 on a fully reared pig, £1.33 on a turkey and just 8.1 pence on a broiler hen, and that is without factoring in rent and other overheads.[48] Milk production habitually operates at a deficit, and 'producers have relied upon income from cull cows, calves and other associated sources to overcome the difference'.[49] These pittances are an insult to the animals themselves, and to the people who look after them.

With meat a scarcer commodity than it is now, prices will rise relative to other goods, and so every scrap of the carcase will be used, rather than boiled up for pet food or incinerated. Farmers will require fewer livestock to make a living, and so will be less dependent on production line techniques and fossil-fuel powered machinery to complete their work load. Most important of all to many people, farmers will have more time to devote to the care of each animal, and to employ labour intensive rather than chemical methods of ensuring that they stay healthy.

Finally, the role played by animals in a low carbon permaculture economy to a large extent revolves around the fact that they can walk. God gave them legs, so they might as well use them. No other industry can be operated quite so easily without the use of fossil fuels, but that does depend upon how everything is arranged. The examples given briefly in earlier chapters – the uses of draught horse and oxen, shepherding etc – may have appeared frivolous or quaintly Luddite in that context, even though they are widely practised around the world, because they are so plainly at odds with the economic circumstances in modern Britain. An urbanized society, with urbanized animals, whose citizens expect just-in-time steak and hamburger, must deliver animal feed, livestock, meat and waste over distances and at speeds that cannot be attained by lumbering quadrupeds. Default livestock, on the other hand, are the natural issue and cohabitants of a permaculture system and so less under pressure to meet targets of economic performance. And they will be destined for a more local market, so there should be no need to pack them into multi-storey cattle trucks.

In a ruralized society, where decentralization prioritizes proximity, and where human settlement is more attuned to the cycles and rhythms of nature, animal automotive power is more likely to be economically viable. And in a low

48   Nix, J (2007), Farm Management Pocketbook, Imperial College Wye Campus.
49   RU Source (2009), *Milk Production Costs Briefing 692*, http://www.arthurrankcentre.org.uk/projects/rusource_briefings/rus08/692.pdf

carbon society where energy is at a premium, it makes no sense to waste it. The energy which animals expend on movement, digestion and reproduction is an unnecessary waste to the vegan, and an unavoidable cost to the industrial farmer – both of whom share a similar reductionist logic. The permaculturist, on the other hand, views animal energy as part of the natural cycle and tries to integrate it into the farming system, rather than using it as a reason for shutting the beast up in a factory, or ejecting it from the system altogether. Farmers have lived and worked like this with plants and animals for centuries, and it is arguable that advocates of permaculture have had to coin a new name only because industrial farmers have brought the term agriculture into disrepute.

It is possible that we will be forced to resort to very low carbon farming, and it is time to consider how we would cope if we had to. The following are some of the adjustments we might have to make to our management of livestock if we did reduce carbon emissions to near zero. Since some of these have already been described in foregoing chapters, they are listed in bullet point form.

• Farms would grow a wider range of livestock and crops, selected with a view to working together. Livestock would graze in rotation with crops, eat residues, bring nutrients onto arable land from outlying areas and keep orchards clean – all activities which are still carried out on many mixed farms. But there are other less well known symbiotic approaches which might come into more widespread use under zero carbon farming conditions. Pigs or chickens can be used for gleaning, clearing and fertilizing land. Chickens following three days after cows will pick the parasites out of cow manure. Ducks are one of the most effective organic means of getting rid of slugs. Horse manure can be used to heat greenhouses or polytunnels for out of season crops, or chickens can be housed there for the same purpose. Pigs can be persuaded to turn muck heaps by burying grain in them. Silkworms and fish are well-established companion animals – the fish eat silkworm excretions, and muck dredged out of the fishponds fertilizes the mulberry trees.[50] Fig 5 depicts the nutrient flows between the different components of a mixed livestock farm, while Fig 6 shows how much less complex a vegan but otherwise similar farming system would be.

• Farm traction could be provided by a biomass powered tractor or cultivator; or, alternatively by draught animals. On larger farms both might be used.

• Most pigs and some chickens would be sited close to farmhouses, residential areas, dairies, breweries, schools, hospitals, restaurants and other food sources, and fed on waste. Bought in feed would contain high quantities of waste from rendering plants and larger scale processing units. The dispersed nature of the pig and poultry industry would help redistribute these waste nutrients evenly on the land.

• More pig-slaughter would take place during winter, to economize on refrigeration and more of the produce would be cured, saving in refrigeration. Pigs reared slowly over an entire year have less need for high quality grains, and their diet can contain a higher percentage of waste products. Under less scientific feeding they are more likely to run to fat, which, if Britain were largely self-sufficient in food, would be welcomed.

• The slaughter industry would become more localized. Many pigs and poultry

---

50   Netting, Robert C (1993), *Farm Families and the Ecology of Intensive, Sustainable Agriculture*, Stanford University Press.

Figure 5. Nitrogen and Nutrient Flows in a Livestock Holding

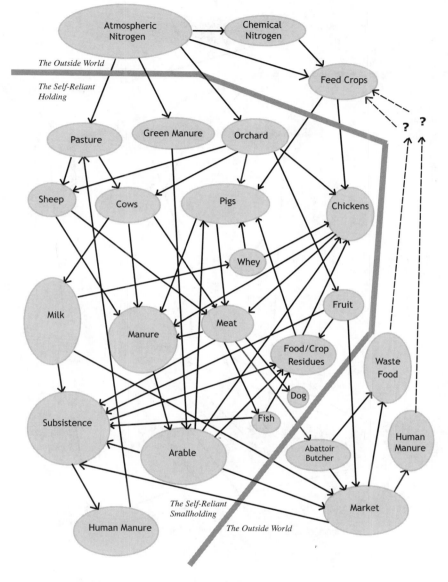

would be slaughtered on the farm. Butchers would become butchers again, rather than meat retailers: that is to say they would store animals on the hoof, behind the shop, and slaughter them as and when they needed it.

• More cows and goats would be hand-milked. This would require smaller dairy cow farms than at present, though many goat herds could remain the size they are at the moment. In the 1970s, I worked on a farm in France where 140 goats were milked by two people twice a day – two litres per goat at a rate of about a litre a minute. The largest number of sheep I have known to be milked by hand by a single person is 35 (in France), and the largest number of dairy cows I have heard of, ten (in Britain, 1940s).

Figure 6. Nitrogen and Nutrient Flows on a Stockless Holding

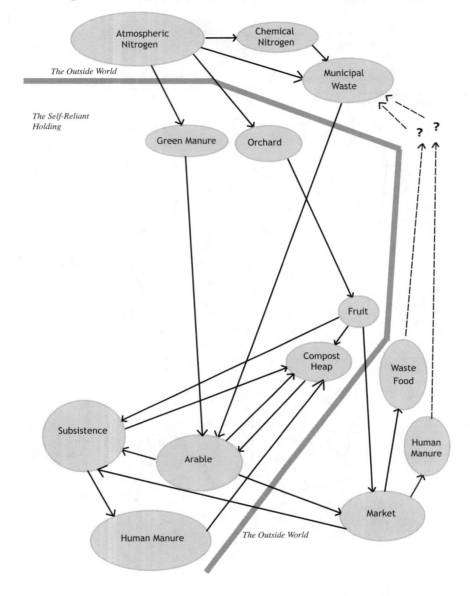

• Pasteurization of milk, which involves heat and then cooling, is widely agreed to be necessary at an industrial scale, for public health reasons; but unpasteurized milk is acceptable at a local level where the source is known, and risks can be assessed by the consumer.

• Milk production would become more localized. Distribution in many areas could best be achieved through temperature controlled milk-dispensing machines which dispense with the need for bottling and packaging by serving straight into the household jug or container. Similar devices operate successfully on farms in Austria. Street corner vending machines, loaded daily by the farmer with churn-sized cartridges, could provide fresh milk in villages and town neighbourhoods only a few hours after milking. Churns can be moved manually (so pumping machinery is not necessary) and are a convenient size for smaller herds.

• Getting fresh milk into larger towns would require sophisticated logistics, perhaps based, as before, upon milk trains. Milk deliveries might make a comeback in towns, and might be horse-drawn, since the milkman can advance the vehicle without mounting, just by calling to the horse, who is familiar with the route. But even in cities, the street corner dispensing machine has the benefit of reducing the amount of packaging and eliminating the burden of bottling, allowing for more speedy delivery.

• Green belt land would be highly valued for agriculture; extensive ruralization would free space in cities that could be used for market gardening. Urban contact with city farms, particularly through school visits, would enhance city-dwellers' understanding about where their food came from.

• Livestock markets would return to local towns, rendering the Market Towns Initiative (which ought really to be called the Marketless Towns Initiative) redundant. The distance between market towns is a reflection of the distance the local populace and traders could manage on foot in a day. In medieval Nottinghamshire the average distance between all neighbouring markets was 3.25 miles and lawyers of the time considered that markets on the same day needed to be at least 7.75 miles apart to avoid harmful competition.[51] In North East China, during late imperial times, markets were 4.9 miles apart, and the most isolated villagers were 2.8 miles from a market.[52] It is common to hear horror stories of people in the 19th century walking distances of up to 12 miles before and after work – but that was in the days when the rural economy was being wrenched apart. You don't have to travel very far in a well furnished local economy. Market days have the advantage of combining sale of produce, purchase of commodities, business meetings, public events and social encounters on a single day which can be served by a frequent public transport service. The disappearance of markets is a classic example of an economy of scale at distribution level sacrificed to obtain economies of scale at production level.

• Over a certain radius (and depending on their value) livestock could be driven to market. The activity of droving and even the profession of drover might return, for example to bring stock from more remote areas for fattening. According to an 18th century observer, driven livestock maintain their weight over the first 50 miles, and then lose it very quickly if they travel any further; while in the US cattle would lose weight if they travelled more than 15 miles in a day.[53] *Transhumances* would be revived and conducted on foot rather than by truck.

---

51   Steane, John (1985), *The Archaeology of Medieval England and Wales*, Taylor and Francis, p 126.
52   Skinner, G William (1964) cited in Frances Berdan, *Aztec Imperial Strategies*, Dumbarton Oaks, 1996.
53   Lisle 1757, p 394, cited in Davis, Simon J (2002), 'British Texts for the Zoo Archaeologist', *Environmental Archaeology* 7, pp 47-60. An article at Wikipedia on *Cattle Drives in the United States* notes: 'While cattle could be driven as far as 25miles (40 km) in a single day, they would lose so much weight that they would be hard to sell when they reached the end of the trail. Usually they were taken shorter distances each day, allowed periods to rest and graze both at midday and at night. On average, a herd could maintain a healthy weight moving about 15 miles per day.' The reference given is Malone, John William (1971), *An Album of the American Cowboy*, Franklin Watts, Inc.

• A highly reruralized society would see a revival of fairs, not unlike the growing number of festivals that embellish the modern English summer, but with stalls that purveyed the practical as well as the fanciful, and plenty of dealing in horses, livestock, cheese, and the like. 'When fairs were frequent shops were not necessary,' says Cobbett with characteristic hyperbole, and he goes on to explain:

> A manufacturer of shoes, of stockings, of hats, of almost anything that man wants, could manufacture at home in an obscure hamlet, with cheap house rent, good air, and plenty of room. He need pay no heavy rent for a shop; and no disadvantages from confined situation; and then by attending three or four or five or six fairs in a year, he sold the work of his hands, unloaded with a heavy expense attending the keeping of a shop. He would get more for ten shillings in a booth at a fair or market than he would get in a shop for twenty pounds.[54]

Anybody who doubts the ability of a society entirely dependent upon animal traction to shift large volumes of material, or who thinks that country life before the motor vehicle must have been boring, should read Defoe's description of Sturbridge fair, an event the size of Glastonbury Festival, but with no entrance fee or security fence, and lasting many days longer.[55]

• In some circumstances it might be more economic to employ a shepherd/cowherd to guard ruminants, rather than to fence them. This has advantages for animal welfare (disease is more readily spotted) and for land management (grazing of different areas can be calibrated by the shepherd, and overgrazed areas such as riverbanks avoided).

• Some herds of ruminants would be returned to barns or folded on fields at night to supply manure.

• There might be a return to the formerly widespread practice – still found in Eastern Europe – of family-owned dairy cows being collected by the community shepherd after milking in the morning and returned for milking in the evening after a day in common pastures. There are great advantages to this system: it uses economies of scale where they exist (in grazing and bull provision), but spreads the burden where scale is a disadvantage (in handmilking and veterinary care); it distributes milk to every family participating in the scheme; and gives every participating family a stake in the management of village lands, and the economic and emotional satisfaction of owning a cow. The same system was used on the Eastern coast of the United States, prior to the days of barbed wire. Defoe describes this system operating in Cheddar where 'the whole village are cowkeepers' and 'before the village is a large green, or common, a piece of ground in which the whole herd of the cows, belonging to the town, do feed'. Milk not consumed at home was turned into a single huge cheese at the co-operative dairy (another operation where economies of scale are a great advantage); but families were paid, when they had contributed enough gallons, with a huge cheese weighing up to a hundredweight, which it was their responsibility to sell: 'Thus every man has equal justice, and though he should have but one cow, he shall, in time, have one whole cheese'. Unfortunately Defoe doesn't tell us what happened to the whey.[56]

• Livestock production would become more localized resulting in a high level of biodiversity amongst breeds, bred to adapt to local conditions.

---

54   Cobbett, William (1826), *Rural Rides*, 22 October 1826.
55   Defoe, Daniel (1724), *A Tour Through England and Wales*, Everyman, Vol 1, 1927, pp 80-5.
56   *Ibid.*, pp 277-8.

• Being predominately grass fed in the case of ruminants, and somewhat so in the case of monogastrics, the quality of meat, milk and eggs would be superior in taste and nutrition, and higher in Omega 3 oils. The fat content, colour and consistency of milk would be less predictable, but this would be of less concern in a non-industrial market.

• Livestock, kept outside and on grass for longer, would become hardier and less dependent on antibiotics. Diseases associated with factory farm conditions would disappear, and when diseases did occur they would endanger far fewer animals. The days of mass slaughter for classical swine fever or foot and mouth would be over. With meat being of higher value than now, and with more farmworkers per beast, animal welfare would be improved.

• Although there would be some risk if cross-country droving were permitted, animal epidemics would spread slower, making their control easier and allowing some relaxation of livestock regulations. Animal movement regulations which, for example, currently make it illegal for a farmer to graze his cow in a neighbour's field and bring it back to milk in the evening, could be abolished.

This may be the point at which some readers, if they have persisted this far, will finally throw this book away in disgust exclaiming that the author is off with the fairies. There is a certain kind of person for whom anything that smacks of Luddism is on the same level of acceptability as Stalinism or paedophilia. The concept of 'progress' is so deeply engraved upon their psyche that the idea that some of the activities which humans carried out for thousands of years and then stopped doing about fifty years ago might actually have been quite sensible is beyond their comprehension.

In fact, all the activities described above are carried on somewhere in the world at this present time, and mostly they are carried on in countries whose per capita carbon emissions are much less than ours, by people who are not necessarily any less content or any less secure than the average British farmer. The measures imposed by a policy of low-carbon livestock husbandry do 'turn the clock back' in some respects, but the result is in many ways attractive, and more in line with the public's expectations of what our farms should be like: a variety of animals, farm horses, healthier stock, pigs you can scratch, tastier food, people actually working in fields and so on. It comes closer to the Arcadian picture of English farming which is depicted so mendaciously on much of our food packaging precisely because it is what the public want to see.

There is also the possibility that livestock farmers might be held accountable for any methane and nitrous oxide emissions significantly over natural and social background levels. Vegans might, reasonably enough, insist upon some sort of reward for their forbearance, and the obvious one is to give them priority to other extravagant crops, notably energy. It is fanciful, but not illogical, to imagine a society dependent upon renewable energy, where vegans are allowed to use biodiesel and biomass fuels to power tractors and private vehicles, while meat and dairy eaters are restricted to horse or bullock power, or else public transport. Well, which would you choose? With such a caste-riven transport system, the sporting highlight of the year would probably be the Boat Race which, instead of pitting dark blue against light blue, would see the reds battling it out against the greens. My money would be on the carnivores, not because of any inherent superiority in their diet, but because they would be more used to exercise than the vegans.

Ruralization

Reliance on renewable energy raises questions about where we live. Firewood is one resource that points towards a change in settlement patterns, but it is by no means the only one.

The sun, the wind and the rain dispense their bounty reasonably equitably across the whole of the British Isles. The atmosphere envelopes all of us with a homogenous blanket of nitrogen, oxygen, carbon and other elements. Soil and minerals have been laid down over the ages less evenly in terms of their usefulness to humanity, but nonetheless, every place has its geology. The only great gift of nature which some places have access to and others lack is the ocean.

These are the elements from which we humans must generate all our energy and food, derive our livelihoods and build our civilizations, if we are not to rely on fossil fuels, nuclear energy or technofix. They are not all concentrated into a few oases (except in deserts), but spread out around the globe. In order to use these resources we have two options, or rather there is a spectrum spread out between two opposing strategies. On one hand people can go out to the resources, which means dispersing our settlements and farming resources *in situ*. Or we can bring these resources to people, which means tapping them at source and conveying them to concentrated settlements. The first option, or end of the spectrum, is what we term rural – the other end is urban.

The reliance over the last 200 years upon fossil fuels has pushed humanity towards urbanization because these sources of energy tend to occur in specific places and in a concentrated form, which can be easily transported. There is no need to go into detail explaining why London and Manchester expanded as a result of the 19th century coal economy in England; or why Mexico City, Bombay, Shanghai and similar other accretions of humanity have occurred as a result of the superior efficiency to be derived from delivering fossil fuel energy and food to people living in concentrated settlements. The oil dependency which herded people into towns pursues a similar dynamic to that which made it more profitable (for some) to export 120 kilos of sheep from the uplands of northern England, than to grow three tonnes of firewood and produce milk, meat and oats for local consumption.

Insofar as the response to global warming may involve a move back to neglected sources of renewable energy, which can now hopefully be captured more efficiently through devices such as wind generators, solar panels, wood burners, tide-mills and biomass gas generators, we are faced with the question: should we bring the energy to people in towns, or should we consider dispersing people?

The matter was broached by George Monbiot in *Heat*,[57] an overview of the UK's energy options in the face of global warming. Monbiot looked at a great many alternatives for supplying energy, either through centralized systems involving large power stations and the existing grid, or through decentralized methods of energy production, such as microgeneration and combined heat and power. His preferred solution was a mixture, including decentralized microgeneration of heat and electricity using solar panels and hydrogen boilers or fuel cells, together with grid-based electricity from fossil fuel power stations where the carbon was extracted and buried.

I have no particular quarrel with Monbiot's conclusions about where the national grid element of our energy supply should come from. My concern is that he reaches these conclusions without any wider consideration of how, in

---

57   Monbiot, *op cit., 10.*

a post fossil fuel world, we are to provide the other things that people need, the most important of which are food, water and a way of life that people find satisfying. The newly developed DC electric grid which Monbiot advocates magically transfers energy around the country without any transmission losses; but the material things of life cannot be moved without transmission losses. You cannot transport food, fibre and building materials around the country without generating expenses which over the last century have been paid for by cheap fossil fuels. If you derive all your water from a small number of sources which offer economies of scale, you not only experience diseconomies of distribution (infrastructure costs, pumping costs and leaks), you also run the risk of running your sources dry. And if you bring all this biomass into the cities to maintain an urban population, then you invite various kinds of congestion (too many vehicles, too much smoke, too much waste, too many animal diseases); and you have to find a way of getting the biomass back out again, once it has been used, so that it can go back to the land. When each of these transmission costs is analysed separately it appears slight. But when you add together all the extra costs that become necessary when we house people so densely that there is not enough local solar energy or water to go round – the transport, the refrigeration, the packaging, the heating, the street lighting, the clothes drying, the waste disposal, the composting, the congestion, etc – they represent a sizable proportion of our CO_2 emissions.

When these factors are taken into account, there is a *prima facie* case for reconsidering our current urban settlement patterns. A more rural population would be closer to the sources of food, and so would its sewage and waste disposal system. Enough rain falls on people's roofs every year to supply virtually all their domestic needs and this, together with local wells, springs, boreholes, and village-scale supplies that the water companies have abandoned over the years, offers a more reliable source of water than the reservoirs and aquifers which even in our rain-drenched country always seem to be on the brink of running out. The small-scale wind generators which Monbiot points out would be impracticable on town houses, could, he agrees, work fairly well in conjunction with solar panels in the countryside. Building materials can be sourced, as they always used to be, only a stone's throw from the site of the building, resulting in a genuine rediscovery of the vernacular which modern architects can only imitate. And, to bring us back to where we began, local biomass – be it timber, SRC, *Miscanthus*, sawdust or hedgeclippings – instead of being turned inefficiently into gigawatts, fed into the grid, and consumed at the flick of a switch, can be harvested, distributed, consumed and managed in cords and faggots or chippings at parish or district council level.

This is permaculture on the grand scale and a similar vision is proposed for the UK by Patrick Whitefield:

> In the long term we need to design truly sustainable settlements. This means cities and towns which are small enough to get most of their resources locally. It also means repopulating the countryside with new hamlets and smallholdings. More people will work where they live, and more food and manufactured goods will be consumed near where they're produced … The key to implementing permaculture in the countryside is repopulation. This includes the breaking up of the present large, mechanized farms into small farms, smallholdings and new hamlets, where energy-intensive production can be replaced with design-intensive and human-attention-intensive production. As well as

farming and gardening there can be small-scale manufacturing in the countryside, mainly for local needs and/or using local resources, and people tele-working from home. Many people will have polycultural incomes, usually involving some food production.[58]

The first proponent of post-industrial decentralization was perhaps Peter Kropotkin, who as early as 1901 noticed that the replacement of steam power by electricity was resulting in lighter machinery more accessible to smaller communities, which ought to have been leading to the reintegration of industry with small highly productive farms. 'The 'concentration' so much spoken of is nothing but an amalgamation of capitalists for the purpose of *dominating the market*, not for cheapening the technical progress'.[59]

This theme was taken up in the 1920s by the American Ralph Borsodi, in this prophetic paragraph:

The coming of cheap power in a form suitable for application to domestic machines may help to redress the present adverse balance between the home and the factory. When we shall have become sufficiently civilized to create a demand for small generating plants driven by windmills and watermills, they will be developed and placed on sale at even lower prices than the very ingenious plants driven by gasoline engines which are now on the market. The domestic producer will then have power, heat and light at no cost in money except for lubricants and maintenance.

Borsodi was ahead of his time in a number of ways, but no more so than in his identification of what we now call 'food miles' in his 1927 book *The Distribution Age*. The thesis of this book is that economies of scale are more than offset by diseconomies of distribution. He provides pie-charts showing that the distribution costs for products such as corn flakes and rolled oats, even in the 1920s, were twice as high as manufacturing costs. And he depicts the working of Say's Law (that 'supply creates its own demand') in the freight industry:

We may have reduced the ton-mile cost of transporting freight, but at the same time we have increased the average miles per ton shipped so as to completely wipe out the saving and to actually increase the transportation cost on each ton of merchandise we consume.[60]

In particular, he lambasts the pointlessness of what he calls 'cross-hauling' (transporting the same item in opposite directions):

If Chicago manufacturers could secure a thousand desirable outlets in New England in addition to those they supply in their own natural territory, they could probably lower their cost of production sufficiently so as to quote prices FOB New England in full competition with Boston. If, however, the Boston manufacturers at the same time secure one thousand desirable outlets in the Chicago manufacturer's territory, isn't the net return a mere exchange of accounts?

Centralized processing units also generate a form of cross-haulage:

It has always seemed absurd to me that Kansas farmers should ship wheat to Minneapolis and then have flour shipped all the way back

58   Whitefield, P (2004), *The Earth Care Manual*, Permanent Publications.
59   Peter Kropotkin (1912), *Fields, Factories and Workshops*, Tho Nelson and Son, 1912 (1898), p 354.
60   Borsodi, Ralph (1927), *The Distribution Age*, D Appleton and Co.

from Minneapolis to Kansas ... . In the milling of flour – which lends itself so admirably to local production – we have drifted into a state of affairs where ten of the national mills are able to supply 50 per cent of the consumption of the country.

The same applies to meat:

The development of the refrigerator car made it possible to ship fresh meats great distances ... As a result we now have the absurd system whereby the raw products of an industry, in this case the livestock, are shipped great distances to the meat packing centres and then, after being slaughtered, shipped equally great distances back to the point of origin.[61]

Borsodi wrote several other books, including one called *Flight from the City*, which was a practical guide to rural self-reliance. He initiated a number of agrarian settlement schemes (some more successful than others), worked in India with Jayaprakash Narayan on the financing of the *gramdan* land resettlement project, founded a 'School for Living' which is still going, and helped introduce the now flourishing concept of Community Land Trusts.[62] However his economic analysis was ignored and the academics and campaigners who initiated the concept of food miles in the early 1990s did not appear to remember him.[63] Now, when even the NFU is talking about food miles and has belatedly launched a campaign in support of farmers' markets, most commentators have not yet caught up with Borsodi, nor will they until they recognize (a) that it is not just a matter of food miles but 'resource miles'; and (b) that we will just be tinkering around with this problem unless we find a way of siting people close to the resources that they use.[64]

The case for ruralization has been advanced recently by Richard Heinberg, as a corollary of his view that oil prices will rise dramatically when global oil reserves reach their predicted peak in the next few years.[65] Heinberg anticipates that in an economy running on limited supplies of oil (and without any substitute form of energy, such as nuclear) the US will need to find another 50 million farmers to grow its food. This figure is based on the experience of Cuba in the 1990s when, deprived of financial support and cheap oil from the Soviet Union, some 15 to 20 per cent of the population became involved in food production.

Another advocate of ruralization, the Swedish academic Folke Günther, has extended this line of thinking by comparing Cuba's performance with

---

61   *Ibid.*
62   Borsodi, R (1933), *Flight from the City*, Harper and Bros; Fogarty, S, 'Ralph Borsodi: Decentralist Theorist and Community Builder', in R Borsodi (1929) *This Ugly Civilization*, Porcupine Press, 1975. On the failure of the Borsodi's first settlements: Issel, W (1967), 'Ralph Borsodi and the Agrarian Response to Modern America' in *Agricultural History*, Vol XLI, No 2, April 1967. On Borsodi in India: Slastrom, P (1975), 'Ralph Borsodi's Vision of Land Reform', *The Green Revolution*, Sept 1975, www.cooperativeindividualism.org/salstrom-paul_on-borsodi-and-land-reform.html. On Community Land Trusts: Loomis, M (1978), 'Ralph Borsodi's Principles for Homesteaders', *Land and Liberty* Nov-Dec 1978 http://www.cooperativeindividualism.org/loomis_borsodi_bio.html. School of Living: http://www.schoolofliving.org/
63   See Lang, T and Raven, H (1994), 'From Market to Hypermarket: Food Retailing in Britain', *The Ecologist*, July/August 1994.
64   *NFU Countryside*, April 2007.
65   Heinberg, Richard (2006), 'Fifty Million Farmers', *Energy Bulletin*, 17 November 2006, http://energybulletin.net/22584.html

that of North Korea.[66] With the collapse of communism, both countries found themselves in a similar situation, but whereas Cuba weathered its difficulties by increasing its farmers and its rural population, North Korea persisted with an industrial mode of production, and the proportion of people living in the countryside decreased. By 1999, 80 per cent of the motorized capacity in North Korea's agricultural sector was inoperable, and grain production had slumped to 40 per cent of previous production levels. The country was eventually bailed out with food aid but according to former Korea Worker's Party Secretary Hwang Jang Yop (a defector) the death toll during the food crisis was three million people.

Günther anticipates eco-units of about 200 inhabitants and 50 hectares, largely self-sufficient in terms of food, animal fodder, water, energy and with a maximum proportion of waste return to the soil; three or four such settlements might share a common social infrastructure, such as primary schools and small service business. The transition to a ruralized social structure would occur gradually:

> The development strategy implies successive replacement of houses in need of extensive restoration or rebuilding. Instead of building new houses in existing areas, small settlements integrated with agriculture can be created in the hinterland of the urban areas.

In a series of four plans, Günther shows how an urban centre of 33,000, with a population in its surrounding countryside of 3000, has declined to 24,000 inhabitants with 12,000 inhabitants living in the periphery after 12 years. After 25 years the town has only 12,000 people, with 24,000 ruralized; and by the time 50 years have passed, the proportions are completely reversed with just 3000 people in the urban centre and 33,000 in outlying areas. The region has become ergonomically decentralized and self-sustaining, with the town centre serving mainly as a cultural and decision-making centre.

Whether Günther's vision for a provincial town in sparsely populated Sweden can be applied over the same time scale to a metropolis like Manchester is perhaps another matter. Ruralization on the scale he proposes – with 92 per cent of the population in the countryside – seems ambitious for a densely populated country such as England. Günther calculates that 0.2 hectares per person are required for food production, which amounts to about half the land available per person across the UK. That accords closely with the projections made in my Livestock Permaculture scenario where the food needs for 60.6 million people are met on 13 million hectares of land. Technically there does not seem to be any geographical impediment to prevent Britain ruralizing its economy to a radical extent, should it wish to do so.

Günther calculates that by ruralizing the economy, and recreating direct local food links between farmers and consumers, payments to farmers could be increased six-fold from the current average in Sweden of 3.7 per cent of the shop price, to 22 per cent (allowing for a 30 per cent increase in production costs). UK farmers who have shifted from producing for corporate distributors to direct sales, for example from farm shops, can testify that increases of this order are normal. 'A rise in farmer income of that magnitude' says Günther, 'can be expected to enforce even larger changes in agricultural practice.'

---

66 Gunther, F, *Ruralization: A Possible Way to Alleviate Our Current Vulnerability Problems*, IV Biennial International Workshop Advances in Energy Studies, Unicamp, Campinas, SP Brazil; 15-19 June 2004. See also: Boys, Tony (2000), *Causes and Lessons of the North Korean Food Crisis*, Ibaraki Christian University, 2000.

On Getting Matter in the Right Place

While Günther recognizes the need to improve accessibility to food, energy and water, he focusses especially on waste disposal and the need to return nutrients back to the soil efficiently. Günther's particular concern is phosphorus, which (in contrast to nitrogen and carbon) is not a volatile element present in the atmosphere, but has to be mined:[67]

> Whereas other non-volatile nutrients [eg silicon] are more abundant in the earth's crust than in biological systems, phosphorus is both essential in biological systems, and is needed in much higher concentration than its existence in the crust. Consequently, phosphorus needs to be concentrated from outside the organism, and a lack of availability is immediately reflected in its vigour.[68]

I have already explained, in Chapter 8, how sustainable agriculture is dependent upon (a) returning phosphorus back to the land from which it came and (b) upon a small but essential input of phosphorus from outlying land or other sources to make up for erosion and leaching. For the last 150 years much of the phosphorus has not been returned to the land in the correct proportions and the shortfall has been made up by mining rocks rich in phosphorus, which are found in relatively few parts of the world. As with many non-renewable resources, the issue is not that phosphorus supplies are about to run out, but that there are diminishing returns from continued exploitation. John Driver, an industry consultant, observes, 'what is patently clear to anyone who has worked in the phosphate industry for a number of years is that the quality of phosphate rock is declining inexorably'.[69]

Moreover, the politics of phosphorus mining have always been fraught with conflict. Britain exhausted its own supplies at the end of the 19th century, so the British Phosphate Commission set about strip mining the tiny Pacific island of Nauru. The island eventually gained control of its resources after World War II, but by the year 2000 these were exhausted, leaving Nauru's population impoverished, and its ecology devastated. The Nauru Phosphates Royalties Development Group, on the other hand, has invested in five luxury condominiums on prime real estate on the Hawaiian island of Oahu, the Hawaiki Tower in Honolulu, and Nauru House, once the tallest building in Melbourne.[70]

The USA still has 11 per cent of known reserves of phosphate, but there have been estimates that production will halve over the first 15 years of this century, and tail off after that.[71] Canada's substantial grain exports are entirely dependent upon phosphate imports from the US and Togo. The world's largest reserves are in Morocco, where in the early 1990s, guerillas from the former Spanish Sahara contested the ownership of disputed phosphate-rich territory.[72] In Sri Lanka, in 1999, a campaign of demonstrations by local farmers fought off an attempt to sell the phosphate deposits at Eppawela to the US company Freeport McMoran,

---

67   See discussion of this in Chapter 8.

68   Günther, *op cit.,* 66.

69   John Driver (1988), 'Phosphates Recovery for Recycling from Sewage and Animal Wastes', *Phosphorus and Potassium*, Issue 216, July–August 1998.

70   *Nauru Phosphate Corporation*, Wikipedia http://en.wikipedia.org/wiki/Nauru_Phosphate_ Corporation.

71   Herring, J and Fantel, R (1993), 'Phosphate Rock Demand into the Next Century: Impact on World Food Supply', *Non-Renewable Resources* (2)3, 1993, 226-46, cited in Grimm, next reference.

72   Grimm, K (1998), 'Phosphorites Feed People: Finite Fertilizer Ores Impact Canadian and Global Food Security', *The Monitor* (Journal of the Canadian centre for Policy Alternatives), http:// www.eos.ubc.ca/personal/grimm/phosphorites.html

which intended to strip mine an area of 56 square kilometres, relocating 12,000 people in 26 villages. Freeport McMoran planned to exhaust the supply over a period of 30 years, whereas campaigners claimed the supply could meet Sri Lanka's needs for the next 200 years.[73] In 2005 the Sri Lankan government again put the mine up for sale, this time attracting a Chinese bid, but again this was fought off by protests.[74]

This economic and political scarcity can be addressed by agricultural and food systems which recycle most of the phosphate in manure, human sewage and animal bones, but there are three problems here. The first is that because the centralization and transportation of livestock makes animal disease such a high risk, we no longer recycle slaughterhouse and food wastes, but incinerate them (see Chapter 5). The second is that the easiest way to transport excessive volumes of animal manure is often to mix it with water, but this makes composting and storage of nitrogen more difficult and encourages leaching into the surrounding environment. The nature of the problem can be seen by a visit to a typical struggling UK dairy farm, where manure from the milking parlour is hosed into a lagoon to sit around bubbling methane, or overflowing into the nettles, until it is pumped out onto the fields.

The third and principal problem is that nutrients accumulate in what Günther calls HEAP traps – HEAP standing for Hampered Effluent Accumulation Process – while other locations are deprived. These HEAPs are caused, first of all by feeding grains to animals, since grains are in themselves a transportable form of nutrient concentration; and second, by trying to obtain economies of scale by feeding a great many animals in a small space. John Driver of Albright and Wilson, a firm who have been manufacturing phosphates for over 150 years' agrees:

> Problems have arisen due to the intensification of livestock production, particularly pigs and poultry. This has resulted in local excesses of manure production, far beyond the capacity of nearby farmland to absorb the output. In such situations, alternative disposal routes have to be found; incineration is amongst those already being employed.[75]

Another industry analyst, Ingrid Steen, estimates that the amount of phosphate excreted by livestock in Western Europe could be some 50 per cent more than the amount currently applied as mineral phosphate fertilizer – but that a large amount of this is wasted – 'perhaps up to around 40 per cent could be more effectively spread onto agricultural land'.[76] Günther estimates that by 're-ruralizing' livestock and people, ie keeping them close to the source of their food, most of the nutrients can be returned to the soil: 'About six persons are in nutrient balance with one hectare of land. This means that about 0.2 hectares of

73   Independent press articles about the Eppawela Issue, listed at http://people.whitman. edu/~walterjs/localpress.html
74   Centre for Environmental Justice (2005), *Selling of Eppawela Phosphate Deposit and Other Public Properties for Rebuilding Sri Lanka is a Crime*, press release, 2 March 2005; Green Movement of Sri Lanka, *The Protest Campaign to Prevent the Phosphate Deposits from being Sold to the Chinese*, internet site no longer on line. Colombo Page (2007) 'Sri Lanka Government to Utilize Eppawala Apatite for Local Fertilizer, press release, *Colombo Page*, 12 January 2007, http://www.colombopage.com/ archive_07/January12143554SL.html
75   John Driver, *op cit.*, 69.
76   Steen, I (1998), 'Phosphate Recovery', *Phosphorus and Potassium*, No 217, Sept–Oct 1998, http://www.nhm.ac.uk/research-curation/projects/phosphate-recovery/p&k217/steen.htm

such agriculture can support one individual without HEAP effects, provided that the nutrient-containing residues are returned to the agriculture'.[77]

HEAPs are also an area of concern for the FAO, who observe:

> There is growing concentration of livestock activities in certain favoured locations ... This concentration is driven by the newly gained independence of industrial livestock from the specific natural endowments of given locations which have previously determined the location of livestock production (as they still do for most of crop agriculture).[78]

The FAO's term for this problem is 'nutrient loading' which, it explains, is due to the 'urbanization of livestock':

> Geographic concentration or what could be called the 'urbanization of livestock' is in many ways a response to the rapid urbanization of human populations ... The separation of livestock production and the growing of feed crops is a defining characteristic of the industrialization of livestock production. Nutrient loading is caused by high animal densities, particularly on the periphery of cities and by inadequate animal waste treatment.[79]

The FAO's solution is to ruralize 'confined animal feeding operations', or factory farms, which it considers should no longer be sited like satellites around megacities, but should be more widely distributed throughout the countryside. Since the SARS and swine flu epidemics, the FAO also cites the risk of disease as another reason for dispersal.[80] This is a move in the right direction, but unless these factories are subdivided back into smaller farms, the nutrients will still end up in a huge methane generating HEAP (though in a more bucolic location) and still require fossil-fuel powered transport to take them back to the land from which they came.

Funnily enough, although the FAO view urbanization and nutrient loading to be a problem for livestock, they do not consider it to be so for humans, nor do they advocate that people's habitations should be determined by the 'specific natural endowments of given locations'. Yet human waste treatment in cities is problematic for precisely the same reason as animal manure is in factory farms; too much of it is gathered in one place, resulting in unmanageable concentrations of phosphorus and nitrogen. And it too is conveyed to one spot by being commingled with vast amounts of water and pumped along antiquated sewage pipes. Leaking and contaminating ground water on the way, it gathers up heavy metals from industrial sources, as well as an extra dose of phosphorus from the detergents used in washing machines and dishwashers. No sooner is this soup all gathered in one place than the water has to be separated off and cleaned up for release into the rivers or sea, bearing with it a permissible proportion of the nutrients and contaminants, and leaving the rest of the phosphorus, together with the chemical precipitate that removed it, in a pile of toxic sludge, which in John Driver's words 'is of dubious agronomic value and presents its own

77 Günther, Folke (2001), 'Ruralization: Integrating Settlements and Agriculture to Provide Sustainability', Proceedings from the NJF-seminar No. 327 'Urban Areas - Rural Areas and Recycling – The organic way forward?' Copenhagen, Denmark, 20–21 August 2001, Darcof Report no. 3, Copenhagen August 2001, http://www.holon.se/folke/lectures/Ruralisation.shtml

78 Steinfeld, H et al (2006), Livestock's Long Shadow, FAO, p 229.

79 Ibid., pp 229 and 261.

80 FAO (2009), Livestock in the Balance, State of Food and Agriculture 2009, draft July 2009, chapter 5.

disposal problems'.[81] When Ralph Borsodi coined the term 'cross-hauling' to describe similar consignments of apples or meat going in opposite directions, he failed to observe that the most pervasive form of cross haulage is the transport of huge volumes of biomass into urban concentrations and then back out again.

In the UK, some of this sludge is landfilled or incinerated, but since dumping at sea has been banned, the majority is now put back to the land – though not where the nutrients are most needed (since its use is not permitted for organic agriculture), nor in the correct proportions. A further problem is that, en route, the sludge tends to lose nitrogen, which is volatile and leaks into the atmosphere (from what I can decipher less than 20 per cent of the N gets back to the land).[82] However it not only retains most of the phosphorus we excrete, but also gains as much again from the addition of detergents and other industrial effluents.[83] The result is that 'the ratio of P to N in the sludge is significantly higher than that required by plants. Hence the practice of applying biosolids at rates based on the N requirements of crops results in an excessive P supply relative to crop needs'.[84] The phosphorus ends up in a HEAP while the remaining land is not seeing its phosphorus recycled.

Admittedly, the volume of domestic sewage is only about a quarter of the volume of livestock dung available in livestock units.[85] But then humans make up for that with all the other organic detritus of civilization: the grey water, the effluents from food, drink and fibre processing plants, the waste food, the hedge clippings and lawnmowings, the unrecyclable scraps of paper and cardboard, the chewing gum, the dog poo, the cat litter, the old clothes and shoes, the human hair and flakings of skin, the ash, and the dust and fluff that seems to arrive from nowhere and fills up hoover bags – all the sort of stuff that over time buried the successive incarnations of Troy under a layer of midden 50 foot deep.

None of this is a problem for people who live dispersed in the countryside. Manure, human sewage and solid organic waste can all be composted; human sewage can also be treated on site through wet water systems, reed beds or composting; liquid waste can be recycled through reed beds, or cured and sprayed onto meadows when the grass is about to grow. In a predominately local economy the quantities of waste are neither too large to cause a pollution problem, nor are they likely to deliver too little phosphorus, because mostly they are going back to the place that they came from. Moreover, in a renewably resourced economy where all disposable items are made of organic material, there is none of the fossil-fuel derived packaging and the wretched scraps of plastic that make sweeping up such a chore for the conscientious: we could forget about waste sorting, and return to the carefree existence of our forefathers

---

81   Driver, *op cit.*, 69.

82   It is hard to get any uniform figures for nitrogen loss, but it looks as though about three quarters of the N are lost. The sums are something like this: humans excrete about 3 kilos of N per year, so the UK excretes about 180,000 tonnes of N. This ends up as 1.5 millions tonnes of sewage sludge (dry matter) with about 2.6% to 3% N content, ie <45,000 tonnes N. Currently about two-thirds is returned to farmland. Bozkurt, M A *et al* (2006), 'Possibilities of Using Sewage Sludge as Nitrogen Fertilizer for Maize', *Acta Agricultura Scandinavica, Section B Plant Soil Science*, Vol 56 2 May 2006, pp 143-9 (abstract); Dove Associates 2002, Sewage Sludge Information, www.dovebugs.co.uk; DEFRA (2005), *eDigest of Statistics, Waste and Recycling*, Tables 11 and 12.

83   'Raw sewage normally contains between 8 and 10 mg/l phosphorus arising from: human waste 45%; detergents 40%; industrial effluent 10%; other 5%. If green detergents were used, raw sewage phosphorus levels could be reduced to about 6 mg/l'. EA West Water Treatment (2002) *Sewage Treatment*, http://www.huntsman.com/pigments/Media/SEWAGE_brochure_Oct_02.pdf

84   Xiap-Lan Huang and Shenker, Moshe (2004), 'Water Solid-State Speciation of Phosphorus in Stabilized Sewage Sludge', *J Environ Qual* 33:1895-1903.

85   Driver, *op cit.*, 69.

who could sweep everything unwanted outdoors onto the midden, or toss it to the pigs, knowing that that was the right place for it.

With animal manure, a move away from slurry and towards dry composting would improve nutrient retention. When manure is composted the methane and nitrous oxide emissions are significantly lower than when manure is applied to the land as slurry, or is simply left in a heap.[86] One Canadian researcher, Gupreet Singh, estimates that by composting manure from beef cattle, Canada could reduce its emissions of greenhouse gases by the equivalent of 16 million tonnes of carbon but adds that 'the slight hurdle' to persuading farmers to adopt composting techniques is the 50,000 Canadian dollars needed to fit a tractor with compost turning equipment.[87] Perhaps he should consider Polyface Farm's 'Pigaerator' system where pigs are persuaded to root through beef cattle bedding by throwing grains into it.[88] Compost is also the method of returning plant biomass to the land that has the best record in accumulating soil carbon. A well stacked and composted muck heap ought be as much an object of pride as a well-turned furrow, and as much a symbol of prosperity in spring as a bulging granary is in autumn.

Another option is using the manure on the farm, including human manure, to produce biogas for the farm and the surrounding community. According to one study, the cow manure from one hectare, in a farm anaerobic digestion unit, has the potential to produce 1,500 kwh and 1,100 kwh of heat per year (this is presumably from cows kept indoors day and night). In Germany there are an estimated 2,500 such units on farms with an average capacity of 200 kw.[89] In India there were 3.67 million domestic and village sized biogas units installed by the end of 2004, most running on cow manure, but a few using human waste. They can be built out of brick by a competent mason, and produce gas for electricity and lighting.[90]

In Britain, however, the main incentive for building anaerobic digesters seems to be to deal with the food waste problem. The UK's largest installation at Holsworthy in Devon, which supplies the grid with enough electricity for 3600 homes, processes food waste from local councils, fish processors, bakers, slaughterhouses etc; but it also trucks in slurry and manure from 25 farmers, and then trucks it back again in the form of wet digested residue (six per cent dry matter), with its nutrient content boosted by the food waste. The biggest cost of the operation is transport, involving 27 deliveries by 20-tonne lorries every weekday. The plant is barely profitable, and would make a substantial loss were it not for the gate fees that can be charged for accepting the food waste that can no longer be recycled profitably to pigs.[91]

Profit aside, a great many solutions are technically possible and these waste problems are also being addressed by people with an urban perspective. The German technological city of Braunschweig (population 245,000), for example,

86   Pattey, E et al (2005), 'Quantifying the Reduction of Greenhouse Gas Emissions as a Result of Composting Dairy and Beef Cattle Manure', Nutrient Cycling in Agroecosystems, Springer Vol 72, 2 June 2005, pp 173-87; Burton, D (2004), Influence of Method of Application on Nitrous Oxide Emissions from Animal Manure, University of Manitoba Project, 1 May 2004.

87   Bond, Sam (2006), Composting Farm Manure Could Slash Gas Emissions, Edie Newsroom, www.edie.net 28 March 2006.

88   Polyface Farm (n.d.), Products, http://www.polyfacefarms.com/products.aspx

89   Frost, Gilkinson and Buick (2006), cited in Easson, Lindsay (n.d.)Energy from Waste, AFBI Renewable Energy Centre, http://www.cafre.ac.uk/leassonafbi.pdf

90   Van Nes, Wim J (2005), 'Asia Hits the Gas', Renewable Energy World, Jan/Feb 2005, http://www.snvworld.org/en/Documents/20060209%20Article%20on%20Biogas%20Asia%20in%20Renewable%20Energy.pdf

91   Geraghty, M et al (2003), Analysis of the Holsworthy Biogas plant, Devon, Biomass Using Aerobic Digestion, http://www.esru.strath.ac.uk/EandE/Web_sites/03-04/biomass//case%20studyhols.htmll; Andigestion (n.d.) Holsworthy, http://www.andigestion.co.uk/content/holsworthy

has plans to grow biomass maize on a dedicated 1500 hectare farm, turn it into gas in an on-site anaerobic digester and pipe it 12 miles into town. In return Braunschweig will pipe back its waste water for irrigation. This sounds like a raw deal for the countryside, which ought to be getting the sewage nutrients back as well. And a distance of 12 miles does not break the bounds of Günther's ruralized regions. But it is easy to see how advocates of large cities and empty countrysides – those who look to rural biomass to pay for the carbon excesses of an urban lifestyle – could set about devising schemes involving tens of thousands of acres of biomass maize or *Miscanthus*, and pipelines ferrying the diluted waste nutrients of cities much bigger than Braunschweig fifty miles or more out into their hinterland.

Meanwhile, chemical and waste disposal firms who recognize 'the problems that have risen as a result of livestock intensification' have no intention of de-intensifying. On the contrary they see these problems as an opportunity to devise even more ingenious technofixes, and for the last decade they have been researching ways to extract chemical fertilizer from biomass. SBN, the owner of the largest sewage incineration plant in Europe, is working on a project euphemistically called SUSAN, designed to extract phosphorus and eliminate heavy metals from the ash of incinerated sewage sludge. Another process, developed by the US Department of Agriculture, removes first nitrogen and then phosphorus from pig manure to produce chemical fertilizers, including bags of calcium phosphate. The technology was developed 'to solve problems of excess nutrient land application due to confined animal production by removing P from animal wastewater'.[92] As another advocate of this technology points out, the recovery of chemical phosphate:

> is likely to be an economic option only in the case of large,
> geographically concentrated waste streams (sewage from urban areas,
> intensive livestock units). In rural areas, agricultural sludge or manure
> spreading will probably always remain the best option for recycling
> nutrients.[93]

In other words, if you don't want your poo turned into chemical fertilizer, move out to the country.

The Future: Urban or Rural?

There is no shortage of high tech proposals for addressing the problems associated with urbanization and fossil fuels. If they are successful, our future may depend less upon technological restraints than upon what kind of life most people want to live. If we choose to use arable land for biogas to help fuel power-hungry cities and develop technologies for turning biomass residues into chemical fertilizers, then our farmed countryside will become a deserted hydroponic factory. The more biomass that is required for energy, the less there will be for livestock, leading to increased pressure for vegan solutions. Under the vegan option, areas of less fertile land could be cordoned off for rewilding, a bargain which would be welcomed by those who, like James Lovelock, Paul

92   Szögi, A A *et al* (2006), 'Innovative Technology for Recycling Manure Phosphorus with Rapid Amorphous Phosphate Precipitation', *12th RAMIRAN International Conference*, USDA-ARS Coastal Plains Research, 12 September 2006; this is just one of several projects described in Reindl, John (2007), *Phosphorus Removal from Wastewater and Manure Through Hydroxylapatite Formation: An Annotated Bibliography*, Dane Co WI, 2007, http://danedocs.countyofdane.com/webdocs/PDF/lwrd/lakes/hydroxylapatite.pdf

93   Driver, *op cit., 69.*

Shepard and the GOOFs, believe that the resurgence of Gaia requires the human race to live a crabbed urban existence, expelled from the Garden for our sins. Following this course would signal our further withdrawal from the natural world, and point us further down the road that leads ultimately towards the spooky scenarios advanced by transhumanists.

This is, I fear, what is more likely to happen. The GOOFs have too much on their side: the pressure of rising population, the increasing numbers of people brought up in cities who have no connection with or understanding of land-based livelihoods, a media with an urban bias, and the financial muscle of capitalist corporations whose sole interest is profit. There are some, such as Richard Heinberg,[94] who think that peak oil will force society to ruralize (as it did in Cuba). But the supporters of headlong economic growth will do their utmost to identify other technologies such as nuclear energy to substitute for fossil fuels, and the most likely thing to scupper them is a miscalculation about the onset of global warming – and that will scupper all of us.

However, my hunch is that there will be rising levels of resistance to increasing urbanization, from people who seek a lifestyle more closely related to the natural world. The natural world is controlled by God, while the technological world is managed by scientists. Both are tyrannical, but as tyrants go, the former has a better record than the latter. Every survey ever conducted in the UK in the last 20 years has concluded that a majority of people would rather live in the countryside than in the town.[95] While the rest of the world lures or forces its peasants into ever-expanding metropolises, Britain, which for two hundred years was at the forefront of urbanization, is now one of the first countries to experience a reversal of the process. The flight from cities – 'counter-urbanization' as sociologists prefer to call it – has become so marked that in 1999 the government formed an Urban Task Force, headed by architect Richard Rogers, to make them more attractive places to live.[96] It wasn't very successful: the population of London was jacked back over the eight million mark, but is still 12.5 per cent smaller than it was in 1951, even though 31 per cent of its current population are people who were born abroad. London's population would be barely half what it was in the 1950s, were it not for the influx of young immigrants from foreign countries, attracted by the wealth derived from dodgy speculations in international finance.

Counter-urbanization tends to be viewed as a middle age and middle class phenomenon, pursued by people with an urban income who are drawn to the country by nothing more profound than the prospect of owning a nice house with a view. However a large proportion of these incomers, even if only as an afterthought, do participate in the local rural economy, they 'eat the view' by buying local food, and they (or their daughters) mess around with horses and grass and manure, because that is a socially acceptable way of maintaining contact with the animal kingdom and the natural world. Biophilia – the innate a love of nature identified by Erich Fromm and EO Wilson – is alive and kicking in British horseyculture.

The trouble is that these prosperous refugees are the only people who can currently afford to buy a house in the countryside, when a land-based economy needs working people (and horses). There is no shortage of people eager to do perform that role, either full or part time. The demand to live and work in the countryside,

---

94   Heinberg, *op cit.*, 65.
95   77 per cent of rural dwellers are 'very satisfied' with their location, as against 35 per cent of urban dwellers. DETR (2000), *Our Towns and Cities: the Future*, Urban White Paper, 2000, p 26.
96   *Ibid.*

as I can attest from my work with the planning campaign, Chapter 7, comes from people at all social and economic levels.[97] However, most can't afford to live in the countryside because the government will not allow them to build houses there.

The English planning system was established in 1947 with the express purpose of preventing people living in the countryside – more specifically working class people who during the inter-war years made their escape from the city by erecting plotland shacks on surplus agricultural land.[98] It still fights a running battle against people who want to establish a rural dwelling, whether it be to enjoy a leafy environment, to make a living, or for environmental reasons. Downsizers who have concluded that the only way to live a low impact renewable lifestyle is by moving close to the plants, animals, building materials, water and energy that they intend to use, inevitably meet with resistance from planning authorities, unless they can afford £400,000 for a rural dwelling with land.

The planners are applying the government policy that 'sustainable patterns of development' require new buildings to be concentrated in towns and key settlements.[99] A rural lifestyle, they argue, will cause people to drive more, a contention which (if it is true at all) only remains a problem as long as the government persists in failing to tax adequately the social and environmental costs of private car use. There is an opposite line of reasoning which suggests that the more people, production and services are spread around the countryside, the less need there will be to travel back and forth and to cart stuff around, and the more viable public transport and delivery services become – in other words that decentralized networks are more efficient and less congested than centralized systems where everything converges on the metropolis (a view already commonly accepted in respect of the transmission of information). Even today, when most development is concentrated around cities, government figures show that the average journey to work is longer, in terms of distance, in peri-urban areas than it is in rural areas.[100]

The scarcity of agricultural infrastructure which results from these urban-biased planning policies is now the chief obstacle to the repopulation of the land-based economy. The soaring value of farmhouses and farm–buildings makes it hard for new entrants to buy into the industry, and tempting for landowners' sons and daughters to sell up and get out, with the result that the average age of farmers is now allegedly over 60. Because agricultural wages won't meet the costs of rural housing, there is a chronic shortage of labour, so casual workers are shipped in from abroad and lodged in caravans. The conclusion so often inferred is that people no longer want to do the work, when the real problem is that they can't afford to. There is no problem finding people to work in the horse industry because its well-heeled clientele (in contrast to the supermarkets that buy our food from farmers) pays people properly.

The myth that nobody wants to engage in the 'drudgery' of land-based work any longer is habitually enlisted in support of a more sweeping assertion that 'we cannot put the clock back, we cannot return to a ruralized peasant economy of small farms'. Why not? We have crop yields and milk yields that are three or four times as great as 250 years ago, an increase which has nothing to do with economies of scale, but results from plant and livestock breeding and better command of nutrient flows. Even if we did cheat and used a dab of

97   Chapter 7: http://www.tlio.org.uk/chapter7/index.html
98   Hardy, D and Ward, C (1984), *Arcadia for All*, Five Leaves, 2004.
99   DCLG (2004), *Planning Policy Statement 7*.
100   Mark Thurstain Goodwin, of Ecofutures Ltd, based in Bath, has produced a map, based on census figures, depicting this phenomenon.

artificial nitrogen here or there, that would be energy well spent, compared to the plasma TVs, outdoor heaters, plastic toys and flights to the Costa del Sol that currently inspire people to fritter away our depleting reserves of fossil fuels. The truth is not that we cannot ruralize, but that powerful elements in society are opposed to it. Their objection is that food will cost more (30 per cent more according to Günther's estimate) and hence there will be correspondingly less wealth available for the purchase of energy intensive consumer goods, and the maintenance of parasitic elites and an overweight 'creative class'.

Two centuries ago, in *The Wealth of Nations*, Adam Smith observed how 'the industry which is carried on in towns is everywhere in Europe more advantageous than in the countryside'. A greater share of 'the whole annual produce of the labour of society ... is given to the inhabitants of the town than would otherwise fall to them, and a less to those of the country'.[101]

The three main reason given by Smith for this inequality are:

(i) The superior ability of urban capitalists and workers to form cartels and unions which make regulations and fix prices in their own favour. 'The inhabitants of a town, being collected into one place, can easily combine together ... The inhabitants of the country, dispersed in distant places cannot combine together.'[102]

(ii) Monopoly control of land, means that little reaches the market and 'what is sold always sells at a monopoly price. The rent never pays the interest of the purchase money ... Such land is in North America to be had for almost nothing, or at a price much below the value of the natural produce'.[103]

and (iii) the construction of roads and canals, which 'break down the monopoly of the country in its neighbourhood' – in other words destroy local food links.[104]

These conditions still obtain today in the UK, almost exactly as Smith describes them. A cartel of supermarkets and agrobusiness corporations plays farmers around the world off against each other to whittle prices down to a minimum. Landowners hold on to land and ask more for it than can be paid off by agriculture, enhancing the competitiveness of imports from countries with fewer people competing for land or where people are paid a pittance. And transport infrastructure is constantly improved and cheapened so as to allow the influx of these imports.

The economic dominance of the town is a drain on the countryside, on its resources and its people. Over time, the terms of trade between agriculture and industry become increasingly disadvantageous to the farmer, the margins for agricultural produce are reduced, economies of scale become necessary, farms become bigger, and peasants desert the country for the town. Nowadays the process can be seen taking place most clearly in developing countries, but not so long ago the UK was at the forefront of urbanization in Europe. Ever since the enclosures of the 18th and early 19th century, Britain has had larger farms and fewer farmworkers than anywhere else in Europe, and it still does today.[105]

---

101 Smith, A (1776), The Wealth of Nations, Book I, Chapter X, part II.
102 *Ibid.*
103 *Ibid.*, Book III, Chapter IV.
104 *Ibid.*, Book I, Chapter XI, Part I.
105 For a comparison of European farm sizes and employment rates in the 1939s, see Warringer, D (1939), *Economics of Peasant Farming*, Oxford.

In the 1950s, in Britain the average sized dairy herd was 18, and it was possible for a man to save money and to buy land by milking just five cows.[106] By 2006 the average herd size was 92 and DEFRA stated, as if it were a matter it had no control over, that 'the trend for herd size is to increase' while the number of dairy farmers 'is forecast to decline'. By 2008 the average had gone up to 112. In Europe in 2006, the average dairy herd was 36, but by 2008 it had gone up to 42.[107] Yet in countries such as Austria and Switzerland a herd of six or eight dairy cows is still normal, and in 2004 the average size herd in Poland was 3.2 cows; two thirds of Polish farms had fewer than three cows, and less than two per cent had more than 20.

The classic view of the economist is that such tiny herds must be inefficient: yet that isn't any bar to productivity, since in 2004 Poland was the fourth largest producer of milk in Europe, and was providing one and a third times as much milk for each of its citizens as Britain. In its wisdom, the EU is reducing Poland's milk quota and forcing farmers and processors who cannot comply with new regulations out of business. The reason the EU gives is that it is necessary 'to increase the competitiveness of Poland's dairy sector';[108] what it is actually doing is reducing competition and introducing urban monopoly control. Poland's productivity is mirrored in India, where the world's largest dairy industry is maintained by millions of peasant producers, and yet it too is under threat from Tesco and Carrefour.[109]

Small scale decentralized dairying is productive because it is a default activity that doesn't demand much more than the ecosystem yields anyway; it is efficient because milk doesn't keep, and decentralized production can more readily meet the demands of a decentralized market. Why put milk into tankers when it has to be distributed by the pint? But it can't compete in a market where a cartel of supermarkets control food distribution and drive down the price of rural goods so that peasants are squeezed off the land and into cities, and economies of scale reign supreme,.

There is nothing inevitable about the UK's large farms and overstretched farmworkers. In a democratic society, we can choose whether or not we pay more for our food, fibre and energy in order that more people can live and work in land-based occupations. More farmers would mean fewer car manufacturers, media workers, or academics, for example, but that might not be such a bad thing.

However, the UK will only make that choice once it has eliminated the deep-rooted prejudice against peasantry that dates back to the industrial revolution. Nearly 200 years ago William Cobbett observed that the term 'peasant' was being adopted as a term of contempt.[110] Yet its equivalents, *paysan* in French and *campesino* in Spanish, are borne with pride in countries that speak those

106 Hoskins, R (1998), *The Perfect Pinta?*, SAFE Alliance, http://www.sustainweb.org/pdf/ff_pinta.pdf; pers com Brian.

107 DEFRA (2006), *UK Dairy Industry*, Nov 2006. www.defra.gov.uk/foodrin/milk/dairyindustry. htm The web page which contained this information has been deleted (along with a lot of other DEFRA material). DEFRA (2008), *Milk and Milk Products: UK Dairy Industry*, http://www.defra.gov. uk/foodfarm/food/industry/sectors/milk/dairyindustry/index.htm The figures for Europe come from these two web pages.

108 Jesse, E V (2005), *The Dairy Industry of Poland*, Dairy Updates World Dairy Industries No 105, Babcock Institute, University of Wisconsin, 2005. Formerly on line, now listed at http://www. babcock.wisc.edu/?q=node/508#2005

109 Wiggerthale, Marita (2009), *To Checkout Please!*, Oxfam Germany, http://www.oxfam.de/download/to_checkout_please.pdf

110 Cobbett, William (n.d.), *Political Register LXXVIII*, 710, cited in Snell, K D M (1985), *Annals of the Labouring Poor*, Cambridge, 1985.

languages. There is a *Confédération Paysanne* in France and an organization representing peasants worldwide called *Via Campesina*. In the UK, the peasant has been the victim of a slur campaign spearheaded by political economists ranging from the 'Scotch philosophers' of the Edinburgh Review whom Cobbett excoriated, to their modern day reincarnation, Sean Ricard, erstwhile economist to the National Farmers Union. The same bias against the peasantry can be seen in *Livestock's Long Shadow*, and the growing tendency to focus the responsibilty for global warming upon extensive livestock farmers 'who often extract marginal livelihoods from dwindling resources', when it is as clear as daylight that, if global warming is a problem, it is primarily caused by fossil fuels.

The restoration of the status of the peasant requires the unravelling of two pernicious delusions. The first is that a peasant economy is necessarily a poor one. The grinding rural poverty of the UK in the 19th century was not intrinsic but imposed. The peasant who has adequate land is wealthy and independent, as long as surplus value is not being creamed off through rent, tax or disadvantageous terms of trade. Cobbett, riding through Milton in the Wiltshire Avon Valley in 1826, calculated that the 100 families living in the village produced enough grain, meat, beer and wool to feed and clothe five hundred families, 'yet those who do the work are half starved … The food and drink and wool are almost all carried away to be heaped on the fund-holders, pensioners, soldiers, dead weight and other swarms of tax eaters'.[111]

Seventy years later Richard Jeffries, not exactly a left-wing radical, spoke of adults and children undernourished 'in the midst of a country teeming with milk.[112] This has become especially the case of late years, now that so much milk is sent to London.' The countryside is now awash with perhaps five times as much milk as in Jeffrey's time, yet the urban economy is prepared to pay for only a sixth the number of agricultural workers, and is driving dairy farmers out of business.[113] Nobody starves, but increasing numbers of people are coming to the conclusion that they are being poisoned by the bleached, homogenized apology for milk that emerges from our understaffed, over–mechanized cow factories.

The situation today in a country such as India, despite the social conscience of the White Revolution, is not so different from that described by Jeffries in 19th century England. In a recent programme on the BBC World Service a reporter commented that peasants continued to flock into cities like Bombay 'because the countryside cannot support them'. She did not explain why the countryside cannot support them, yet it is a reasonable question. India is self-sufficient in grain most years (whereas the UK wasn't when Jeffries was writing), it is the world's largest dairy producer and it is an exporter of numerous agricultural goods. If it can feed and clothe its citizens in town, why can't it feed and clothe them in the country? The answer is that India's farmers, like most farmers in most parts of the world, are not paid enough to compete with the urban economy. The wealth of the land, which is the source of all wealth before labour is applied to it, is not distributed in due portion to the peasantry, but finds its way into the pockets of India's middle class elite, who drive and fly and enjoy air conditioning and generally gobble up fossil fuels at the same rate that we do in the North – the modern equivalent of Cobbett's 'fund-holders, pensioners, soldiers, dead weight and other swarms of tax eaters'.

---

111 Cobbett, William (1826), *Rural Rides*, Everyman, 1912, 30 August 1826.
112 Jeffries, Richard (1998) The Toilers of the Field, Longmans, p 113.
113 There are 1.66 times the number of cows today as in 1890s, with maximum yields today three times as high: Mitchell, B R (1971), *Abstract of British Historical Statistics*, Cambridge; Smith, Robert Trow (1959), *A History of British Livestock Husbandry, 1700-1900*, Routledge and Kegan Paul.

The other delusion that needs to be challenged is that there is something demeaning about using our muscles, and those of our domestic animals, to perform useful work, while it is regarded as heroic to exercise them to no purpose at all in the stadium, or on the racecourse. The more fanatical advocates of progress look upon manual labour with horror as something unbearably primitive. Within 25 years of the replacement of a hand tool by a machine, a generation grows up deprived of that manual skill, and convinced, by a few abortive experiments, that the manual activity is far more arduous than it need be. Yet the medical profession is constantly urging people to engage in physical exercise, and large numbers of town-dwellers pay for the privilege of exhausting themselves in a health and fitness club. Working as a shepherd, or a coppice worker, or a field worker, providing you are paid decently, is no more gruelling than working at a supermarket till, minding an injection moulding machine, or turning a treadmill in a gym, and a good deal more skilful and stimulating.

To champion the virtues of a rural lifestyle against the arguments of today's political economists, who better to cite than Adam Smith, the original Scotch philosopher, who famously began *The Wealth of Nations* by praising the division of labour involved in the manufacture of a pin?

> Not only the art of the farmer, the general operations of husbandry, but many inferior branches of country labour, require much more skill and experience than the greater part of mechanic trades … The man who ploughs the ground with a team of oxen, works with instruments of which the health, strength and temper, are very different upon different occasions. The condition of the materials which he works are as variable as that of the instruments he works with and both require to be managed with much discretion. The common ploughman, though generally regarded as the pattern of stupidity and ignorance, is seldom defective in his judgment and discretion. His voice and language are more uncouth and difficult to be understood by those who are not used to them. His understanding, however, being accustomed to consider a greater variety of objects is generally much superior to that of the other, whose whole attention from morning till night is commonly occupied in performing one or two very simple operations.[114]

And again:

> The pleasures of a country life, the tranquility of mind which it promises, and wherever the injustice of human laws does not disturb it, the independence which it really affords, have charms that more or less attract everybody; and as to cultivate the ground was the original destination of man, so in every stage of his existence he seems to retain a predilection for this primitive employment.[115]

That was published in 1776, when agricultural yields were a fraction of what they are today. Now that we are wealthier, instead of emptying the countryside of its farmers, and its animals, should we not allow more people to enjoy this 'primitive employment'?

---

114 Smith, *op cit*. 101, Book I, Chapter X, Part II.
115 Smith, *ibid*, Book III, Chapter 1.

INDEX

compiled by Auriol Griffith-Jones

Note: Page numbers in **bold** refer to tables; those in *italics* refer to figures

ABOUT THE AUTHOR

Simon Fairlie is editor of *The Land* magazine and runs Chapter 7, a UK organization that provides planning advice to smallholders and other low-income people in the countryside. Previously he was co-editor of *The Ecologist* magazine. For ten years, Fairlie lived on a community farm where he managed cows, pigs, and a working horse. Fairlie lives in Dorset, England, and sells scythes for a living.

www.permaculture.co.uk

Gene Logsdon farms in Upper Sandusky, Ohio, and has written more than two dozen books, including *The Contrary Farmer* and *Holy Shit: Managing Manure to Save Mankind*. Logsdon writes a popular blog at OrganicToBe.org, is a regular contributor to *Farming* magazine and *The Draft Horse Journal*, and writes an award-winning weekly column in the Carey, Ohio, *Progressor Times*.

Positive Inspiration
for Making
a Better World

Permaculture Magazine **offers tried and tested ways of creating flexible, low cost approaches to sustainable living.**

It enables you, your family and community to:

- Discover the inspiring and exciting world of permaculture

- Meet the world's most creative and innovative green pioneers

- Learn practical, cutting edge methods for self reliance

- Find interesting courses, contacts and opportunities

- Save money and have fun

Permaculture Magazine is published quarterly for enquiring minds and original thinkers worldwide. Each issue gives you practical, thought provoking articles written by leading experts as well as fantastic eco-friendly tips from readers!

permaculture, organic gardening, agroforestry, eco-building, renewable technology, transition towns, sustainable agriculture, ecovillages,community activism, human-scale economy... and much more!

Permaculture Magazine gives you access to a unique network of people and introduces you to pioneering projects from around the world. Join the 100,000+ readers in 77 countries and share their knowledge!

Each issue of *Permaculture Magazine* brings you the best ideas, advice and inspiration from people who are working towards a more sustainable world.

Available from all good outlets

or by subscription from

DISTICOR DIRECT   1-877-474-3321   www.magamall.com

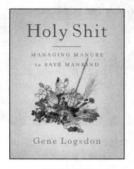

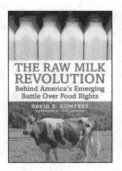

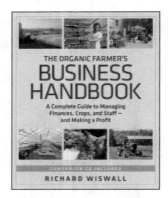